STUDENT'S SOLUTIONS MANUAL

JAMES LAPP

MATHEMATICS ALL AROUND

FIFTH EDITION

THOMAS PIRNOT
Kutztown University of Pennsylvania

PEARSON

Boston Columbus Indianapolis New York San Francisco Upper Saddle River
Amsterdam Cape Town Dubai London Madrid Milan Munich Paris Montreal Toronto
Delhi Mexico City São Paulo Sydney Hong Kong Seoul Singapore Taipei Tokyo

Reproduced by Pearson from electronic files supplied by the author.

Publishing as Pearson, 75 Arlington Street, Boston, MA 02116.

ISBN-13: 978-0-321-83737-0
ISBN-10: 0-321-83737-1

1 2 3 4 5 6 EBM 16 15 14 13

www.pearsonhighered.com

PEARSON

CONTENTS

Chapter 1

Problem Solving: Strategies and Principles

Section 1.1 Problem Solving

1. Drawings may vary.

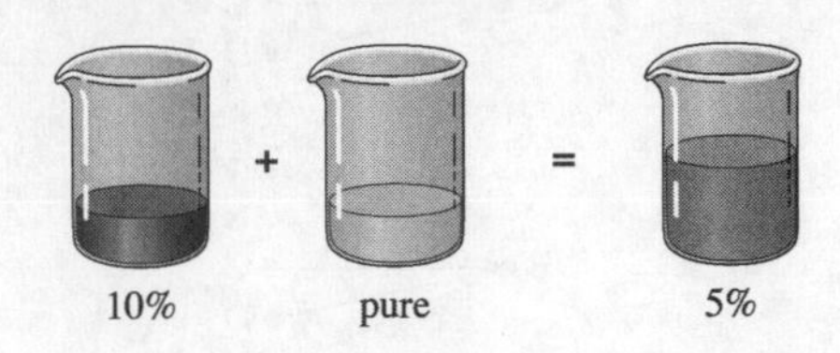

3. Drawings may vary.

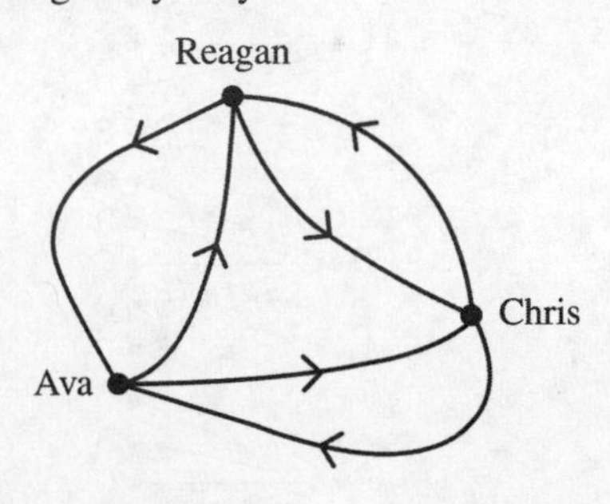

5. Answers may vary.
   Let H be the hybrid automobiles, W be the windmill turbines, and S be solar energy.

7. Answers may vary.
   Let s be the dollar amount invested in stocks and b be the dollar amount invested in bonds.

9. Answers (order) may vary.

Penny	Nickel
Heads	Heads
Heads	Tails
Tails	Heads
Tails	Tails

   Combinations would be *HH*, *HT*, *TH*, and *TT*.

11. $2\times2\times2\times2\times2=32$

13. Answers (order) may vary.
    Using the "Be Systematic" strategy, first list all pairs that begin with *B*, next all new pairs that begin with *R*, etc. Pairs would be *BR*, *BU*, *BE*, *BL*, *RU*, *RE*, *RL*, *UE*, *UL*, and *EL*.

15. (1,1), (1,2), (1,3), (1,4), (2,1), (2,2), (2,3), (2,4), (3,1), (3,2), (3,3), (3,4), (4,1), (4,2), (4,3), (4,4)

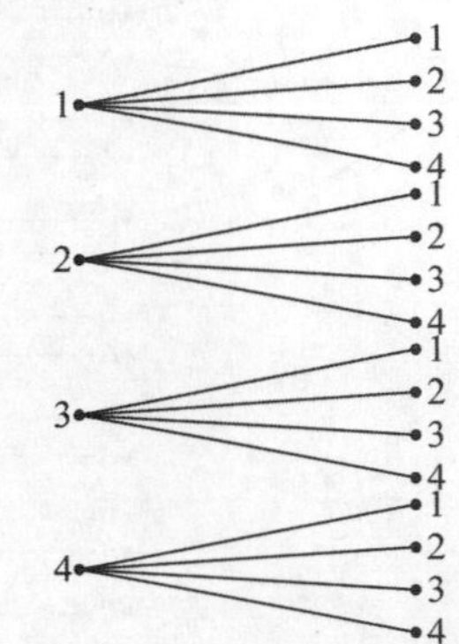

17. r_3 is the set of people who are good singers and appeared on "American Idol". r_4 is the set of people who have appeared on "American Idol" and are not good singers.

19. 7, 14, 21, 28, **35, 42, 49, 56, 63,...**

21. Answers may vary.
    ab, *ac*, *ad*, *ae*, *bc*, *bd*, *be*, ***bf*, *cd*, *ce*, *cf*, *cg*,...**

23. 1, 1, 2, 3, 5, 8, 13, **21, 34, 55, 89, 144,...**

25. Answers may vary.
    In how many ways can we line up three people for a picture?

    Let the people be labeled A, B, and C.

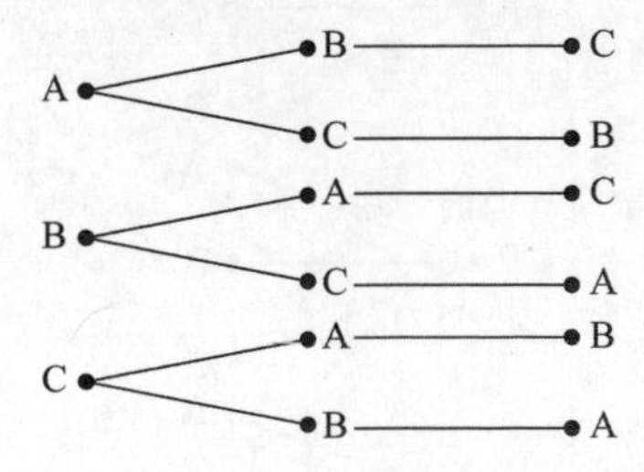

The possible orders are ABC, ACB, BAC, BCA, CAB, and CBA.
6 different ways

25. (continued)

In how many ways can we line up four people for a picture?

Let the people be labeled A, B, C, and D.

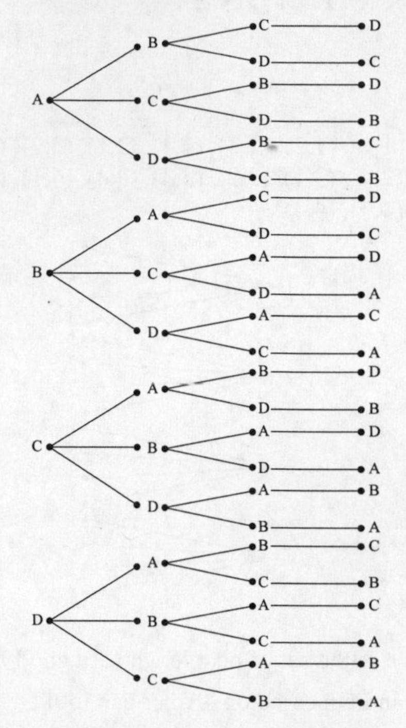

The possible orders are ABCD, ABDC, ACBD, ACDB, ADBC, ADCB, BACD, BADC, BCAD, BCDA, BDAC, BDCA, CABD, CADB, CBAD, CBDA, CDAB, CDBA, DABC, DACB, DBAC, DBCA, DCAB, and DCBA.

24 different ways

27. Answers may vary.
Using the first three letters of the alphabet, how many two-letter codes can we form if we are allowed to use the same letter twice?

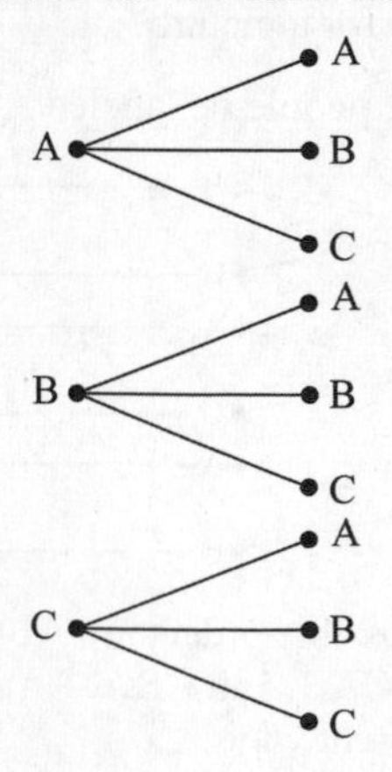

The possible codes are AA, AB, AC, BA, BB, BC, CA, CB, and CC.
9 different codes

Using the first five letters of the alphabet, how many two-letter codes can we form if we are allowed to use the same letter twice?

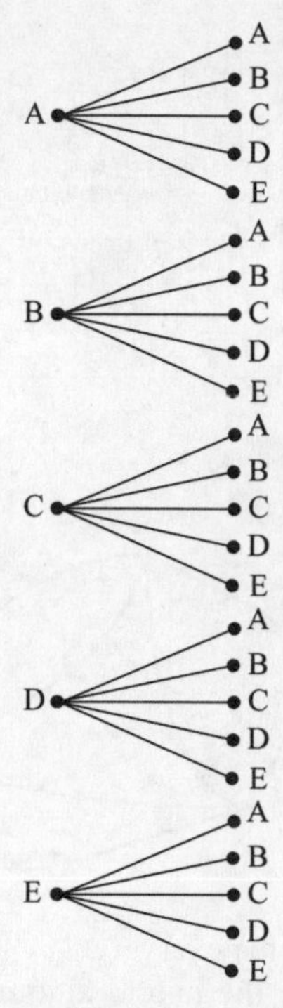

The possible codes are AA, AB, AC, AD, AE, BA, BB, BC, BD, BE, CA, CB, CC, CD, CE, DA, DB, DC, DD, DE, EA, EB, EC, ED, and EE.
25 different codes

29. Answers may vary.
An electric-blue Ferrari comes with two options: air conditioning and a CD player. You may buy the car with any combination of the options (including none). How many different choices do you have?

Let A be air conditioning and C be CD player. If a feature is not included, it is indicated by a "0". If it is included, it is indicated by a "1".

A	C
0	0
0	1
1	0
1	1

4 different choices

29. (continued)

An electric-blue Ferrari comes with three options: air conditioning, CD player, and air bags. You may buy the car with any combination of the options (including none). How many different choices do you have?

Let A be air conditioning, C be CD player, and B be air bags. If a feature is not included, it is indicated by a "0". If it is included, it is indicated by a "1".

A	C	B
0	0	0
0	0	1
0	1	0
0	1	1
1	0	0
1	0	1
1	1	0
1	1	1

8 different choices

31. False, counterexamples may vary.
September has only 30 days.

33. False, counterexamples may vary.

$$\frac{1}{2}+\frac{3}{4}=\frac{2}{4}+\frac{3}{4}=\frac{5}{4}$$

$$\frac{1+3}{2+4}=\frac{4}{6}=\frac{2}{3},\quad \frac{5}{4}\neq\frac{2}{3}$$

35. False, A is the grandfather of C.

37. False, counterexamples may vary.
If the price of a \$10.00 item is increased by 10%, its new price is \$11.00. If the \$11.00 item is then decreased by 10%, the new price would be \$9.90, not \$10.00.

39. Explanations may vary.
These two sequences do not give the same results. The question here is equivalent to asking if the algebraic statement, $x^2+5=(x+5)^2$, is true.

If we let $x=1$, we have a counterexample.

$$1^2+5\overset{?}{=}(1+5)^2$$

$$1+5\overset{?}{=}6^2$$

$$6\neq 36$$

Hence, the statement is false.

41. Explanations may vary.
These two sequences do give the same results. The question here is equivalent to asking if the algebraic statement, $\frac{x+y}{3}=\frac{x}{3}+\frac{y}{3}$, is true.

If you think of dividing by 3 equivalent to multiplying by $\frac{1}{3}$, then you can use the distributive property to prove this statement.

$$\frac{x+y}{3}=\frac{1}{3}(x+y)=\frac{1}{3}x+\frac{1}{3}y=\frac{x}{3}+\frac{y}{3}$$

43. Answers may vary.
5 is a number; {5} is a number with braces around it. Moreover, 5 is a singly listed element, while {5} is a set that contains the single element 5.

45. Answers may vary.
One is uppercase and the other is lowercase. Moreover, U usually denotes the universal set, and the lower case letters are usually elements that appear in some set.

47. Answers may vary.
The order of the numbers is different. Moreover, if you interpret (2,3) and (3,2) as intervals, they are both expressing the same set of numbers. The standard, however, is to express intervals (a,b) in such a way that $a<b$ (assuming a and b are real numbers).

Note: Another interpretation of (2,3) and (3,2) is as points in the plane, and these two points would be different. They would be points that are reflected over the line $y=x$.

The context of this question, however, is relative to intervals.

49. – 51. No solution provided.

53. Answers may vary.
Let O be the age of the older building and Y be the age of the younger building.

Guesses for O and Y	Good points	Weak Points
100, 221	Sum is 321.	The older is more than twice the younger.
110, 211	Sum is 321.	The older is less than twice the older.
107, 214	All conditions satisfied. We have the solution.	

55. Answers may vary.
Let S be the number of hours Janine worked in the sporting goods store and P be the number of hours Janine gave piano lessons.

Guesses for S and P	Good points	Weak Points
11 and 4	Sum is 15.	Amount earned is less than $137.25.
8 and 7	Sum is 15.	Amount earned is more than $137.25.
9 and 6	All conditions satisfied. We have the solution	

57. Answers may vary.
Let T be the number of times Tom Brady threw a touchdown pass, P be the number of times Phillip Rivers threw a touchdown pass, and A be the number of times Aaron Rodgers threw a touchdown pass.

Guesses for T, P and A	Good points	Weak Points
30, 24, and 22	P is 2 more than A and 6 less than T.	Sum is less than 94.
40, 34, and 32	P is 2 more than A and 6 less than T.	Sum is more than 94.
36, 30, and 28	All conditions satisfied. We have the solution.	

Brady had 36 touchdowns.

59. Answers may vary.
Let A be the amount invested at 8%and B be the amount invested at 6%.

Guesses for A and B	Good points	Weak Points
$3,000 and $5,000	Sum is $8,000.	Return is less than $550.
$4,000 and $4,000	Sum is $8,000.	Return is more than $550.
$3,500 and $4,500	All conditions satisfied. We have the solution	

61. Answers may vary.
Let A be the number of administrators, S be the number of students, and F be the number of faculty members.

Guesses for A, S and F	Good points	Weak Points
10, 2, and 7	A is 5 times S and F is 5 more than S.	There are less than 26 people.
20, 4, and 9	A is 5 times S and F is 5 more than S.	There are more than 26 people.
15, 3, and 8	All conditions satisfied. We have the solution.	

There are 3 students.

63. LCHPL, LCPHL, LHCPL, LHPCL, LPCHL, LPHCL

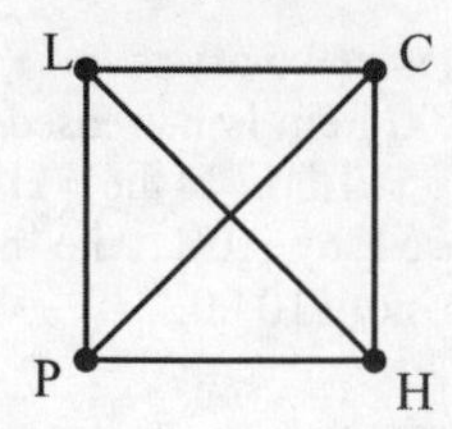

65. The possible schedules are given in the table below.

Math	English	Sociology	Art History
9	11	12	10
9	12	10	11
10	9	12	11
10	11	12	9
12	9	10	11
12	11	10	9

67. The possible schedules are given in the table below.

Math	English	Art History
9	11	10
9	12	10
9	12	11
10	9	11
10	11	9
10	12	9
10	12	11
12	9	10
12	9	11
12	11	9
12	11	10

69. 79; The top and bottom rows will both have 21 tiles, the middle row will have one tile, and the remaining 18 rows will have 2 tiles each. $2\times21+1\times1+18\times2=79$

71. You will pay more in the second year if the 8% raise occurs first. As an example, if tuition is \$100, you will pay $100+0.08(100)=\$108$ in the second year if the 8% raise occurs first, but only $100+0.05(100)=\$105$ in the second year if the 5% raise occurs first. Note, in either case, you would pay the same amount at the end of the two years.

73. – 75. Answers will vary.

77. $6\times6\times6=216$

79. (3, 5), (5, 7), (11, 13), (17, 19), (29, 31), **(41, 43), (59, 61), (71, 73)**; pairs of sequential primes that differ by 2.

81. There are a total of 55 rectangles.

1×1	25
2×2	16
3×3	9
4×4	4
5×5	1

83. As you look at the intersections, you'll see that there is a pattern as to how many routes can be created as you leave the Hard Rock Cafe on the way to The Cheesecake Factory. To choose direct routes, you must always be traveling down and/or to the right. The numbers indicate how many ways there are to get to the intersection below and to the right of the number. For example, the "2" below and to the right of the Hard Rock Café indicates there are two ways to arrive at that intersection, one by going right, then down, and another by going down, and then right. There are a total of 252 possible routes.

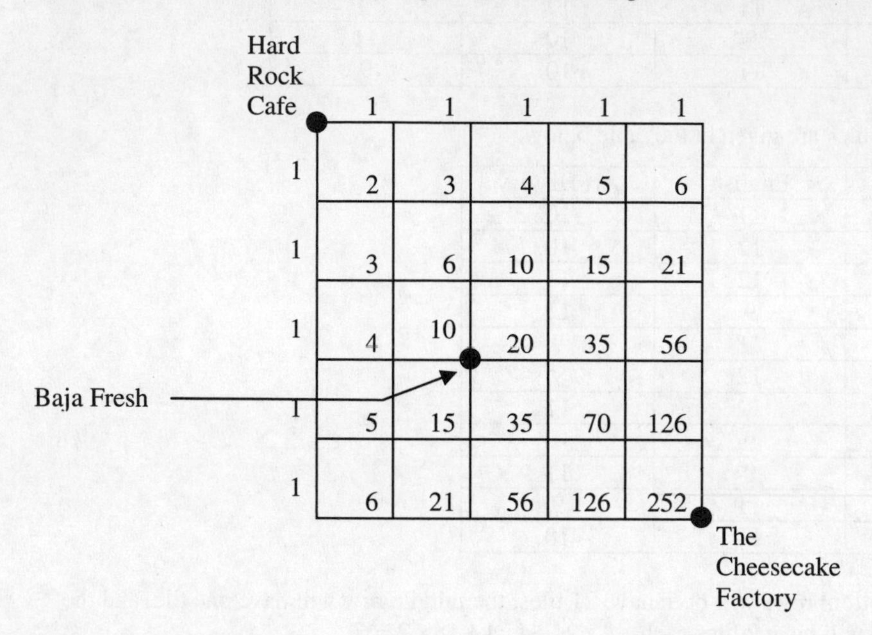

Section 1.2 Inductive and Deductive Reasoning

1. inductive

3. deductive

5. inductive

7. deductive

9. inductive

11. 16

13. 96

15. 21

17.

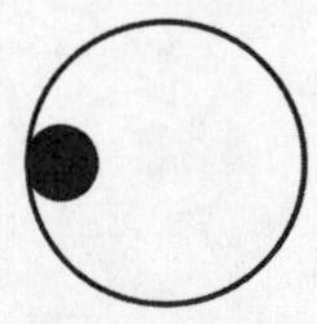

19.

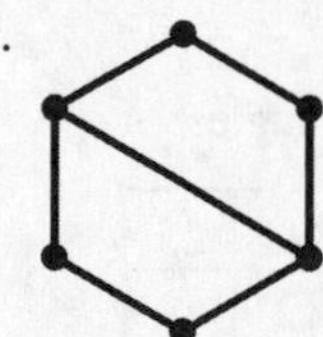

21. Hint: Think of prime numbers.

X X X X X X
2 3 5 7 11 13

23. 3+13 or 5+11

25. 3+17 or 7+13

27. $1+2+3+4+5=\frac{5\times6}{2}$, $1+2+3+4+5+6=\frac{6\times7}{2}$

29. $1+3+5+7+9=25$, $1+3+5+7+9+11=36$

31. $1+4+9+16=30$

33. a) The total of all the numbers in the square is $1+2+3+...+16=136$.

    b) The total of the numbers for each row, column, and diagonal would be $\frac{136}{4}=34$.

    c) One can deduce the missing numbers to yield the following.

7	6	12	9
10	11	5	8
13	16	2	3
4	1	15	14

35. In this trick, you will always get the result three.

    a) Call the number n.

    b) $3n$

    c) $3n+9$

    d) $\frac{3n+9}{3}=\frac{3(n+3)}{3}=n+3$

    e) $(n+3)-n=n+3-n=3$

37. In this trick, you will always get a result that is twice the number that you started with.

    a) Call the number n.

    b) $8n$

    c) $8n+12$

    d) $\frac{8n+12}{4}=\frac{4(2n+3)}{4}=2n+3$

    e) $(2n+3)-3=2n+3-3=2n$

39. Adriana (political issues), Caleb (solar power), Ethan (water conservation), and Julia (recycling)

41. 36644633; the letters represent the numbers in reverse order.

43. 986763; reverse the numbers and delete one number from any pair of numbers.

45. Answers may vary. Possible responses include that Sharifa would begin by visiting one of the six branches and then visit the five original cities in 120 ways. The total number of ways she could make her visits is $6\times120=720$ ways. The same reasoning would lead to $7\times720=5040$ ways for seven cities.

47.

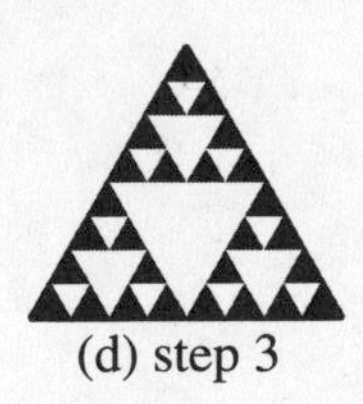

(d) step 3

49. By looking at examples, inductive reasoning leads us to make conjectures which we then try to prove.

51. – 53. Answers will vary.

55. 53, 107, 213; Next term is the previous term plus twice the term before the previous term.

57. 47, 76, 123; Next term is sum of two previous terms.

59. There are a total of 20 squares.
2 3×3 squares; 6 2×2 squares; 12 1×1 squares

61. a) There are a total of 60 rectangles.

1×1	12
1×2	9
1×3	6
1×4	3
2×1	8
2×2	6
2×3	4
2×4	2
3×1	4
3×2	3
3×3	2
3×4	1

b) There are a total of 210 rectangles.

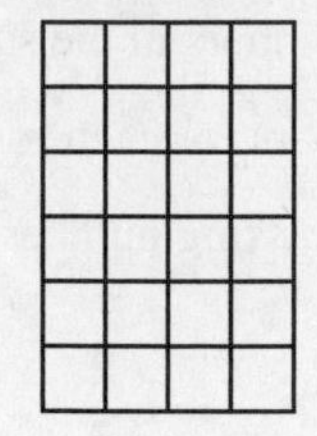

1×1	24
1×2	18
1×3	12
1×4	6
2×1	20
2×2	15
2×3	10
2×4	5
3×1	16
3×2	12
3×3	8
3×4	4

4×1	12
4×2	9
4×3	6
4×4	3
5×1	8
5×2	6
5×3	4
5×4	2
6×1	4
6×2	3
6×3	2
6×4	1

63. The base is a 6×4 rectangle. If you consider the diagram below as the base, we have $6\times4=24$ baseballs. In order to build the next level, we are looking for the number of places in which four baseballs (squares) meet. There are $5\times3=15$ such places.

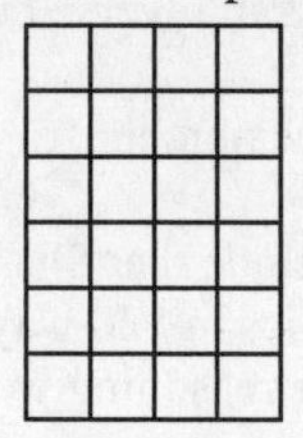

For the next level we would have $4\times2=8$ meeting places.

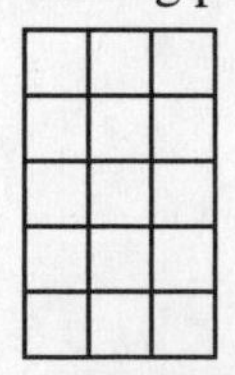

For the last level we would have $3\times1=3$ meeting places.

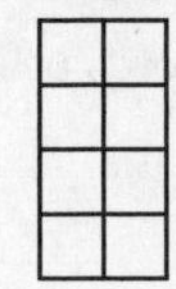

This yields a total of $6\times4+5\times3+4\times2+3\times1=24+15+8+3=50$ baseballs.

65. No solution provided.

67. If you expand the expression as follows,

$(2n+5)\cdot 50+1763-1995=100n+250+1763-1995=100n+2013-1995=100n+18$,

you see that the $1763+250$ gives a multiple of 100 plus 2013 – 1995, which is your age. If you have already had your birthday, you need to add the extra year, which is why we would then add 1764.

Section 1.3 Estimation

1. Answers may vary.

   $20+40+190+40=290$

3. Answers may vary.

   $35-15=20$

5. Answers may vary.

   $5\times 16=80$

7. Answers may vary.

   $18\div 3=6$

9. Answers may vary.

   $0.1\times 800=80$

11. Answers may vary.

    $9\%\times 1000=0.09\times 1000=90$

13. Answers may vary.

    $4\times 5\times 6=120$ miles
    The estimate is larger than the exact answer.
    Exact answer is 111 miles.

15. Answers may vary.

    $325\div 50=6.5$ more hours, 7:30PM
    The estimate is earlier than the actual time.
    Actual answer is 7:50PM

17. Answers may vary.

    $\$120.00\times 15\%=\$120.00\times 0.15=\$18.00$
    The estimate is more than the exact answer.
    Exact answer is \$17.77.

19. Answers may vary.

    $(3\times \$3.00)+(4\times \$1.50)+\$3.00=$

    $\$9.00+\$6.00+\$3.00=\$18.00$
    The estimate is larger than the exact answer.
    Exact answer is \$17.20.

21. Answers may vary.

    It seems safe. Alicia probably weighs less than 200 pounds, so that leaves $2,300-200=2,100$ pounds for the 21 students. They probably each weigh less than 100 pounds.

23. Answers may vary.

    $\$40,000\times 4\%=\$40,000\times 0.04=\$1,600$
    The estimate is larger than the exact answer.
    Exact answer is \$1324.40.

25. Answers may vary.

    $\frac{1000}{1}=1000$ times greater
    The estimate is very close to the actual answer.
    Actual answer is 996.67 times greater.

27. Answers may vary.

    Her total expenses are about \$100 per month; $100\div 7\approx 14$ $12\times \$14=\$168$
    The estimate is larger than the actual answer.
    Actual answer is \$163 (you round to nearest dollar on deductions).

29. Answers will vary. Actual answers are:

    Male high school graduates; \$42,000
    Females with associates degrees; \$41,400

31. College graduates; those with associate degrees

33. category 2.

35. response 2.

37. Estimated answers may vary.

    Exact answer is 17.3 million.

39. Estimated answers may vary.

    Exact answer is 24.7 million.

41. Estimated answers may vary.
Exact answer is $874.66 billion
$2165 \times 0.404 = 874.66$

43. Estimated answers may vary.
Exact answer is $155.88 billion
$2165 \times 0.072 = 155.88$

45. $705,361 \times 0.302 \approx 213,019$
The actual number of immigrants was 213,019.

47. $130,661 \div 705,361 \approx 0.185$
18.5%

49. – 53. Answers will vary.

55. Answers may vary.
The amount of lawn that needs to be fertilized is represented by the size of the lot, less the non-grassy areas such as the garden, driveway and house.
$96 \cdot 169 - 96 \cdot 30 - 65 \cdot 28 - 18 \cdot 65 =$
$16,224 - 2,880 - 1,820 - 1,170 = 10,354$
If you divide the grassy area into rectangles, you get 10,354 square feet. They need slightly over two bags of fertilizer.

57. – 59. No solution provided.

Chapter Review Exercises

1. Understand the problem, devise a plan, carry out your plan, check your answer.

2. An example that shows a conjecture is false.

3. 10.
Let *A*, *C*, *L*, *R*, and *T* represent Amber, Chris, Lawrence, "Thirteen," and Travis.

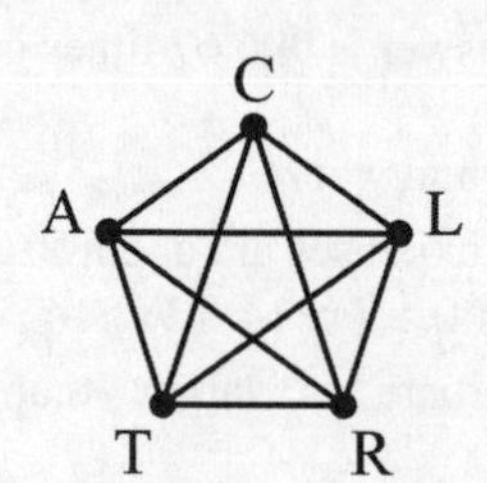

The pairs are AC, AL, AR, AT, CL, CR, CT, LR, LT, and RT.

4. Answers may vary.
At a restaurant, you have 2 appetizers, 3 entrees, and 2 desserts. How many different meals can you choose if you select one appetizer, one entrée, and one desert?

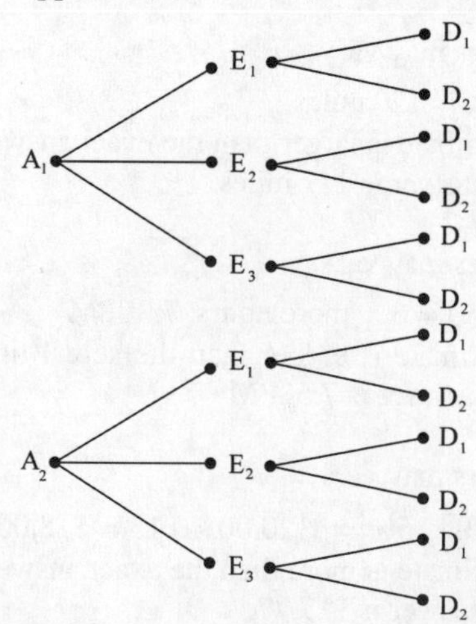

There are twelve different meals possible.

5. Answers may vary.
Let P be the number of hours Picaboo worked as a stock person and I be the number of hours she worked as a ski instructor.

Guesses for P and I	Good points	Weak Points
9 and 11	Sum is 20.	Amount earned is less than \$141.20.
7 and 13	Sum is 20.	Amount earned is more than \$141.20.
8 and 12	All conditions satisfied. We have the solution.	

6. false
Counterexamples may vary.
$\frac{1}{2}+\frac{3}{4}=\frac{2}{4}+\frac{3}{4}=\frac{5}{4}$
$\frac{1+3}{2+4}=\frac{4}{6}=\frac{2}{3}, \quad \frac{5}{4}\neq\frac{2}{3}$

7. No solution provided.

8. Answers may vary. Possible answers include: Inductive reasoning is the process of drawing a general conclusion by observing a pattern in specific instances. In deductive reasoning, we use accepted facts and general principles to arrive at a specific conclusion.

9. a) deductive
b) inductive

10. a) 27
b) 47

11. 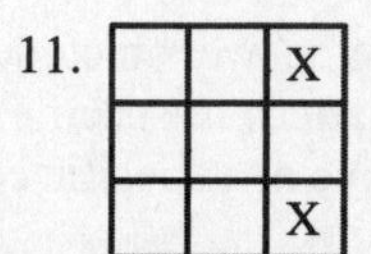

12. $5+43$, $7+41$, $11+37$, $17+31$, or $19+29$

13. In this trick, you will always get twice the number you started with.

a) Call the number n.

b) $8n$

c) $8n+12$

d) $\frac{8n+12}{4}=\frac{4(2n+3)}{4}=2n+3$

e) $(2n+3)-3=2n+3-3=2n$

14. a) 46,000
b) 28,000

15. a) $210-60=150$
b) $6\times15=90$

16. Estimated time left to travel would be $150\div50=3$ hours, arriving at 7:00 PM.

17. Estimated answers may vary.
a) Private college \$21,000; public college \$8,000
b) Statements ii is true.
c) from 1980–81 to 1990–91

Chapter Test

1. Answers will vary.

2. a) true b) false; $\frac{3}{4+5}\neq\frac{3}{4}+\frac{3}{5}$

3. Answers may vary.
Let W be the number of Wii's sold, S be the number of Super Mario Brothers' sold, and P be the number of Pokémon's sold.

Guesses for W, S, and P	Good points	Weak Points
36, 29, and 20	Super Mario Brothers sold 9 more than Pokémon and 7 less than Wii Sports.	Total number is less than 118.
56, 49, and 40	Super Mario Brothers sold 9 more than Pokémon and 7 less than Wii Sports.	Total number is more than 118.
47, 40, and 31	All conditions satisfied. We have the solution.	

4. a) 4320
b) 2820

5. a) 36,000
b) 36,500

6. If two terms are similar but sounds slightly different, they usually do not mean exactly the same thing.

7. Answers may vary. Possible answers include: Inductive reasoning is the process of drawing a general conclusion by observing a pattern in specific instances. In deductive reasoning, we use accepted facts and general principles to arrive at a specific conclusion.

8. a) inductive b) deductive

9. Mathematical ideas can be understood verbally, graphically, and through examples.

10. Answers will vary. One estimate is given below. (True value is $3,220)

$$\left(\frac{\$600+\$200}{3}\right)\cdot 12=(\$800)\cdot 4=\$3,200$$

11. $1 + 1^2 = 2$, $2 + 2^2 = 6$, $6 + 3^2 = 15$, $15 + 4^2 = 31$, $31 + 5^2 = 56$, $56 + 6^2 = 92$, $92 + 7^2 = \mathbf{141}$

12. cde, cdf, cdg

13.

	X	X	
		X	X

14. $7+53$, $13+47$, $19+41$, or $29+31$

15. False; suppose the laptop costs $1,000. After the 10% discount, the laptop would cost $900. If that price is increased by 10%, the laptop would cost $990, not $1,000.

16. In this trick, you will always get twice the number you started with.

a) Call the number n.
b) $4n$
c) $4n+40$
d) $\dfrac{4n+40}{2}=\dfrac{2(2n+20)}{2}=2n+20$
e) $(2n+20)-20=2n+20-20=2n$

Chapter 2
Set Theory: Using Mathematics to Classify Objects

Section 2.1 The Language of Sets

1. $\{10,11,12,13,14,15\}$

3. $\{17, 18, 19, 20, 21, 22, 23, 24, 25\}$

5. $\{4, 8, 12, 16, 20, 24, 28\}$

7. {Sunday, Monday, Tuesday, Wednesday, Thursday, Friday, Saturday}

9. $\varnothing$

11. $\varnothing$

13. Answers may vary. Possible answers include $\{x : x$ is a multiple of 3 between 3 and 12 inclusive$\}$.

15. {28, 29, 30, 31}

17. {January, February, March, April, May, June, July, August, September, October, November, December}

19. Answers may vary. Possible answers include $\{101, 102, 103, \ldots\}$.

21. Answers may vary. Possible answers include $\{x : x$ is an even natural number between 1 and 101$\}$.

23. well defined

25. not well defined

27. not well defined

29. well defined

31. $\notin$

33. $\in$

35. $\in$

37. $\in$

39. $\notin$

41. $\in$

43. 6

45. 0

47. 4

49. 2 elements; {1, 2}, {1, 2, 3}

51. 1 element; $\{\{\varnothing\}\}$

53. finite

55. infinite

57. Answers may vary. Possible answers include 4.5.

59. Answers may vary. Possible answers include Sony.

61. Answers may vary. Possible answers include Angela Merkel.

63. Answers may vary. Possible answers include Sunday.

65. $\{x : x$ is a humanities elective$\}$

67. {History012, History223, Geography115, Anthropology111}

69. {AZ, FL, GA, LA, NJ, NM, TX, VA}

71. $\{x : x$ had an average price of gasoline above 380$\}$

73. {Amazon, Apple, eBay, Facebook, Interactive Corp, News Corp }

75. $\{x : x$ had an audience of less than 50 million$\}$

77. Answers will vary.

79. $\varnothing$ is the empty set, it contains no elements. $\{\varnothing\}$ is not empty, it contains 1 element, $\varnothing$.

81. Precise definitions are important, not only in mathematics, but in everyday life.

83. Answers will vary.

85. If the barber shaves himself, then he (the barber) does not shave himself. If the barber does not shave himself, then he (the barber) does shave himself.
Conclusion: This is a paradox.

87. If $S \in S$, then $S \notin S$. If $S \notin S$, then $S \in S$.
Conclusion: This is a paradox.

Section 2.2 Comparing Sets

1. These two sets are equal. They have the same elements arranged in a different order.

3. These two sets are not equal. The second set contains (infinitely many) elements that don't appear in the first set.

5. These two sets are equal. They are both $\{1, 3, 5, \ldots, 99\}$.

7. These two sets are equal. Common sense dictates that nobody born before 1800 should be living.

9. true; All the elements of the first set are understood to be elements of the second set and moreover the first set is not equal to the second set.

11. false; The letter "y" is an element of the first set and not an element of the second set.

13. true; The null set is a subset of all sets.

15. The first set is equivalent to the second set because they both have the same number of elements.

17. The first set is equivalent to the second set because they both have that same number of elements, namely 4.

19. The first set is not equivalent to the second set. The first set has 0 elements while the second set has 1 element.

21. The first set is equivalent to the second set. They both have 8 elements.

23. The first set is not equivalent to the second set. The first set has 366 elements while the second set has 365 elements.

25. $\{1, 2\}, \{1, 3\}, \{2, 3\}$

27. $\{1,2,3\}, \{1,2,4\}, \{1,3,4\}, \{2,3,4\}$

29. There are $2^5 = 32$ subsets and $2^5 - 1 = 31$ proper subsets.

31. T; V = {Carmen, Frank, Ivana} = T

33. The set of lowerclassmen = L; Note: the **bolded** value indicates the set with the largest cardinality.
$n(U)$= 4, $\mathbf{n(L)}$**= 6**, $n(S)$= 2, $n(V)$= 3,
$n(A)$= 2, $n(T)$= 3, and $n(D)$=2

35. $2^6 = 64$

37. $2^4 = 16$

39. $2^7 = 128$

41. $2^8 = 256$

43. 7

45. {5P, 10P, 25D}

47. {5P, 10P, 25D} or (5P, 10P, 25S}

49. a) $\{1\}$ is a subset, not an element of the set.

d) 3 is an element, not a subset of $\{1,2,3\}$.

51. Answers may vary.
At the first branching, the "yes" or "no" indicates that in forming a subset of $\{1, 2\}$, we will either take the 1 or omit it. The second branchings in the tree indicate whether we are going to take the 2 as a member of the subset that we are forming. The tree shows all possible ways that we can decide to take 1 and 2 in forming a subset of $\{1, 2\}$. The top branch corresponds to the subset $\{1, 2\}$, the second branch corresponds to the subset $\{1\}$, the third branch corresponds to the subset $\{2\}$, and the bottom branch corresponds to the subset $\varnothing$.

53. a) 25 is not a power of 2.; b) He confused 5^2 with 2^5.; c) $2^5 = 32$.

55. Answers will vary.

57. $k = 31$; $k = 32$ would take longer than 100 years.

59. The fifth line counts the number of subsets of sizes 0, 1, 2, 3, 4, and 5 of a five-element set.

61. The sixth, seventh, eighth, and ninth lines are:

1 6 15 20 15 6 1

1 7 21 35 35 21 7 1

1 8 28 56 70 56 28 8 1

1 9 36 **84** 126 126 84 36 9 1

So Tyra Banks can choose the three contestants in 84 different ways.

63.

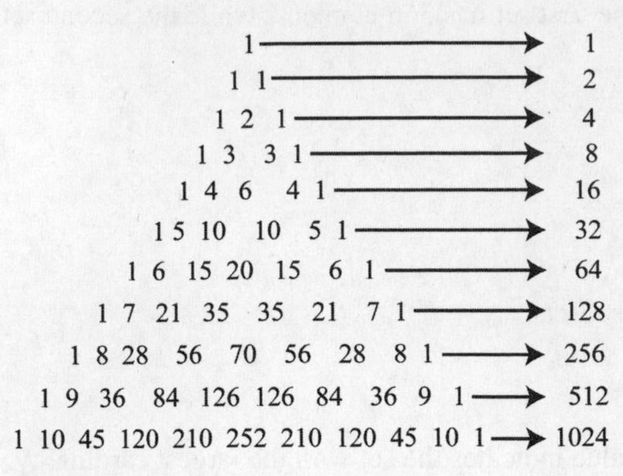

The sum across the rows is always a power of 2, specifically 2^n, where n is the number of the row that is being summed. Note: Recall we start counting these lines with 0, not 1.

65. 16; We are choosing 3, 4, or 5 senior partners from the five possible, so add the last three elements of the 5th row of Pascal's triangle. (Remember we begin numbering the rows with 0.)

67. Corresponding property: If $A \subseteq B$ and $B \subseteq C$, then $A \subseteq C$. If $x \in A$, then $x \in B$. Since $x \in B$, then $x \in C$. Therefore, if we have $x \in A$, we must also have $x \in C$. Note: We use capital letters for sets and lower case letters for elements.

69. Examples will vary. A three-element set has $3! = 6$ correspondences, a four element set has $4! = 24$, and so on. So, a set with n elements will have $n!$ one-to-one correspondences.

Section 2.3 Set Operations

1. $A \cap B = \{1, 3, 5\}$

3. $B \cup C = \{1, 2, 3, 4, 5, 6, 7, 8\}$

5. $A \cup \varnothing = \{1, 3, 5, 7, 9\} = A$

7. $A \cup U = \{1, 2, 3, 4, 5, 6, 7, 8, 9, 10\} = U$

9. $A \cap (B \cup C) = A \cap \{1, 2, 3, 4, 5, 6, 7, 8\} = \{1, 3, 5, 7\}$

11. $(A - B) \cap (A - C) = \{7, 9\} \cap \{1, 3, 5, 9\} = \{9\}$

13. $M \cap E = \{\text{potato chip, bread, pizza}\}$

15. $E - M = \{\text{apple, fish, banana}\}$

17. $M' \cap G' = \{\text{fish}\}$

19. 7

21. 5

23. $A - (B \cup C)$

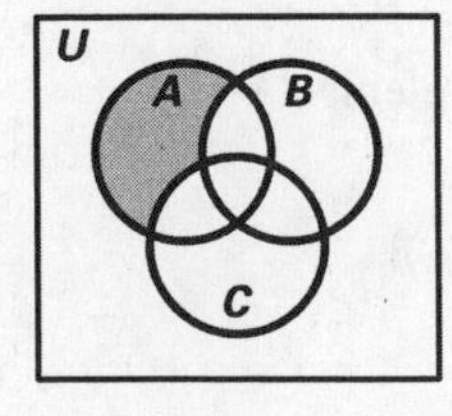

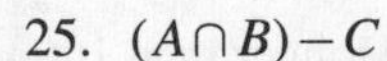

25. $(A \cap B) - C$

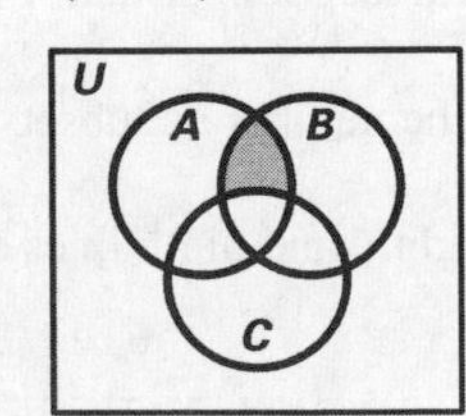

27. 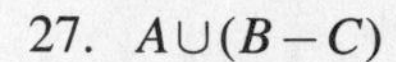
$A \cup (B - C)$

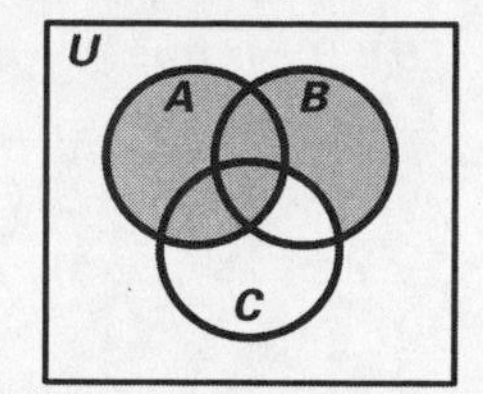

29. $(A \cup (B \cup C))'$

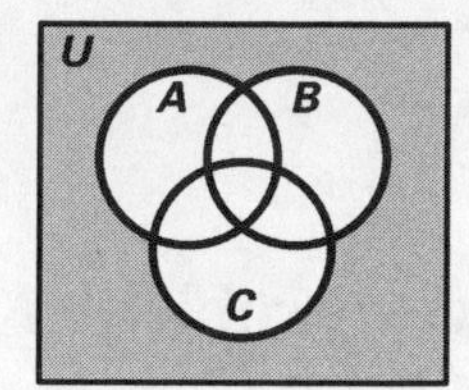

31. $B - A$

33. $(A \cup B)'$

35. $A \cap B \cap C$

37. $(A \cup C) - B$

39. equal; Using the diagrams from Example 2, $(A \cup B')'$ consists of region r_4. $A' \cap B$ also consists of region r_4. so $(A \cup B')' = A' \cap B$.

41. 30

43. 28

45. 20

47. 27

49. $P \cap C$ = the set of cars whose price is above \$20,000 and is compact = $\{d, f, g\}$

51. $W \cap G'$ = the set of cars that have a warranty of at least three years and don't have a good safety rating = $\{c, d, f, g\}$

53. $P \cap (G \cup W)$ = the set of cars that have a price above \$20,000, and a good safety rating or a warranty of at least three years = $\{b, d, f, g, h\} = P$

55. $P - (G \cup A)$ = the set of cars that have a price above \$20,000 and don't have a good safety rating nor have an antitheft package = $\{d, f\}$

57. $P \cap (B \cup A) = P \cap \{m, mc, hc\} = \{m, mc, hc\}$

59. $P \cup C \cup B = \{m, mc, bc, c, hc\}$

61. {FL, NJ, TX}

63. {CA, NY}

65. $E \cap B$ = the set of movies that earned more than \$900 million and were made before 1970 = $\{a, c, e\}$

67. $O \cap B'$ = the set of movies that won an Oscar and were made after 1970 = $\{f\}$

69. $E \cap B \cap O$ = set of movies that earned more than \$900 million, were made before 1970, and won an Oscar = $\{a, c\}$

71. "Union" implies joining together . "Intersection" implies overlapping.

73. Answers will vary. Possible answers include confusing DeMorgan's laws with the distributive property.

75. Answers will vary.

77. false; It is possible that $A = B$ and hence $n(A) = n(B)$. Counterexamples may vary.

79. true

81. A

83. B

85. a) $A \cap B = B \cap A$ is true.

b) $A \cup (B \cup C) = (A \cup B) \cup C$ is true.

Section 2.4 Survey Problems

1. r_2, r_3

3. r_2, r_3, r_4

5. r_2, r_3, r_5, r_6

7. r_4, r_7

9. r_7; $\{r_6, r_7\} \cap \{r_1, r_4, r_7, r_8\} = \{r_7\}$

11. 25

13. 18

15. 16

17. 20

19. 19

21. $n(A) = 7, n(B) = 3$, and $n(C) = 12$

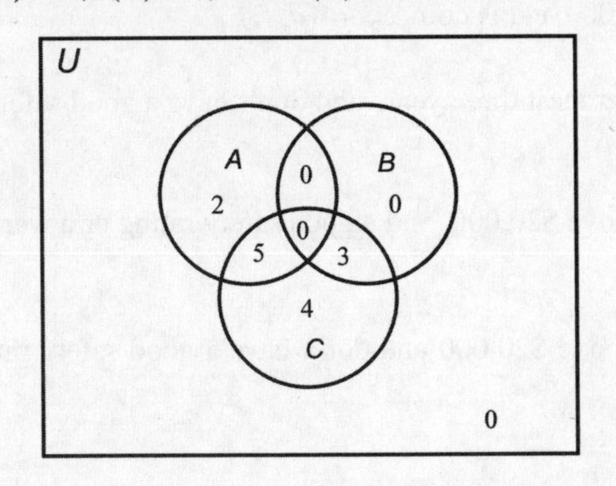

or

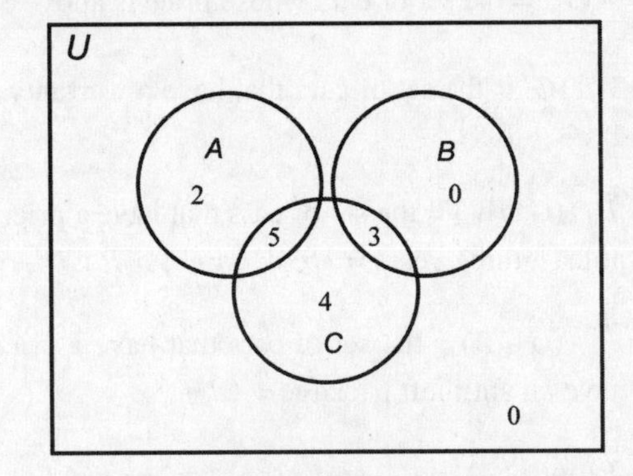

23. $n(A) = 18, n(B) = 15$, and $n(C) = 14$

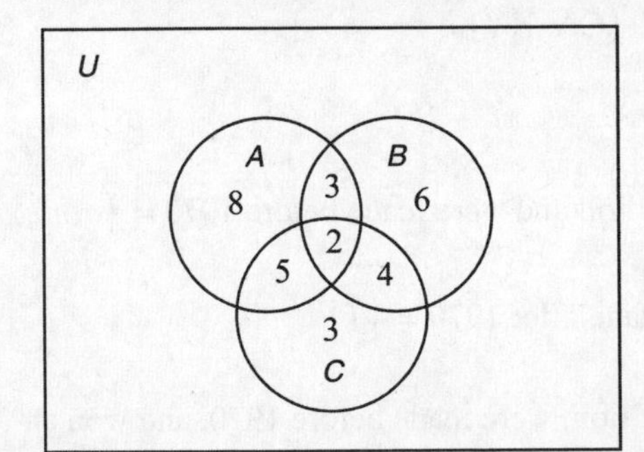

25. $n(A) = 5, n(B) = 14$, and $n(C) = 9$

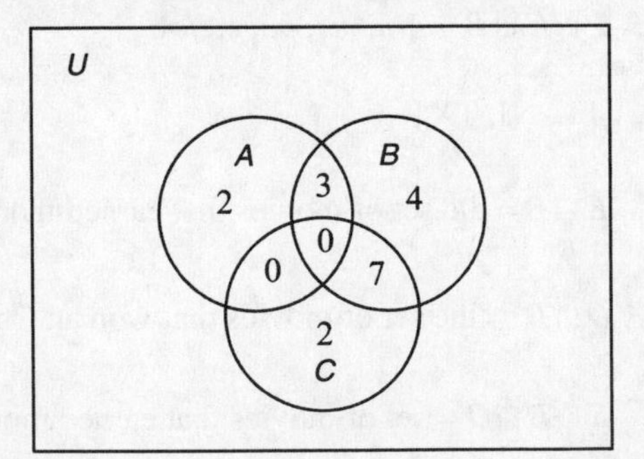

27. 59

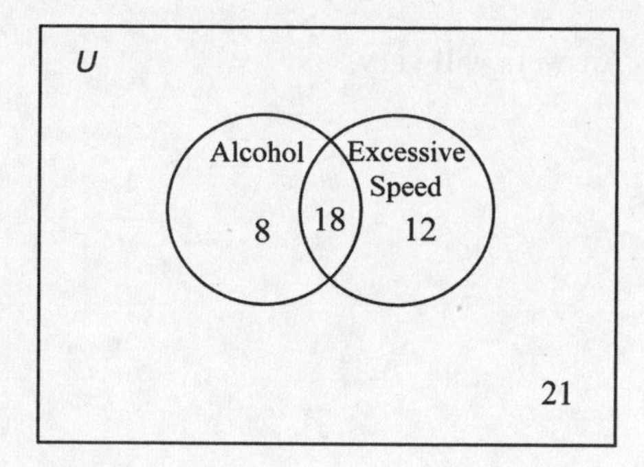

29. 28

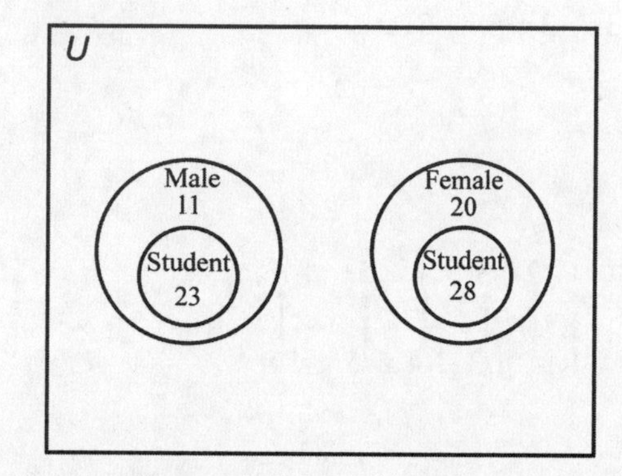

31. 30 attend the barbecue and 49 purchased a tour guide.

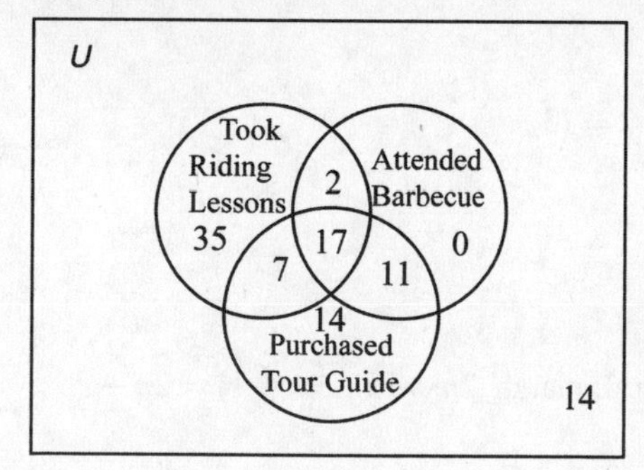

33.

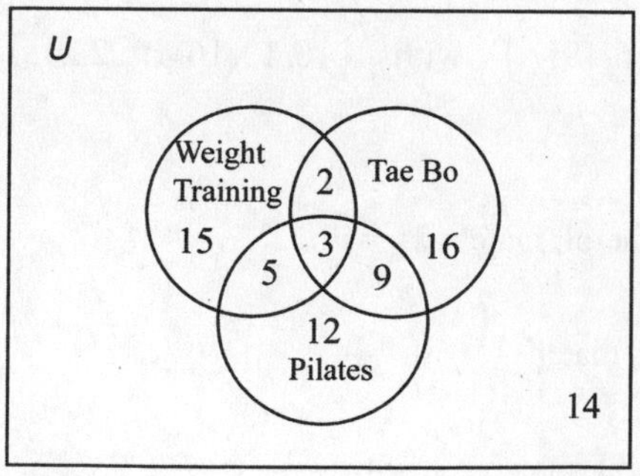

a) 76
b) 16
c) 21

35.

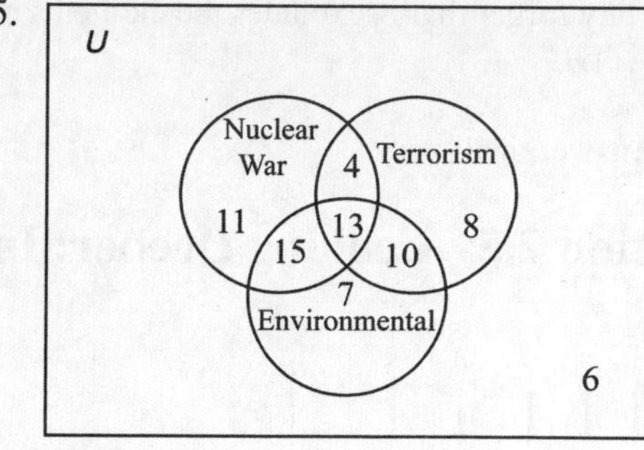

a) 74
b) 45
c) 8

37. Answers may vary.
If none use both the bus and the train, then $68+59=127$ use either the bus or train. If we add the 44 who use only the subway and the 83 that use none, then this total exceeds 200. Moreover, from the given information, we can set up the following Venn diagram:

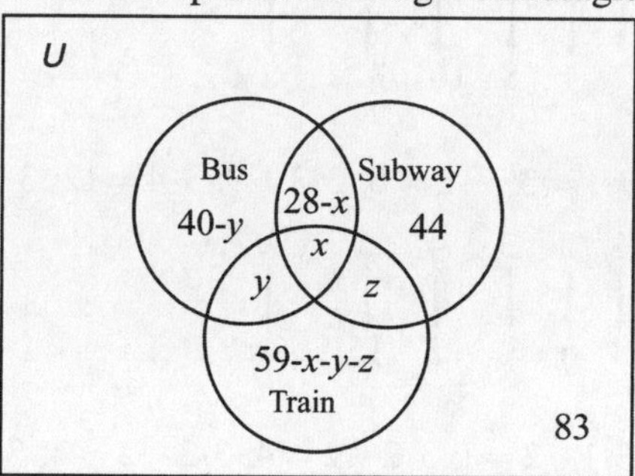

$(40-y)+(28-x)+44+x+y+z+$
$(59-x-y-z)+83=200$
$254-(x+y)=200 \Rightarrow -(x+y)=-54 \Rightarrow$
$x+y=54$

Thus, there are 54 people that use both the bus and the train, so the intersection of bus and train is not empty.

39. $n(A \cup P)=95+63=158$

41. $n(A-(M \cup S))=95-(44+31)$
$=95-75=20$

43. A^-

45. A^-, B^-, or AB^-,

47. $A \cap B \cap Rh$

49. B^-, O^-

51. They forget that r_2 excludes the elements in $A \cap B$.

53. Answers will vary.

55. 8

57. Answers will vary.

Section 2.5 Looking Deeper: Infinite Sets

1.

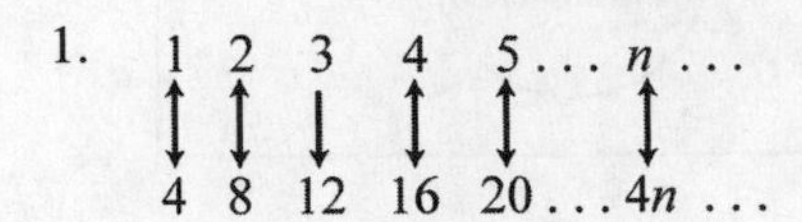

3. Each term is 3 more than the previous term. The general term is represented by 8 plus 3, added $n-1$ times. So the general term would be $8+(n-1)\cdot 3=8+3n-3=3n+5$.

1 2 3 4 5 . . . n . . .
↕ ↕ ↕ ↕ ↕ ↕
8 11 14 17 20 . . . $3n+5$ . . .

5. 1 2 3 4 5 . . . n . . .
↕ ↕ ↕ ↕ ↕ ↕
2 4 8 16 32 . . . 2^n . . .

7.

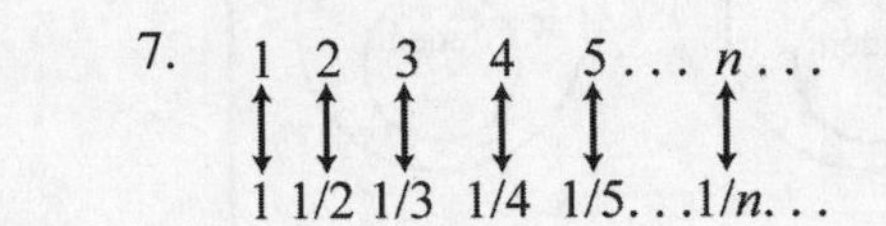

9. $3\cdot 1=3$; $3\cdot 2=6$; $3\cdot 3=9$; $3\cdot 4=12$; $3\cdot 5=15$

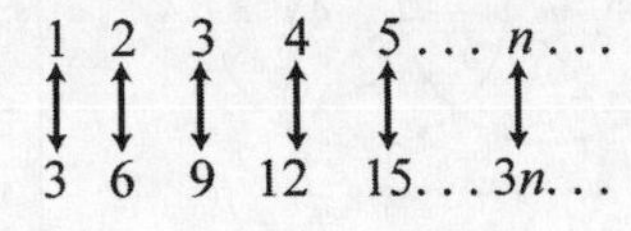

11. $3\cdot 1-2=1$; $3\cdot 2-2=4$; $3\cdot 3-2=7$; $3\cdot 4-2=10$; $3\cdot 5-2=13$

1 2 3 4 5 . . . n . . .
↕ ↕ ↕ ↕ ↕ ↕
1 4 7 10 13 . . . $3n-2$ . . .

13. Match $\{2, 4, 6, 8, 10, ...\}$ with $\{4, 6, 8, 10, 12, ...\}$; in general, match $2n$ with $2(n+1)=2n+2$.

15. Since each term of $\{7, 10, 13, 16, 19, ...\}$ is 3 more than the previous, the general term is $7+3(n-1)=7+3n-3=3n+4$. Match $\{7, 10, 13, 16, 19, ...\}$ with $\{10, 13, 16, 19, 22, ...\}$; in general, match $3n+4$ with $3(n+1)+4=3n+3+4=3n+7$.

17. Match $\{2, 4, 8, 16, 32, ...\}$ with $\{4, 8, 16, 32, 64, ...\}$; in general, match 2^n with 2^{n+1}.

19. Match $\{1, \frac{1}{2}, \frac{1}{3}, \frac{1}{4}, \frac{1}{5}, ...\}$ with $\{\frac{1}{2}, \frac{1}{3}, \frac{1}{4}, \frac{1}{5}, \frac{1}{6}, ...\}$; in general, match $\frac{1}{n}$ with $\frac{1}{n+1}$.

21. Match $\{\frac{1}{2}, \frac{1}{4}, \frac{1}{6}, \frac{1}{8}, \frac{1}{10}, ...\}$ with $\{\frac{1}{4}, \frac{1}{6}, \frac{1}{8}, \frac{1}{10}, \frac{1}{12}, ...\}$; in general, match $\frac{1}{2n}$ with $\frac{1}{2(n+1)}=\frac{1}{2n+2}$.

For Exercises 23 – 25 refer to the following diagram.

1/1	2/1	3/1	4/1	5/1	6/1	7/1	8/1	9/1	10/1	11/1 ...
1/2	2/2	3/2	4/2	5/2	6/2	7/2	8/2	9/2	10/2	11/2 ...
1/3	2/3	3/3	4/3	5/3	6/3	7/3	8/3	9/3	10/3	11/3 ...
1/4	2/4	3/4	4/4	5/4	6/4	7/4	8/4	9/4	10/4	11/4 ...
1/5	2/5	3/5	4/5	5/5	6/5	7/5	8/5	9/5	10/5	11/5 ..
1/6	2/6	3/6	4/6	5/6	6/6	7/6	8/6	9/6	10/6	11/6 ...
1/7	2/7	3/7	4/7	5/7	6/7	7/7	8/7	9/7	10/7	11/7 ...
1/8	2/8	3/8	4/8	5/8	6/8	7/8	8/8	9/8	10/8	11/8 ...
1/9	2/9	3/9	4/9	5/9	6/9	7/9	8/9	9/9	10/9	11/9 ...
1/10	2/10	3/10	4/10	5/10	6/10	7/10	8/10	9/10	10/10	11/10 ...
1/11	2/11	3/11	4/11	5/11	6/11	7/11	8/11	9/11	10/11	11/11 ...

23. 6; We skip 2/2, 2/4, 3/3, and 4/2.

25. 25; We skip 2/2, 2/4, 3/3, 4/2, 2/6, 4/4, 6/2, and 6/3.

27. It was shown that the set of positive rational numbers has a one-to-one correspondence with the natural numbers.

29. {1,2,3,4,5} does not have a one-to-one correspondence with any of its 31 proper subsets.

31. They had already been matched as $1, \frac{1}{2}, 1,$ and 2.

33. 6

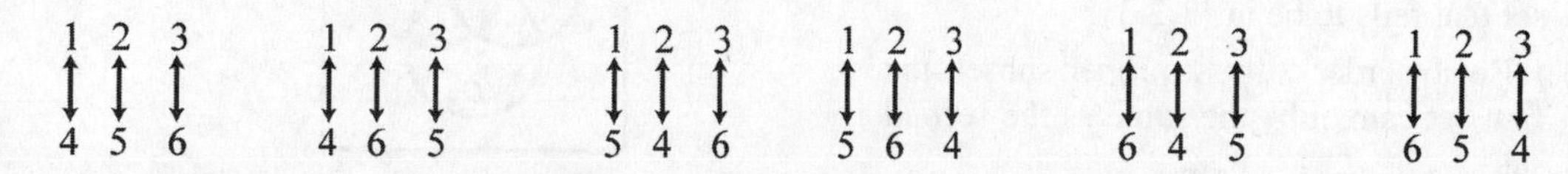

35. Answers may vary.
If we take the union of $\{1\}$, which has cardinal number 1, and $\{2, 3, 4, \ldots\}$, which has cardinal number $\aleph_0$, we get $\{1, 2, 3, 4, \ldots\}$, which has cardinal number $\aleph_0$.

37. From this figure we see that every point on the semi-circle matches with exactly one point on the line and vice versa.

Chapter Review Exercises

1. a) Answers may vary. Possible answers include: $\{x : x$ is an even natural number between 1 and 19$\}$.
   b) $\{x : x$ a month of the year$\}$
   c) {New Hampshire, New Jersey, New Mexico, New York}.
   d) $\varnothing$

2. $\varnothing$ is the empty set; it contains no elements. $\{\varnothing\}$ is not empty; it contains $\varnothing$.

3.

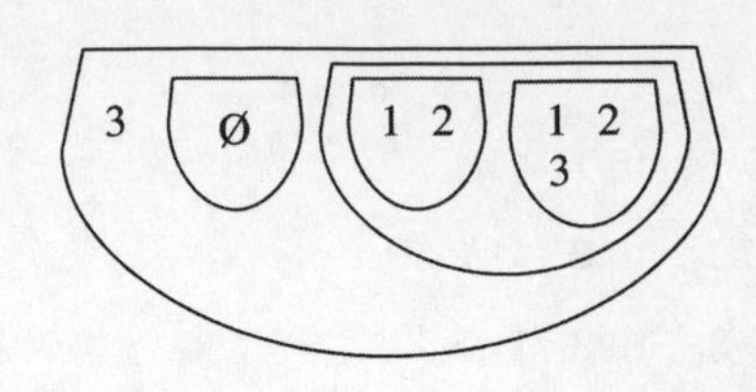

4. a) 7 b) 0 c) 4

5. a) Yes, these two sets are equal. They have the same elements arranged in a different order.

   b) Yes, these two sets are equal. Duplicate elements do not count as distinct (different) elements.

   c) No, these two sets are not equal. The first set has elements, such as 1002, that are not elements in the second set.

6. a) true; It is understood that all of the elements of the first set are also elements of the second set.

   b) true; It is understood that all of the elements of the first set are also elements of the second set.

   c) true; It is understood that all of the elements of the first set are also elements of the second set.

   d) true; The null set is a subset of all sets. We cannot find an element of the empty set that fails to be in $\{1,2,3\}$.

   e) false; In order to be a proper subset, the first set cannot be the same as the second set.

7. a) not equivalent; b) equivalent; c) equivalent; d) equivalent

8. a) There are $2^3 = 8$ subsets. They are $\varnothing$, $\{a\}$, $\{b\}$, $\{c\}$, $\{a,b\}$, $\{a,c\}$, $\{b,c\}$, and $\{a,b,c\}$.

   b) $2^7 = 128$

9. a) $A \cap B = \{5,7,9\}$

   b) $B \cup C = \{2,3,4,5,7,8,9,10\}$

   c) $C' = \{1,4,6,7,10\}$

   d) $A - C = \{7\}$

10. a) $(A \cup B)' = \{1,6\}$

   b) $(A - C) \cup (A - B) = \{2,7,8\}$

   c) $A' \cap (B' \cup C) = \{1,3,6\}$

11. a) $A \cup B$

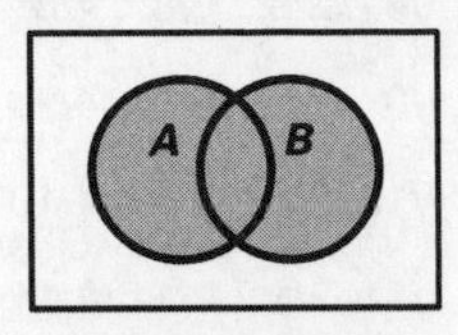

   b) $B \cap C$

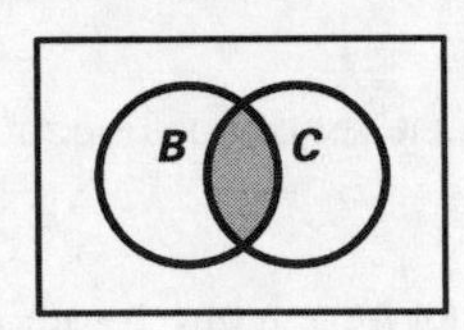

   c) $A' \cap B \cap C$

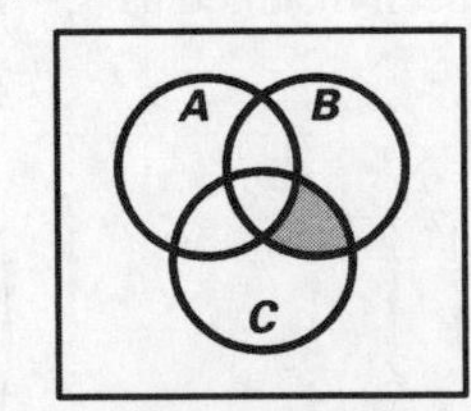

   d) $(B \cup C) - A$

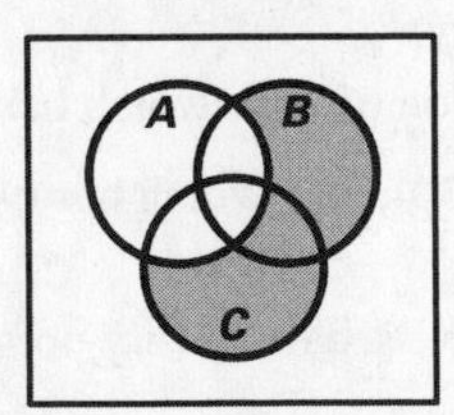

12. $A' \cap B'$

13. a) Any three of the following four are valid responses.
   1) Closure
   2) Commutativity; $A \cup B = B \cup A$
   3) Associativity; $(A \cup B) \cup C = A \cup (B \cup C)$
   4) Identity; $A \cap U = A$

13. (continued)

b) Any three of the following four are valid responses.

1) Closure
2) Commutativity; $A \cap B = B \cap A$
3) Associativity; $(A \cap B) \cap C = A \cap (B \cap C)$
4) Identity; $A \cup \varnothing = A$

c) Union distributes over intersection; $A \cup (B \cap C) = (A \cup B) \cap (A \cup C)$ and intersection distributes over union; $A \cap (B \cup C) = (A \cap B) \cup (A \cap C)$

14. $n(A \cup B) = n(A) + n(B) - n(A \cap B)$; We sometimes forget to subtract $n(A \cap B)$.

15.

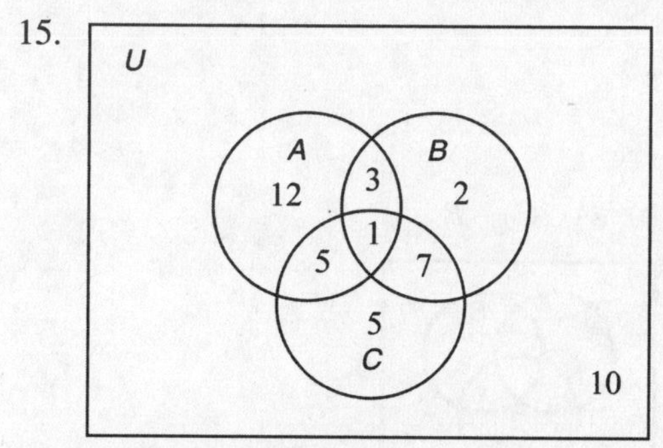

a) $C - B = 5 + 5 = 10$

b) $10 + 2 + 7 + 5 = 24$

16. 83 said academics. 58 said both academics and social life.

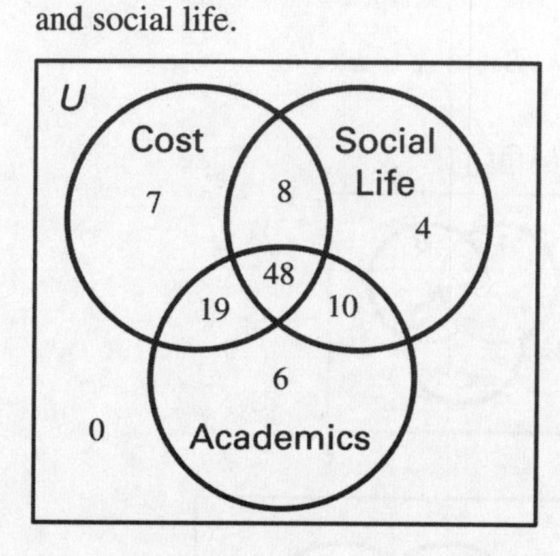

17. a) B^-

b) O^-

18. An infinite set can be put in a one-to-one correspondence with one of its proper subsets.

19. $\{1,2,3,4,\ldots\}$ can be put in a one-to-one correspondence with $\{2,4,6,8,\ldots\}$

20. $\frac{3}{2}$

21. We chose a digit that was different from the third decimal place of the third number in the list.

Chapter Test

1. a) $\{x : x \text{ is a natural number greater than } 100\}$
b) {January,February,March,…,December}
c) $\varnothing$

2. a) equal; order of elements does not matter.
b) not equal; the set $\{1\}$ is not the same as the number 1.
c) equal; the sets contains the same elements.

3. a) not equivalent
b) equivalent;
c) equivalent

4. a) $C' = \{1,3,5,7,9\}$

   b) $B - C = \{3,5\}$

   c) $(A \cap B)' = \{1,3,4,6,7,8,9,10\}$

   d) $(A' \cap B') \cup C = \{2,4,6,7,8,10\}$

5. $\varnothing$ contains no elements, but $\{\varnothing\}$ contains one element, namely $\varnothing$.

6. a) 8; b) 1

7. $2^8 = 256$

8.

9. a) True; every element of the first set is a member of the second.
   b) True; every element of the first set is a member of the second
   c) False; $\varnothing$ is not one of the numbers 1,2,3.

10. $(A \cap B)' = A' \cup B'$

11. a) $A \cap C$

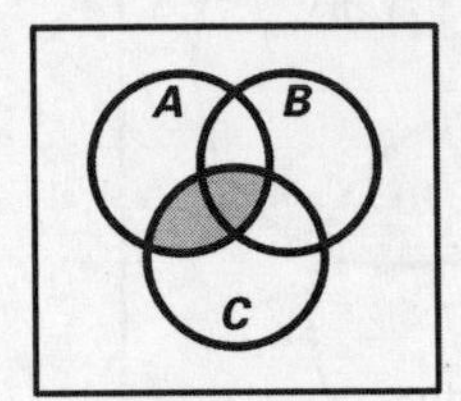

b) $(A \cup C) - B$

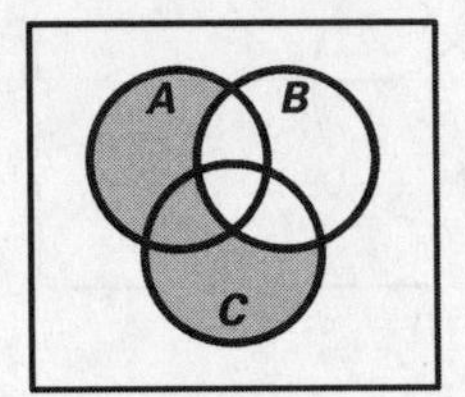

12.

a) 10
b) 64

13.

gas mileage: 89
both gas mileage and safety: 64

14. An infinite set can be put in a one-to-one correspondence with one of its proper subsets.

15. $\{1,2,3,4,\ldots\}$ can be put in a one-to-one correspondence with $\{2,4,6,8,\ldots\}$

16. $\frac{1}{4}$

17. We chose a digit that was different from the fifth decimal place of the fifth number in the list.

18. a) $A \cap B \cap Rh'$

    b) $A' \cap B' \cap Rh$

Chapter 3

Logic: The Study of What's True or False or Somewhere in Between

Section 3.1 Statements, Connectives, and Quantifiers

1. statement

3. not a statement

5. not a statement

7. statement

9. not a statement

11. compound; if … then, and, or

13. compound; or

15. simple

17. compound; if … then, or

19. compound; if … then

21. $g \vee \sim c$

23. $(\sim c) \rightarrow (g \vee a)$

25. The radial tires are included or the sunroof is not extra.

27. If the radial tires are included, then the sunroof is extra or the power windows are not optional.

29. There exists at least one snake that is not poisonous.

31. All personal items are covered by your renter's insurance.

33. All scientists believe that it is not true that an asteroid collision led to the extinction of the dinosaurs.

35. false; Nadia is not a commuter.

37. false; All sophomores are athletes.

39. false; Lennox is a freshman that is on a scholarship.

41. true

43. false; All green happy faces are not in yellow boxes. (No green happy faces are in yellow boxes)

45. – 53. Answers for these exercises have not been provided. These statements are complex and sophisticated. You will probably find that you, your classmates, and your instructor do not always agree as to which connectives are present. Nevertheless, it is interesting to try to determine the form of these statements.

55. $p \wedge (q \vee r)$

57. $(p \vee q) \wedge (p \vee r)$

59. true

61. false; There exists a real number that is not a rational number.

63. a) Some are not; Examples will vary.
    b) None are; Examples will vary.

65. – 71. Answers will vary.

Section 3.2 Truth Tables

1.

p	q	3 2 1 $\sim(p\vee\sim q)$

3.

p	q	4 3 2 1 $p\wedge\sim(p\vee\sim q)$

5. T

7. T

9. F

11. T

13. yes; $64 = 2^6$

15. no; 72 cannot be represented as a power of 2.

17.

p	q	2 1 $p\wedge(\sim q)$
T	T	T **F** F
T	F	T **T** T
F	T	F **F** F
F	F	F **F** T

19.

p	q	3 2 1 $\sim(p\vee\sim q)$
T	T	**F** T TF
T	F	**F** T TT
F	T	**T** F FF
F	F	**F** F TT

21.

p	q	2 1 5 4 3 $\sim(p\wedge q)\ \vee\sim(p\vee q)$
T	T	F TTT **F** F TTT
T	F	T TFF **T** F TTF
F	T	T FFT **T** F FTT
F	F	T FFF **T** T FFF

23.

p	q	r	1 4 3 2 $(p\wedge r)\ \vee\ (p\wedge\sim q)$
T	T	T	TTT **T** TFF
T	T	F	TFF **F** TFF
T	F	T	TTT **T** TTT
T	F	F	TFF **T** TTT
F	T	T	FFT **F** FFF
F	T	F	FFF **F** FFF
F	F	T	FFT **F** FFT
F	F	F	FFF **F** FFT

25.

p	q	r	3 2 1 4 $\sim(p\vee\sim q)\ \wedge\ r$
T	T	T	F TTF **F** T
T	T	F	F TTF **F** F
T	F	T	F TTT **F** T
T	F	F	F TTT **F** F
F	T	T	T FFF **T** T
F	T	F	T FFF **F** F
F	F	T	F FTT **F** T
F	F	F	F FTT **F** F

27. exclusive or

29. inclusive or

31. *p*: Bill is tall.
q: Bill is thin
It is not true that Bill is tall and thin.
$\sim(p\wedge q)\Leftrightarrow\sim p\vee\sim q$
Negation: Bill is not tall or Bill is not thin.

33. *p*: Christian will apply for a loan.
q: Christian will apply for work study.
It is not true that Christian will apply for either a loan or work study.
$\sim(p\vee q)\Leftrightarrow\sim p\wedge\sim q$
Negation: Christian will not apply for a loan and he will not apply for work study.

35. p: Ken qualifies for a rebate.
q: Ken qualifies for a reduced interest rate.
It is not true that Ken qualifies for a rebate or a reduced interest rate.
$\sim(p \vee q) \Leftrightarrow \sim p \wedge \sim q$
Negation: Ken does not qualify for a rebate and he does not qualify for a reduced interest rate.

37. p: The number x is equal to five.
q: The number s is odd.
It is not true that the number x is not equal to five and s is not odd.
$\sim(\sim p \wedge \sim q) \Leftrightarrow p \vee q$
Negation: The number x is equal to five or s is odd.

39. Yes; $\sim(p \wedge \sim q)$ is logically equivalent to $\sim p \vee q$ by DeMorgan's laws.

41. No; $\sim(p \vee \sim q) \wedge \sim(p \vee q)$ is not logically equivalent to $p \vee (p \wedge q)$.

		3 2 1 6 5 4
p	q	$\sim(p \vee \sim q) \wedge \sim(p \vee q)$
T	T	F T T F **F** F TTT
T	F	F T T T **F** F TTF
F	T	T F F F **F** F FTT
F	F	F F T T **F** T FFF

		2 1
p	q	$p \vee (p \wedge q)$
T	T	T **T** TTT
T	F	T **T** TFF
F	T	F **F** FFT
F	F	F **F** FFF

43. Yes; $\sim(p \vee \sim q) \wedge \sim(p \vee q)$ is logically equivalent to $(\sim p \wedge q) \wedge (\sim p \wedge \sim q)$ by DeMorgan's laws.

45. Yes; $p \vee (\sim q \wedge r)$ is logically equivalent to $(p \vee (\sim q)) \wedge (p \vee r)$.

			3 1 2
p	q	r	$p \vee (\sim q \wedge r)$
T	T	T	T **T** FFT
T	T	F	T **T** FFF
T	F	T	T **T** TTT
T	F	F	T **T** TFF
F	T	T	F **F** FFT
F	T	F	F **F** FFF
F	F	T	F **T** TTT
F	F	F	F **F** TFF

			2 1 4 3
p	q	r	$(p \vee (\sim q)) \wedge (p \vee r)$
T	T	T	T T F **T** TTT
T	T	F	T T F **T** TTF
T	F	T	T T T **T** TTT
T	F	F	T T T **T** TTF
F	T	T	F F F **F** FTT
F	T	F	F F F **F** FFF
F	F	T	F T T **T** FTT
F	F	F	F T T **F** FFF

47. true

p	q	$\sim p$	$\vee$	$\sim q$
T	F	F	**T**	T

49. false

p	q	r	$(\sim p \vee q)$	$\wedge$	r
T	F	F	FFF	**F**	F

51. true

p	q	r	p	$\vee$	$(\sim q \wedge r)$
T	F	F	T	**T**	TFF

53. p: The earned income did reduce the tax you owe.
q: The earned income gave you a refund.
$\sim p \lor \sim q \Leftrightarrow \sim (p \land q)$
It is not true that the earned income did reduce the tax you owe and gave you a refund.

55. p: You are single.
q: You are the head of a household.
$\sim p \land \sim q \Leftrightarrow \sim (p \lor q)$
It is not true that: you are single or the head of a household.

57. no

59. yes

61. false

63. false

65. $(p \lor q) \lor (p \lor r)$ is logically equivalent to $p \lor (q \lor r)$.

			1	3	2
p	q	r	$(p \lor q)$	$\lor$	$(p \lor r)$
T	T	T	TTT	**T**	TTT
T	T	F	TTT	**T**	TTF
T	F	T	TTF	**T**	TTT
T	F	F	TTF	**T**	TTF
F	T	T	FTT	**T**	FTT
F	T	F	FTT	**T**	FFF
F	F	T	FFF	**T**	FTT
F	F	F	FFF	**F**	FFF

				2	1
p	q	r	p	$\lor$	$(q \lor r)$
T	T	T	T	**T**	TTT
T	T	F	T	**T**	TTF
T	F	T	T	**T**	FTT
T	F	F	T	**T**	FFF
F	T	T	F	**T**	TTT
F	T	F	F	**T**	TTF
F	F	T	F	**T**	FTT
F	F	F	F	**F**	FFF

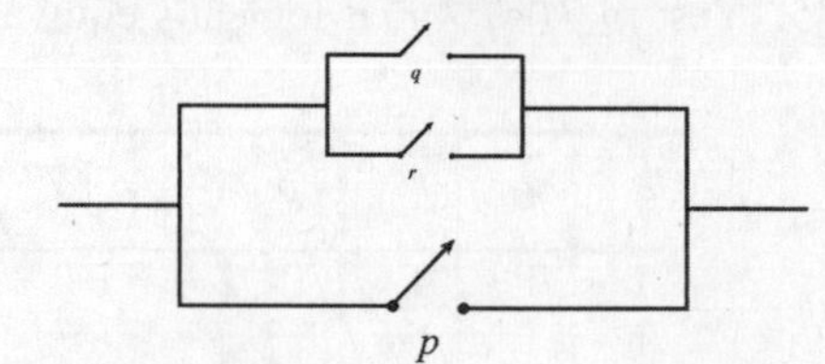

67. $(p \land q) \lor \sim q$ is logically equivalent to $p \lor \sim q$.

		1	3	2
p	q	$(p \land q)$	$\lor$	$\sim q$
T	T	TTT	**T**	F
T	F	TFF	**T**	T
F	T	FFT	**F**	F
F	F	FFF	**T**	T

			2	1
p	q	p	$\lor$	$\sim q$
T	T	T	**T**	F
T	F	T	**T**	T
F	T	F	**F**	F
F	F	F	**T**	T

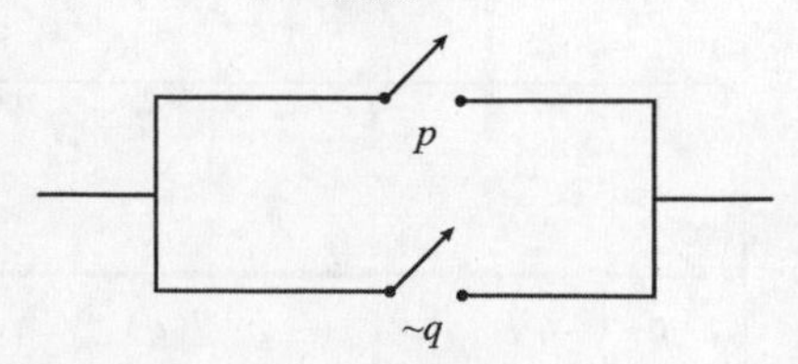

69. We use the inclusive *or*. When filling in the tables in the standard fashion, the first line of the inclusive *or* table would have a *true* whereas the first line of the exclusive *or* table would have a *false*.

71. A "T" would imply that the element is in the subset, and an "F" would imply that it is not an element of the subset.

p	q	r	Set of Variables That Have True Values
T	T	T	$\{p, q, r\}$
T	T	F	$\{p, q\}$
T	F	T	$\{p, r\}$
T	F	F	$\{p\}$
F	T	T	$\{q, r\}$
F	T	F	$\{q\}$
F	F	T	$\{r\}$
F	F	F	$\varnothing$

73. $p \mid q$ is logically equivalent to $\sim(p \wedge q)$.

		2	1	
p	q	$\sim$	$(p \wedge q)$	$p \mid q$
T	T	**F**	T T T	**F**
T	F	**T**	T F F	**T**
F	T	**T**	F F T	**T**
F	F	**T**	F F F	**T**

75.

		1	3	2	
p	q	$(p \mid p)$	$\mid$	$(q \mid q)$	$p \vee q$
T	T	F	**T**	F	**T**
T	F	F	**T**	T	**T**
F	T	T	**T**	F	**T**
F	F	T	**F**	T	**F**

Section 3.3 The Conditional and Biconditional

1. true

		2	1	4	3
p	q	$\sim$	$(p \vee q)$	$\rightarrow$	$\sim p$
T	F	F	TTF	**T**	F

3. true

			1	3	2
p	q	r	$(p \wedge q)$	$\rightarrow$	$(q \vee r)$
T	F	T	TFF	**T**	FTT

5. true

			1 3 2	4	
p	q	r	$(\sim p \vee \sim q)$	$\rightarrow$	r
T	F	T	F T T	**T**	T

7. false

			3	1 2	5	4
p	q	r	$\sim$	$(\sim p \wedge q)$	$\rightarrow$	$\sim r$
T	F	T	T	F FF	**F**	F

9.

			2	1
p	q	p	$\rightarrow$	$\sim q$
T	T	T	**F**	F
T	F	T	**T**	T
F	T	F	**T**	F
F	F	F	**T**	T

11.

		2	1
p	q	$\sim$	$(p \rightarrow q)$
T	T	**F**	TTT
T	F	**T**	TFF
F	T	**F**	FTT
F	F	**F**	FTF

13.

			1	4	3 2
p	q	r	$(p \vee r)$	$\rightarrow$	$(p \wedge \sim q)$
T	T	T	TTT	**F**	TFF
T	T	F	TTF	**F**	TFF
T	F	T	TTT	**T**	TTT
T	F	F	TTF	**T**	TTT
F	T	T	FTT	**F**	FFF
F	T	F	FFF	**T**	FFF
F	F	T	FTT	**F**	FFT
F	F	F	FFF	**T**	FFT

15.

p	q	r	$\sim (p \lor r) \rightarrow \sim (p \land q)$		
			2 1	5	4 3
T	T	T	F TTT	**T**	F TTT
T	T	F	F TTF	**T**	F TTT
T	F	T	F TTT	**T**	T TFF
T	F	F	F TTF	**T**	T TFF
F	T	T	F FTT	**T**	T FFT
F	T	F	T FFF	**T**	T FFT
F	F	T	F FTT	**T**	T FFF
F	F	F	T FFF	**T**	T FFF

17.

p	q	r	$(p \lor q) \leftrightarrow (p \lor r)$		
			1	3	2
T	T	T	TTT	**T**	TTT
T	T	F	TTT	**T**	TTF
T	F	T	TTF	**T**	TTT
T	F	F	TTF	**T**	TTF
F	T	T	FTT	**T**	FTT
F	T	F	FTT	**F**	FFF
F	F	T	FFF	**F**	FTT
F	F	F	FFF	**T**	FFF

19.

p	q	$(\sim p \rightarrow q) \leftrightarrow (\sim q \rightarrow p)$		
		1 2	5	3 4
T	T	FTT	**T**	FTT
T	F	FTF	**T**	TTT
F	T	TTT	**T**	FTF
F	F	TFF	**T**	TFF

21. If it pours, then it rains.

23. If you do not buy the all-weather radial tires, then they will not last for 80,000 miles.

25. If the sides of a geometric figure are not all equal in length, then it is not an equilateral triangle.

27. If x does not evenly divide 6, then x does not evenly divide 9.

29. contrapositive

31. converse

33. $q \rightarrow \sim p$

35. $q \rightarrow \sim p$

37. p is true, so $p \land q$ must be false. Since p is true, q must be false.

39. $\sim p$ is false, so p is true. Since $p \land \sim q$ is true while p is true, $\sim q$ must be true, so q is false.

41. The second statement is the contrapositive of the first. They are logically equivalent.

43. The second statement is the converse of the first. They are not logically equivalent.

45. If I finish my workout, then I'll take a break.

47. If you qualify for this deduction, then you complete Form 3093.

49. If you receive a free cell phone, then you sign up before March 1.

51. If you remain accident free for three years, then you get a reduction on your auto insurance.

53. true; $F \rightarrow T$

55. true; $F \rightarrow F$

57. true; $F \rightarrow F$

59. false; $T \rightarrow F$

61. true

w	f	w	$\rightarrow$	f
T	T	T	**T**	T

63. false

s	w	i	s	$\rightarrow$	$(w \wedge i)$
T	T	F	T	**F**	TFF

65. If you can be claimed by someone else as a dependent, then your gross income is not more than $2,250.

67. If you decrease the amount being withheld from your pay, then the amount you overpaid is large.

69. $(p \leftrightarrow q) \Leftrightarrow (p \rightarrow q) \wedge (q \rightarrow p)$

p	q	p	$\leftrightarrow$	q	$(p \rightarrow q)$ 1	$\wedge$ 3	$(q \rightarrow p)$ 2
T	T	T	**T**	T	TTT	**T**	TTT
T	F	T	**F**	F	TFF	**F**	FTT
F	T	F	**F**	T	FTT	**F**	TFF
F	F	F	**T**	F	FTF	**T**	FTF

71. $F \rightarrow T$; Inverse is $\sim F \rightarrow \sim T$ or $T \rightarrow F$.

73. $(p \rightarrow q) \Leftrightarrow (\sim p \vee q)$

p	q	$(p \rightarrow q)$ 1	$\leftrightarrow$ 4	$(\sim p \vee q)$ 3 2
T	T	TTT	**T**	FTT
T	F	TFF	**T**	FFF
F	T	FTT	**T**	TTT
F	F	FTF	**T**	TTF

75. Jamie has not been a member for at least ten years.

77. Answers will vary.

79. Using the contrapositive and DeMorgan's laws, $r \rightarrow (p \wedge q)$.

81. $p \vee (r \wedge (p \wedge q))$ is logically equivalent to p.

p	q	r	p	$\vee$ 3	$(r \wedge (p \wedge q))$ 2 1
T	T	T	T	**T**	TTTTT
T	T	F	T	**T**	FFTTT
T	F	T	T	**T**	TFTFF
T	F	F	T	**T**	FFTFF
F	T	T	F	**F**	TFFFT
F	T	F	F	**F**	FFFFT
F	F	T	F	**F**	TFFFF
F	F	F	F	**F**	FFFFF

83. $(p \rightarrow q) \Leftrightarrow (\sim p \vee q)$

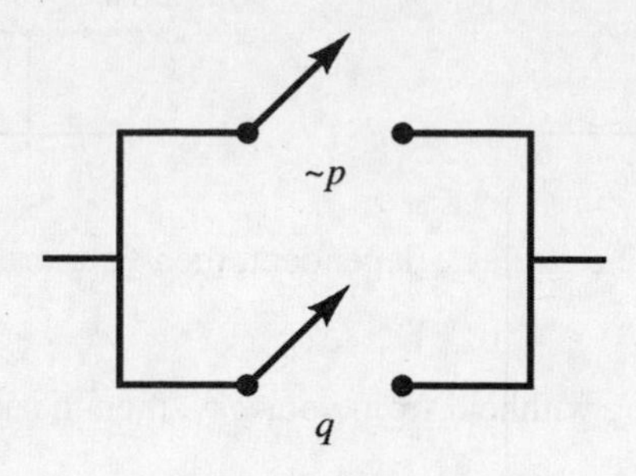

85. $(\sim p) \rightarrow \sim (q \wedge r) \Leftrightarrow p \vee \sim (q \wedge r) \Leftrightarrow p \vee (\sim q \vee \sim r)$

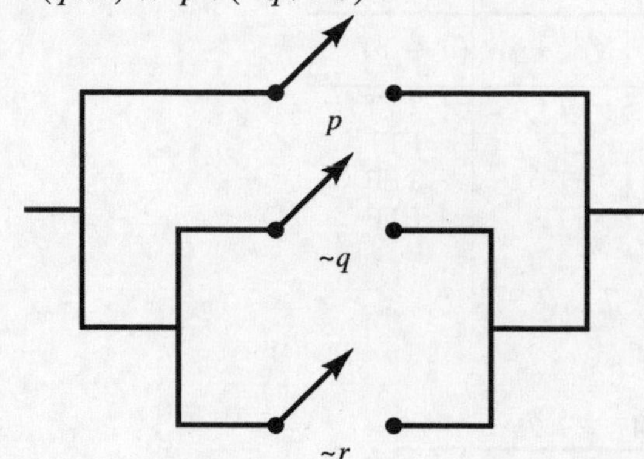

Section 3.4 Verifying Arguments

1. p: The car has air bags.
   q: The car is safe.
   $p \rightarrow q$
   p
   – – – – –
   $\therefore q$
   valid argument
   Law of Detachment

3. p: The movie is exciting.
   q: The movie will gross a lot of money.
   $p \rightarrow q$
   q
   – – – – –
   $\therefore p$
   invalid argument
   Fallacy of the Converse

5. p: The laptop has the enhanced video card.
   q: The computer has an optical disc drive.
   $p \vee q$
   $\sim p$
   – – – – –
   $\therefore q$
   valid argument
   Law of Disjunctive Syllogism

7. p: You pay your tuition late.
   q: You will pay a late payment fee.
   $p \rightarrow q$
   $\sim p$
   – – – – –
   $\therefore \sim q$
   invalid argument
   Fallacy of the Inverse

9. *p:* You watch *The Apprentice*.
   q: You will succeed in business.
   r: A skyscraper will be named after you.
   $p \to q$
   $q \to r$
   — — — — —
   $\therefore p \to r$
   valid argument
   Law of Syllogism

11. *p:* You buy the sports package.
    q: You get the leather seats.
    $\sim p \to \sim q$
    q
    — — — — —
    $\therefore p$
    valid argument
    Law of Contraposition

13. *p:* Phillipe joins the basketball team.
    q: Phillipe will be able to work part time.
    $p \to \sim q$
    $\sim p$
    — — — — —
    $\therefore q$
    invalid argument
    Fallacy of the Inverse

15. *p:* January has 28 days.
    q: February has 31 days.
    r: September has 31 days.
    $p \to q$
    $q \to r$
    — — — — —
    $\therefore p \to r$
    valid argument
    Law of Syllogism

17. valid argument

		3 2 1 5 4
p	q	$[p \wedge (q \to \sim p)] \to \sim q$
T	T	T F T F F **T** F
T	F	T T F T F **T** T
F	T	F F T T T **T** F
F	F	F F F T T **T** T

19. invalid argument

		1 3 2 6 4 5
q	r	$[\sim r \wedge (r \to q)] \to (\sim q \wedge r)$
T	T	F F T T T **T** F F T
T	F	T T F T T **F** F F F
F	T	F F T F F **T** T T T
F	F	T T F T F **F** T F F

21. invalid argument

			1 2 5 4 3 8 6 7
p	q	r	$[(\sim q \to p) \wedge (r \to \sim q)] \to (\sim p \to r)$
T	T	T	F T T F T F F **T** F T T
T	T	F	F T T T F T F **T** F T F
T	F	T	T T T T T T T **T** F T T
T	F	F	T T T T F T T **T** F T F
F	T	T	F T F F T F F **T** T T T
F	T	F	F T F T F T F **F** T F F
F	F	T	T F F F T T T **T** T T T
F	F	F	T F F F F T T **T** T F F

23. invalid argument

3 2 1 7 4 6 5 8

p	q	r	$\{p \wedge (q \to \sim p) \wedge [\sim q \to (r \vee p)]\}$	$\to$	r
T	T	T	T F TF F F F T TTT	T	T
T	T	F	T F TF F F F T FTT	T	F
T	F	T	T T FT F T T T TTT	T	T
T	F	F	T T FT F T T T FTT	F	F
F	T	T	F F TT T F F T TTF	T	T
F	T	F	F F TT T F F T FFF	T	F
F	F	T	F F FT T F T T TTF	T	T
F	F	F	F F FT T F T F FFF	T	F

25. valid argument

3 2 1 5 4 6

p	q	r	$[r \wedge (r \to \sim q) \wedge (p \vee q)]$	$\to$	p
T	T	T	T F TFF F TTT	T	T
T	T	F	F F FTF F TTT	T	T
T	F	T	T T TTT T TTF	T	T
T	F	F	F F FTT F TTF	T	T
F	T	T	T F TFF F FTT	T	F
F	T	F	F F FTF F FTT	T	F
F	F	T	T T TTT F FFF	T	F
F	F	F	F F FTT F FTF	T	F

27. invalid argument

2 1 5 4 3 8 6 7

p	q	r	$[(q \to \sim p) \wedge (r \to \sim q)]$	$\to$	$(\sim p \to r)$
T	T	T	T F F F T F F	T	F T T
T	T	F	T F F F F T F	T	F T F
T	F	T	F T F T T T T	T	F T T
T	F	F	F T F T F T T	T	F T F
F	T	T	T T T F T F F	T	T T T
F	T	F	T T T T F T F	F	T F F
F	F	T	F T T T T T T	T	T T T
F	F	F	F T T T F T T	F	T F F

29. Law of Contraposition: Malik does not have the most expensive Dish TV package.

31. Law of Disjunctive Syllogism: Minxia will attend school in Hawaii.

33. invalid argument

    p: The product has a lower price.
    q: The product has quality.
    r: The product is less reliable.

			2 1 8 3 5 4 7 6 9	10	
p	q	r	$\{(p \to \sim q) \wedge [(\sim p \vee \sim q) \to \sim r] \wedge p\}$	$\to$	r
T	T	T	TFF F FFF T F FT	T	T
T	T	F	TFF F FFF T T FT	T	F
T	F	T	TTT F FTT F F FT	T	T
T	F	F	TTT T FTT T T TT	F	F
F	T	T	FTF F TTF F F FF	T	T
F	T	F	FTF T TTF T T FF	T	F
F	F	T	FTT F TTT F F FF	T	T
F	F	F	FTT T TTT T T FF	T	F

35. valid argument

    p: You buy from a reputable breeder.
    q: Your labradoodle requires shots.
    r: It will cost you extra.

			2 1 5 3 4 7 6	9	8
p	q	r	$\{(p \to \sim q) \wedge (\sim p \to r) \wedge \sim r\}$	$\to$	$\sim q$
T	T	T	TF F F FTT FF	T	F
T	T	F	TF F F FTF FT	T	F
T	F	T	TT T T FTT FF	T	T
T	F	F	TT T T FTF TT	T	T
F	T	T	FT F T TTT FF	T	F
F	T	F	FT F F TFF FT	T	F
F	F	T	FT T T TTT FF	T	T
F	F	F	FT T F TFF FT	T	T

37. valid argument

    p: Health care is improved.
    q: Quality of life will be high
    r: The incumbents will be reelected.

			1 3 2 4 6 9 8 11	10	
p	q	r	$\{(\sim p \to \sim q) \wedge [(p \wedge q) \to r] \wedge q\}$	$\to$	r
T	T	T	F T F T T T T T T T T	T	T
T	T	F	F T F F T T T F F F T	T	F
T	F	T	F T T F T F F T T F F	T	T
T	F	F	F T T T T F F T F F F	T	F
F	T	T	T F F F F F T T T F T	T	T
F	T	F	T F F F F F T T F F T	T	F
F	F	T	T T T T F F F T T F F	T	T
F	F	F	T T T T F F F T F F F	T	F

39. invalid argument
p: Jamie is fluent in Spanish.
q: Jamie will work in Madrid.
r: Jamie will visit Mexico.

2 1 6 3 5 4 8

p	q	r	$[p \wedge (p \rightarrow q) \wedge (\sim r \vee \sim q)] \rightarrow r$
T	T	T	T T TTT F FFF **T** T
T	T	F	T T TTT T TTF **F** F
T	F	T	T F TFF F FTT **T** T
T	F	F	T F TFF F TTT **T** F
F	T	T	F F FTT F FFF **T** T
F	T	F	F F FTT F TTF **T** F
F	F	T	F F FTF F FTT **T** T
F	F	F	F F FTF F TTT **T** F

41. With the hypotheses $p \rightarrow q$ and p, we "detach" the p from the first conditional to get the conclusion, q; "Disjunctive" reminds us of the word "disjunction," which contains an "or."

43. – 45. Answers will vary.

47. Voldemort cannot be a knight since his statement would be a lie. Since he is a knave, Dumbledore must be a knight, otherwise Voldemort would be telling the truth. So, Voldemort is a knave and Dumbledore is a knight.
This problem could also be solved using a method similar to truth tables.

The shaded region gives the only case that would work.

Voldemort	Dumbledore
Knight	Knight
Knight	Knave
Knave	Knight
Knave	Knave

49. They cannot both be knights since one of them would have to be lying. The cannot also both be knaves since one would be telling the truth. Since Rubeus' statement is therefore false, Rubeus is a knave and Bellatrix is a knight.
This problem could also be solved using a method similar to truth tables.

The shaded region gives the only case that would work.

Rebeus	Bellatrix
Knight	Knight
Knight	Knave
Knave	Knight
Knave	Knave

51. $a \wedge b$
$b \rightarrow c$
$d \rightarrow \sim c$
―――――
$\therefore \sim d$

(1) If we assume that $a \wedge b$ is true, then b must be true.
(2) By the law of detachment if b and $b \rightarrow c$ are true, then c is true.
(3) Because $d \rightarrow \sim c$ is equivalent to its contrapositive, we know $c \rightarrow \sim d$ is true.
(4) By the law of detachment again, with c true and $c \rightarrow \sim d$ true we conclude $\sim d$ is true.

53. – 55. Answers will vary.

Section 3.5 Using Euler Diagrams to Verify Syllogisms

1. valid

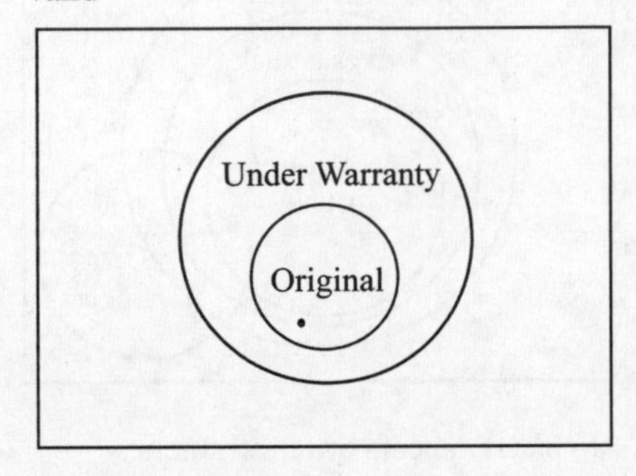

3. invalid

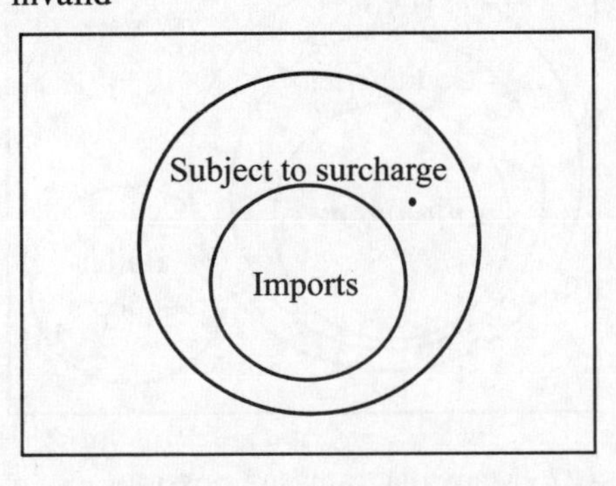

5. invalid

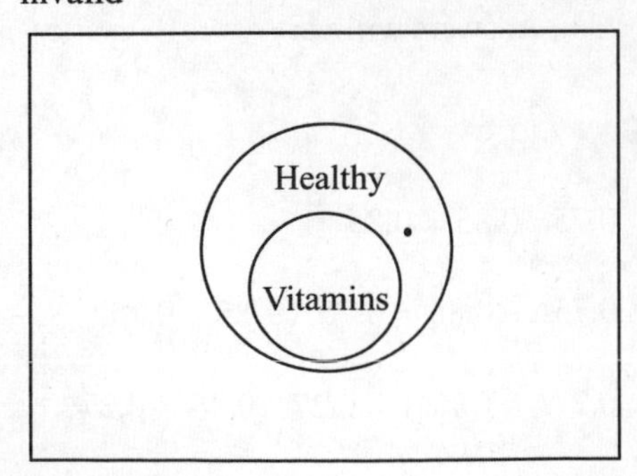

7. invalid

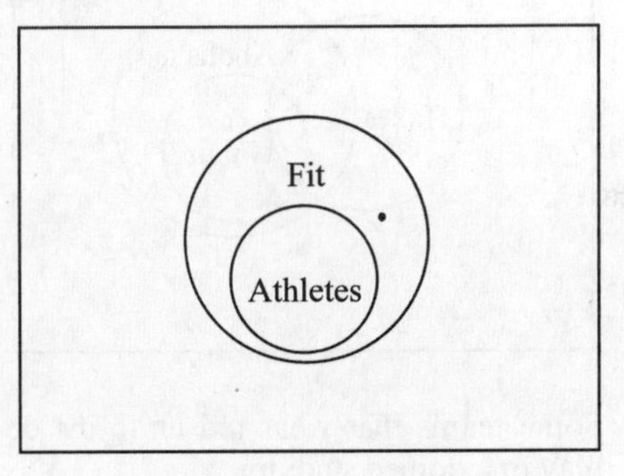

9. valid

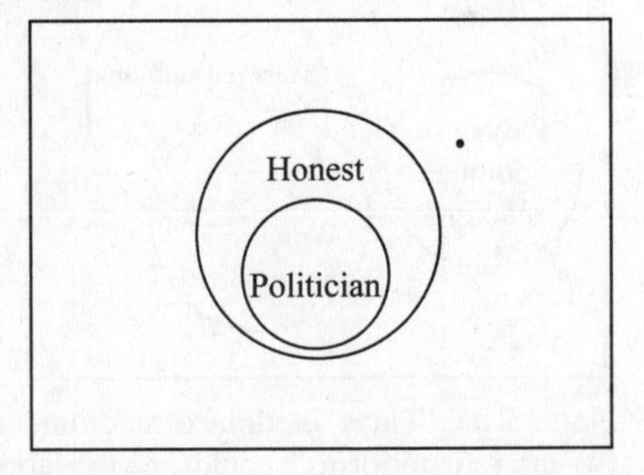

11. invalid

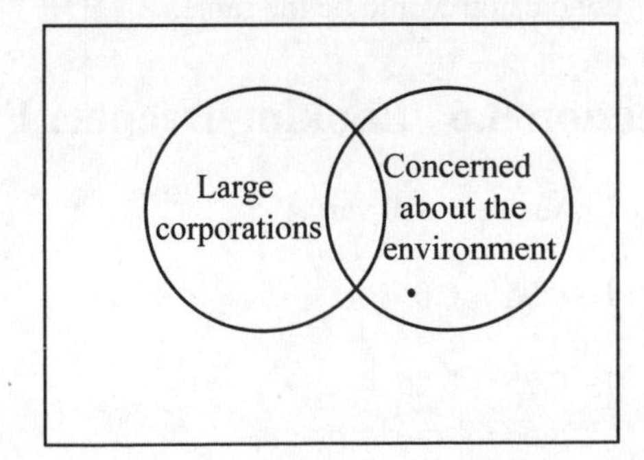

13. invalid

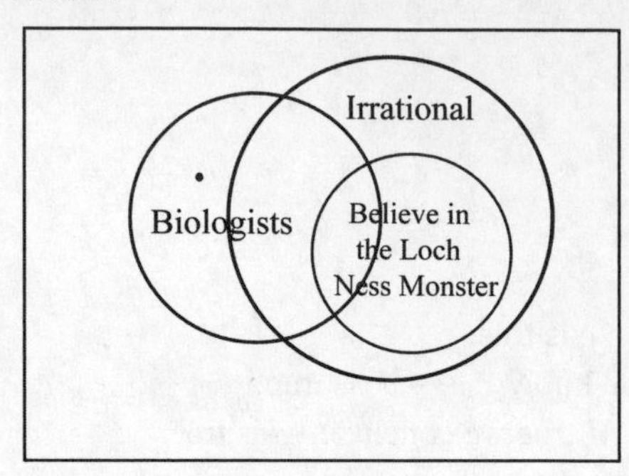

15. valid

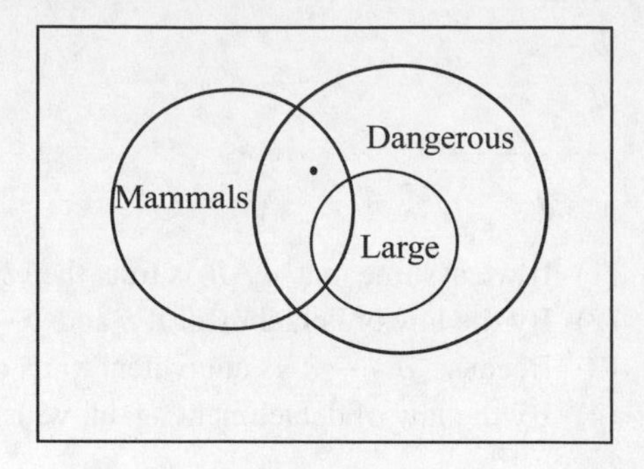

For Exercises 17 – 23, many valid answers are possible.

17. Some taxes should be abolished.

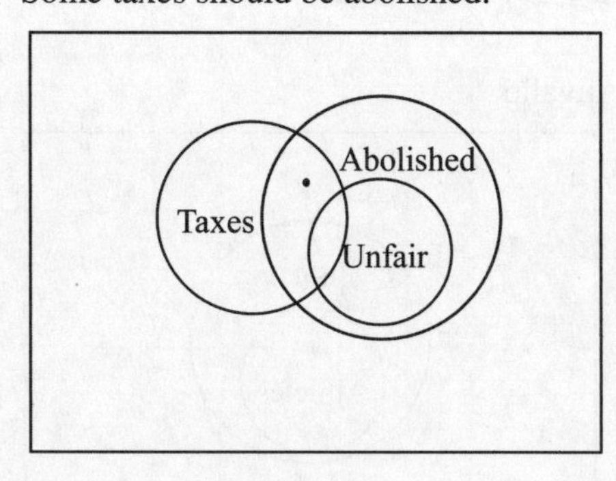

19. Some teams that wear red uniforms do not play in a domed stadium.

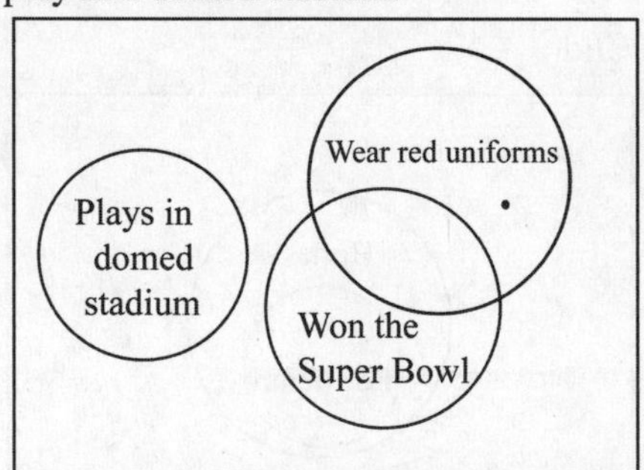

Note: The "Plays in domed stadium" and "Wear red uniforms" could be overlapping (like logic would suggest), but the conclusion would be the same.

21. Some opera singers have dogs.

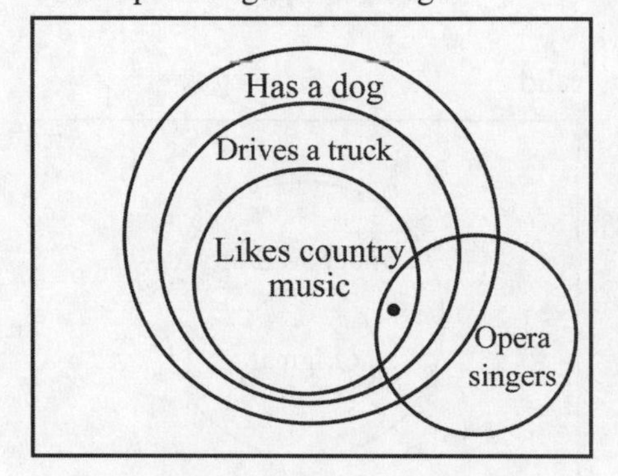

23. No ballet dancers are firefighters.

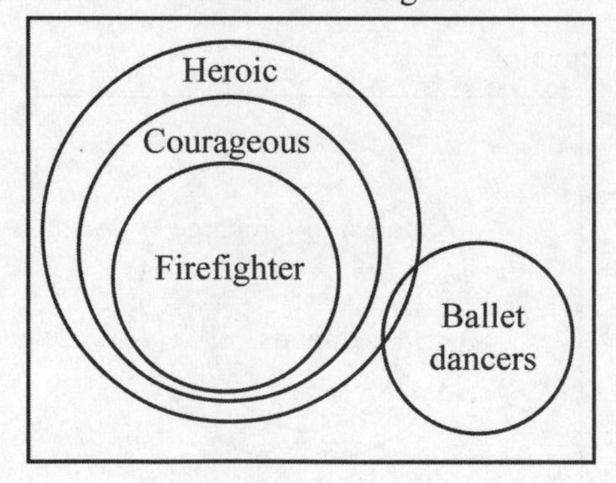

25. – 27. Many diagrams are possible.

29. – 33. Answers will vary.

Section 3.6 Looking Deeper: Fuzzy Logic

1. – 7. Answers will vary.

9. $1-0.95=0.05$

11. $1-0.75=0.25$

13. $0.75\vee 0.85=0.85$

15. $0.75\wedge\sim 0.85=0.75\wedge 0.15=0.15$

17. $0.27\wedge\sim(0.64)=0.27\wedge 0.36=0.27$

19. $\sim(0.27\vee\sim 0.64)=\ \sim(0.27\vee 0.36)=\ \sim 0.36=0.64$

21. $\sim (0.27 \vee \sim 0.64) \wedge \sim 0.71 = \sim (0.27 \vee 0.36) \wedge 0.29 = \sim 0.36 \wedge 0.29 = 0.64 \wedge 0.29 = 0.29$

23. $0.64 \rightarrow (0.27 \vee 0.71) = \sim 0.64 \vee (0.27 \vee 0.71) = 0.36 \vee 0.71 = 0.71$

25.

	i	$\sim i$	Retail Sales Trainee c	Retail Sales Trainee $\sim i \vee c$	Data Analyst c	Data Analyst $\sim i \vee c$
Salary	0.70	0.30	0.60	0.60	0.65	0.65
Interesting	0.80	0.20	0.50	0.50	0.80	0.80
Work with People	0.60	0.40	0.90	0.90	0.30	0.40
Flexible Hours	0.75	0.25	0.80	0.80	0.60	0.60
Conjunction of all $\sim i \vee c$				**0.50**		**0.40**

According to this rating, it is better for you to take the Marketing Trainee position.

27.

	i	$\sim i$	LSU c	LSU $\sim i \vee c$	GOS c	GOS $\sim i \vee c$	MFU c	MFU $\sim i \vee c$
Size (not too large)	0.60	0.40	0.95	0.95	0.60	0.60	0.40	0.40
Cost	0.80	0.20	0.40	0.40	0.80	0.80	0.30	0.30
Academics	0.75	0.25	0.80	0.80	0.70	0.70	0.65	0.65
Social Life	0.90	0.10	0.55	0.55	0.75	0.75	0.70	0.70
Close to Home	0.30	0.70	0.70	0.70	0.65	0.70	0.95	0.95
Conjunction of all $\sim i \vee c$				**0.40**		**0.60**		**0.30**

According to this rating, it is better for her to attend Good Old State (GOS).

29. If we only allowed the values 1 (for true) and 0 (for false), then the rules for computing truth values in fuzzy logic are exactly the same as the rules for computing truth tables.

31. No solution provided.

Chapter Review Exercises

1. a) not a statement
   b) statement
   c) not a statement

2. a) $\sim v \vee s$
   b) $\sim (v \wedge \sim s)$

3. a) It is not true that: Antonio is fluent in Spanish and he has not lived in Spain for a semester.
   b) Antonio is not fluent in Spanish or he has not lived in Spain for a semester.

4. a) It is not true that all writers are passionate.
   There is at least one writer who is not passionate.
   b) It is not true that some graduates received several job offers.
   No graduates received several job offers. –or – All graduates do not receive several job offers.

5. a) true

		2 1
p	q	$p \wedge (\sim q)$
T	F	T **T** T

b) false

			2 1
p	q	r	$r \vee (\sim p \wedge q)$
T	F	F	F **F** F F F

c) false

			3 3 4 1
p	q	r	$\sim(p \vee q) \wedge \sim r$
T	F	F	F TTF **F** T

6. a) $2^3 = 8$ b) $2^5 = 32$

7. a)

		3 2 1
p	q	$\sim(p \vee \sim q)$
T	T	**F** T T F
T	F	**F** T T T
F	T	**T** F F F
F	F	**F** F T T

b)

			3 2 1 5 4
p	q	r	$\sim(p \vee \sim q) \wedge \sim r$
T	T	T	F TTF **F** F
T	T	F	F TTF **F** T
T	F	T	F TTT **F** F
T	F	F	F TTT **F** T
F	T	T	T FFF **F** F
F	T	F	T FFF **T** T
F	F	T	F FTT **F** F
F	F	F	F FTT **F** T

8. a) p: I will take Pilates.
q: I will take Zumba.
$\sim(p \vee q) \Leftrightarrow \sim p \wedge \sim q$
I will not take Pilates and I will not take Zumba.

b) p: I will sign the lease.
q: I will accept the housing agreement.
$\sim(\sim p \vee \sim q) \Leftrightarrow p \wedge q$
I will sign the lease and I will accept the housing agreement.

9. a) $\sim(p \wedge \sim q)$ is logically equivalent to $\sim p \vee q$ by DeMorgan's laws.

b) $\sim(p \vee \sim q) \wedge \sim(p \vee q)$ is not logically equivalent to $p \vee (p \wedge q)$.

		3 2 1 6 5 4
p	q	$\sim(p \vee \sim q) \wedge \sim(p \vee q)$
T	T	F T T F **F** F TTT
T	F	F T T T **F** F TTF
F	T	T F F F **F** F FTT
F	F	F F T T **F** T FFF

		2 1
p	q	$p \vee (p \wedge q)$
T	T	T **T** TTT
T	F	T **T** TFF
F	T	F **F** FFT
F	F	F **F** FFF

10. a) true

		2 1 4 3
p	q	$\sim(p \vee q) \rightarrow \sim p$
T	F	F TTF **T** F

b) false

			1 3 2
p	q	r	$(p \wedge q) \leftrightarrow (q \vee r)$
T	F	T	TFF **F** FTT

c) true

			1 3 2 4
p	q	r	$(\sim p \vee \sim q) \rightarrow r$
T	F	T	FTT **T** T

11. a)

		1 2
p	q	$\sim p \rightarrow q$
T	T	F **T** T
T	F	F **T** F
F	T	T **T** T
F	F	T **F** F

b)

			2 1 5 4 3
p	q	r	$\sim(p \wedge r) \leftrightarrow \sim(p \vee q)$
T	T	T	F TTT **T** F TTT
T	T	F	T TFF **F** F TTT
T	F	T	F TTT **T** F TTF
T	F	F	T TFF **F** F TTF
F	T	T	T FFT **F** F FTT
F	T	F	T FFF **F** F FTT
F	F	T	T FFT **T** T FFF
F	F	F	T FFF **T** T FFF

12. Converse: If I have the new Mocha, then I go to Starbucks.
Inverse: If I don't go to Starbucks, then I won't have the new Mocha.
Contrapositive: If I don't have the new Mocha, then I don't go to Starbucks.

13. a) If the Heat get to the finals, then they beat the Lakers.
b) If you are an astronaut, then you have a pilot's license.

14. a) Law of Detachment (valid argument)
p: You make lots of money.
q: You will be happy.

$p \rightarrow q$
$\sim p$
\-\-\-\-\-
$\therefore \sim q$

b) Law of Contraposition (valid argument)
p: Felicia enjoys spicy food.
q: Felicia enjoys Cajun chicken.

$p \rightarrow q$
$\sim q$
\-\-\-\-\-
$\therefore \sim p$

15. invalid argument

			1 3 2 6 4 5 7
p	q	r	$\{\sim p \wedge (q \rightarrow p) \wedge [(p \vee q) \rightarrow r]\} \rightarrow r$
T	T	T	F F TTT F TTT T T **T** T
T	T	F	F F TTT F TTT F F **T** F
T	F	T	F F FTT F TTF T T **T** T
T	F	F	F F FTT F TTF F F **T** F
F	T	T	T F TFF F FTT T T **T** T
F	T	F	T F TFF F FTT F F **T** F
F	F	T	T T FTF T FFF T T **T** T
F	F	F	T T FTF T FFF T F **F** F

16. valid argument

p: You pay more for your phone plan.
q: You have more calling minutes.
r: You call your mother more often.

p	q	r	$\{(p \to q)$ (1)	$\wedge$ (4)	$[(p \vee q)$ (2)	$\to r]$ (3)	$\wedge p\}$ (5)	(6)	$\to r$
T	T	T	TTT	T	TTT	TT	TT	T	T
T	T	F	TTT	F	TTT	FF	FT	T	F
T	F	T	TFF	F	TTF	TT	FT	T	T
T	F	F	TFF	F	TTF	FF	FT	T	F
F	T	T	FTT	T	FTT	TT	FF	T	T
F	T	F	FTT	F	FTT	FF	FF	T	F
F	F	T	FTF	T	FFF	TT	FF	T	T
F	F	F	FTF	T	FFF	TF	FF	T	F

17. invalid argument

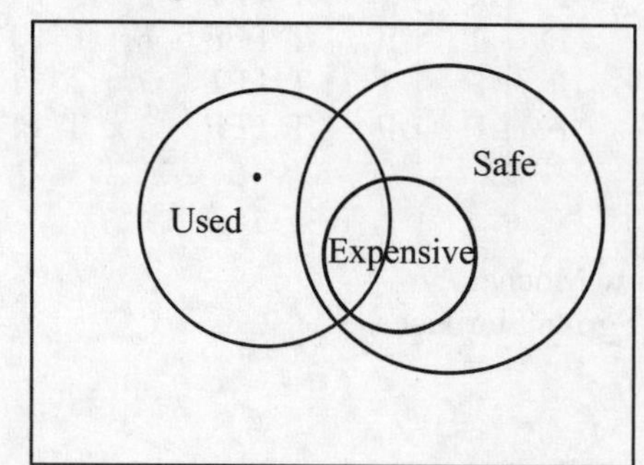

18. Valid

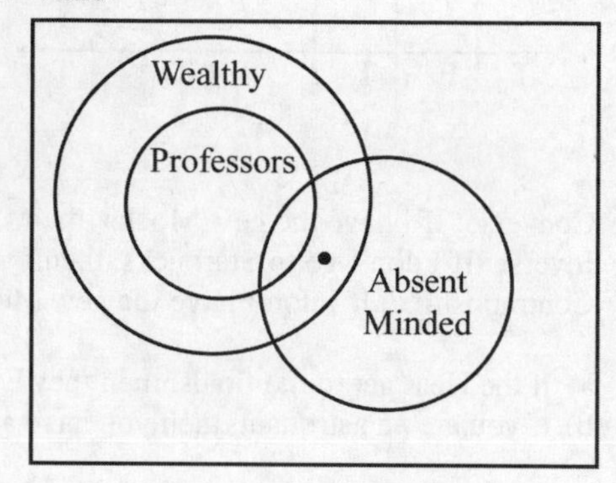

19. a) $0.47 \wedge \sim 0.82 = 0.47 \wedge 0.18 = 0.18$

b) $\sim(0.47 \vee \sim 0.82) = \sim(0.47 \vee 0.18) = \sim 0.47 = 0.53$

c) $0.47 \to \sim 0.82 = \sim 0.47 \vee \sim 0.82 = 0.53 \vee 0.18 = 0.53$

Chapter Test

1. a) statement b) not a statement

2. a) It is not true that all rock starts are fine musicians.
There is at least one rock star who is not a fine musician.
b) It is not true that some dogs are aggressive.
No dogs are aggressive. –or – All dogs are not aggressive.

3. a) $p \vee \sim f$
b) $\sim(\sim p \wedge f)$

4. a) It is not true that: the Tigers will win the series or Verlander will not win the Cy Young Award.
b) The Tigers will not win the series and Verlander will not win the Cy Young Award.

5. $2^6 = 64$

6. a) true

3 2 1

p	q	$\sim(p \vee \sim q)$
F	T	**T** FFF

b) false

3 2 4 1

p	q	r	$\sim(p \vee q) \wedge \sim r$
F	T	F	F FTT **F** T

c) false

4 1 3 2 5

p	q	r	$\sim(\sim r \vee \sim p) \wedge r$
F	T	F	F TTT **F** F

7. a) $\sim(0.65 \vee 0.75) = \sim .075 = 0.25$

b) $\sim 0.75 \vee \sim(0.65 \wedge 0.38) =$
$0.25 \vee \sim 0.38 = 0.25 \vee 0.62 = 0.62$

8. a) 3 2 1

p	q	$\sim(p \wedge \sim q)$
T	T	**T** TF F
T	F	**F** TT T
F	T	**T** FF F
F	F	**T** FF T

b) 1 2 3 5 4

p	q	r	$(\sim p \vee \sim q) \wedge r$
T	T	T	F F F **F** T
T	T	F	F F F **F** F
T	F	T	F T T **T** T
T	F	F	F T T **F** F
F	T	T	T T F **T** T
F	T	F	T T F **F** F
F	F	T	T T T **T** T
F	F	F	T T T **F** F

9. a) If you go to Wikipedia, then you will get enough sources for your term paper.
b) If Ticketmaster will mail the concert tickets, then you will pay a fee.

10. a) p: You can take the final.
q: You can write a term paper.
$\sim(p \vee q) \Leftrightarrow \sim p \wedge \sim q$
You cannot take the final exam and you cannot write a term paper.

b) p: I will finish the painting.
q: I will show it at the gallery.
$\sim(\sim p \vee \sim q) \Leftrightarrow p \wedge q$
I will finish the painting and I will show it at the gallery.

11. a) $\sim(p \vee \sim q)$ is logically equivalent to $\sim p \wedge q$ by DeMorgan's laws.
b) $(\sim p \vee \sim q) \wedge (\sim p \vee q)$ is not logically equivalent to $\sim p \wedge q$.

1 3 2 6 4 5

p	q	$(\sim p \vee \sim q) \wedge (\sim p \vee q)$
T	T	F F F **F** FTT
T	F	F T T **F** FFF
F	T	T T F **T** TTT
F	F	T T T **T** TTF

1 2

p	q	$\sim p \wedge q$
T	T	F **F** T
T	F	F **F** F
F	T	T **T** T
F	F	T **F** F

12. Converse: If it is gold, then it glitters.
Inverse: If it does not glitter, then it is not gold.
Contrapositive: If it is not gold, then it does not glitter.

13. a) true

		2 1 4 3
p	q	$(p \vee \sim q) \rightarrow \sim q$
T	F	T T T **T** T

b) true

			1 2 4 3
p	q	r	$(\sim p \wedge q) \rightarrow \sim r$
T	F	T	FFF **T** F

c) true

		2 1 5 4 3
p	q	$(p \vee \sim q) \leftrightarrow \ \sim (p \wedge q)$
T	F	T T T **T** T TFF

14. a)

		1 2 5 4 3
p	q	$(\sim p \vee q) \rightarrow \ \sim (p \wedge q)$
T	T	F TT **F** F TTT
T	F	F FF **T** T TFF
F	T	T TT **T** T FFT
F	F	T TF **T** T FFF

b)

			1 3 2 5 4
p	q	r	$(\sim p \vee \sim q) \leftrightarrow r$
T	T	T	F F F **F** T
T	T	F	F F F **T** F
T	F	T	F T T **T** T
T	F	F	F T T **F** F
F	T	T	T T F **T** T
F	T	F	T T F **F** F
F	F	T	T T T **T** T
F	F	F	T T T **F** F

15. valid argument

			4 1 3 2 8 5 7 6 9
p	q	r	$\{p \wedge [\sim q \rightarrow \sim r] \wedge [(q \vee r) \rightarrow \sim p]\} \rightarrow \sim r$
T	T	T	T T FTF F TTT F F **T** F
T	T	F	T T FTT F TTF F F **T** T
T	F	T	T F TFF F FTT F F **T** F
T	F	F	T T TTT T FFF T F **T** T
F	T	T	F F FTF F TTT T T **T** F
F	T	F	F F FTT F TTF T T **T** T
F	F	T	F F TFF F FTT T T **T** F
F	F	F	F F TTT F FFF T T **T** T

16. a) Fallacy of the Inverse (invalid argument)
p: It ain't broke.
q don't fix it.

$p \rightarrow q$
$\sim p$
_____
$\therefore \sim q$

b) Disjunctive Syllogism (valid argument)
p: I'll major in music.
q: I'll major in art history.

$p \vee q$
$\sim p$
_____
$\therefore q$

17. $\sim p \vee q$

18. invalid argument
    p: You go to eBay.
    q: You will find a bargain.
    r: You will waste your money.

				2		1		8	3	4		9	5	7	6	10	
p	q	r	$\{[p$	$\to$	$(q$	$\vee$	$r)]$	$\wedge$	$[\sim r$	$\to$	$q]$	$\wedge$	$[\sim p$	$\vee$	$\sim q]\}$	$\to$	r
T	T	T	T	T	T	T	T	T	F	T	T	F	F	F	F	**T**	T
T	T	F	T	T	T	T	F	T	T	T	T	F	F	F	F	**T**	F
T	F	T	T	F	F	T	T	F	F	T	F	F	F	T	T	**T**	T
T	F	F	T	F	F	F	F	F	T	F	F	F	F	T	T	**T**	F
F	T	T	F	T	T	T	T	T	F	T	T	T	T	T	F	**T**	T
F	T	F	F	T	T	T	F	T	T	T	T	T	T	T	F	**F**	F
F	F	T	F	T	F	T	T	T	F	T	F	T	T	T	T	**T**	T
F	F	F	F	T	F	F	F	F	T	F	F	F	T	T	T	**T**	F

19. invalid argument

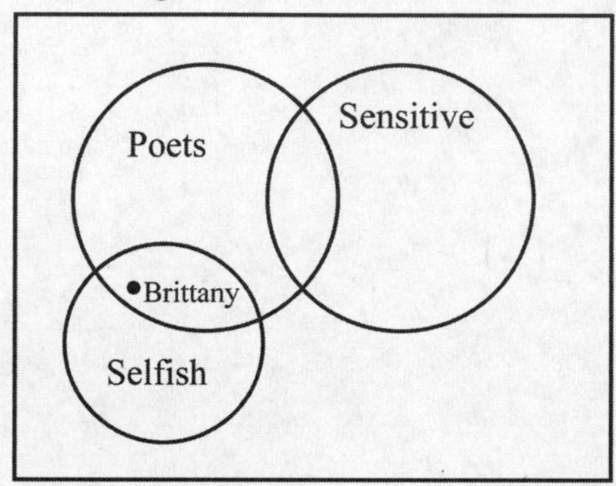

Chapter 4

Graph Theory (Networks): The Mathematics of Relationships

Section 4.1 Graphs, Puzzles and Map Coloring

1. Connected; A and B are odd vertices; C and D are even vertices.

3. Connected; A, B, E, and F are odd vertices; C and D are even vertices.

5. Not connected; All vertices are even.

7. Connected; A, B, E, and F are odd vertices; C and D are even vertices.

9. Yes, this graph can be traced.

11. No, this graph cannot be traced because it has four odd vertices.

13. No, this graph cannot be traced because it is not connected.

15. No, this graph cannot be traced because it has four odd vertices.

17. This is a Eulerian graph because all vertices are even. One possible Euler circuit is: ABFHI GCBEF GECDA

19. This is not an Eulerian graph. Duplicate edges AJ, IJ, DE, and EF.

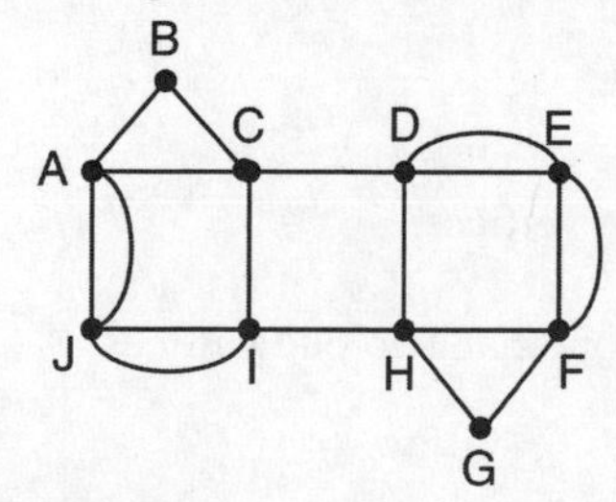

A possible path is
ABCDEFEDHFGHICAJIJA.

21. Graphs may vary.

23. not possible

25. Graphs may vary.

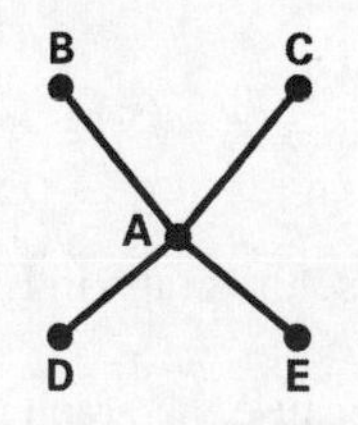

27. Graphs may vary.

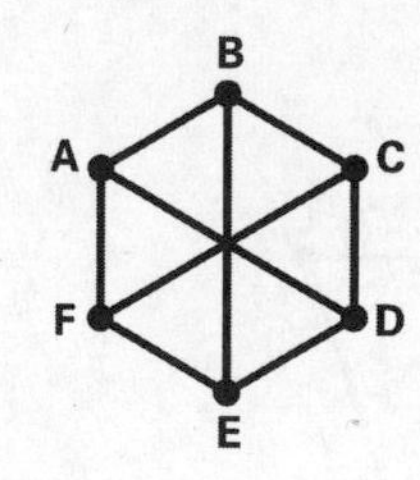

29. Remove *AD*.

31. Remove *AB*.

33. No. If you consider the intersection of two streets as a vertex, then there are four odd vertices. The graph can therefore not be traced, and the taxi driver must travel over parts of the route more than once.

35. If you label the intersections that represent odd vertices as follows you can duplicate edges AB, DE, CF, and GH. Follow the possible path indicated in numerical order.

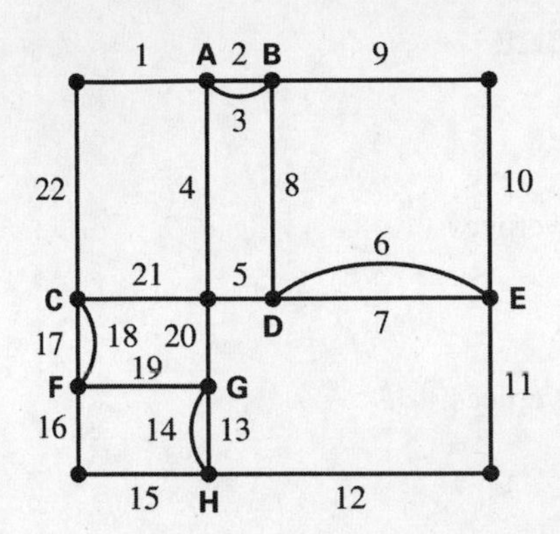

37.

WA MT

OR ID WY

39.

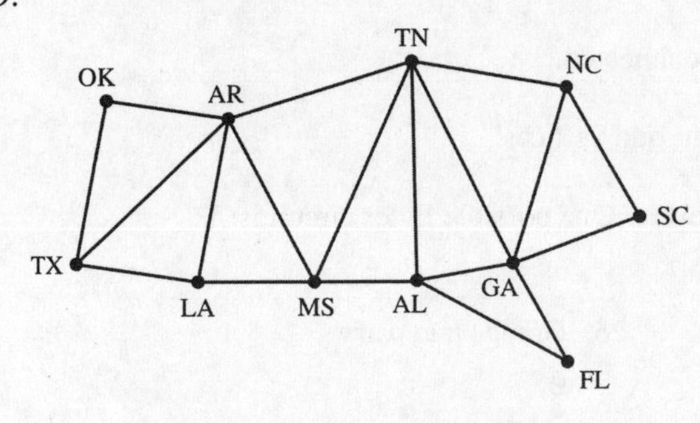

41. 3

43. 3

45. The trip is possible because the graph in Exercise 37 has no odd vertices.

47. The trip is not possible because the graph in Exercise 39 has more than two odd vertices.

49. It is not possible. The room can be graphed as follows:

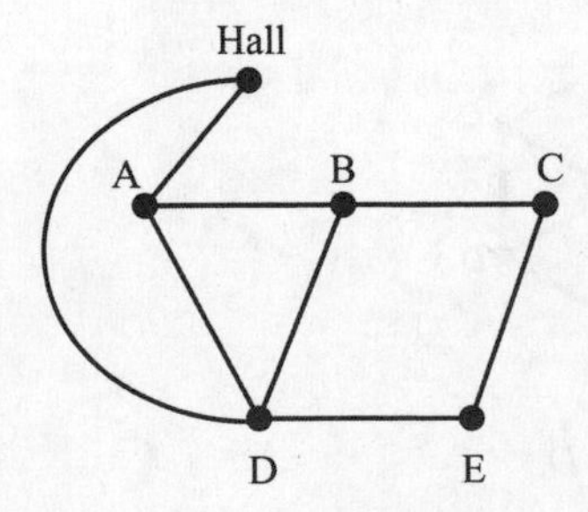

Since there are two odd vertices, it is possible to trace this graph. The tracing (locking of doors) must begin at one odd vertex and terminate at the other. Here this would imply that the guard starts in either Room B and ends in Room A or starts in Room A and ends in Room B. Either way, he would lock himself in the building because he cannot enter or exit through the hallway.

51. Answers may vary. The edges in the following graph represent family members that are unfriendly to each other.

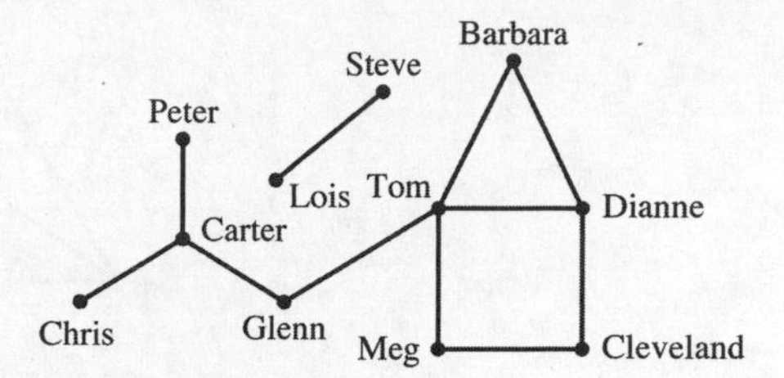

It is possible to color this graph with three colors (indicated by 1, 2, and 3 on the graph). This tells us that three tables will be satisfactory. A satisfactory seating arrangement would be:

Table 1) Peter, Chris, Glenn, and Dianne,

Table 2) Carter, Lois, Tom, and Cleveland,

Table 3) Barbara, Steve, and Meg.

53. Answers may vary.
The edges in the following graph represent committees that would have a conflict because they have common members.

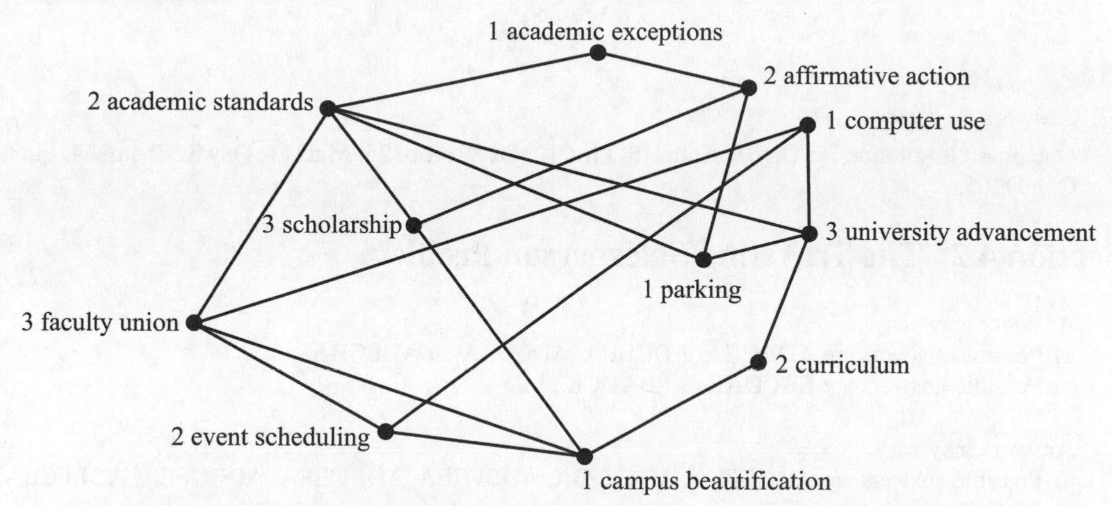

It is possible to color this graph with three colors (indicated by 1,2, and 3). This tells us that three times will be satisfactory to schedule meetings. A satisfactory schedule would be:

Time 1) academic exceptions, computer use, campus beautification, and parking,

Time 2) academic standards , affirmative action, event scheduling, and curriculum,

Time 3) university advancement, faculty union, and scholarship.

55. Every time we go into a vertex we must also leave it. The only possible vertices that we don't both enter and leave are the beginning and ending vertices.

57. No. Since there is a circuit in an Eulerian graph, removing one edge will not disconnect it.

59. Answers may vary.

61. Answers may vary.

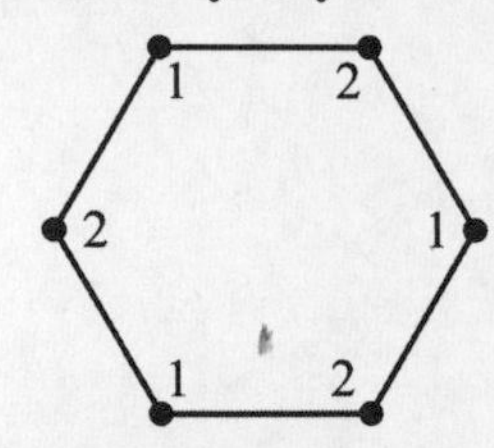

63. Answers may vary.

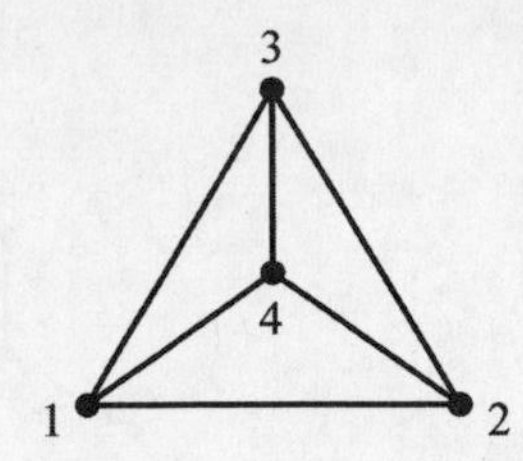

65. One possible sequence is C, F, E, C#, D, A, E, B.

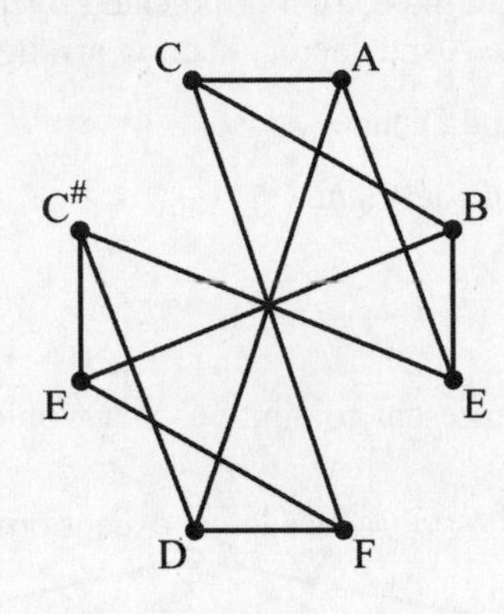

67. Answers may vary.
One possible solution is: Day 1: Anth215, Phy212; Day 2: Bio325, Mat311; Day 3: Chem264, Fin323, ComD265.

Section 4.2 The Traveling Salesperson Problem

1. Answers may vary.
   a) Possible answers are ADBCEA, ADCBEA, ADCEBA, or ADECBA.
   b) Possible answers are EBCDAE or EBADCE.

3. Answers may vary.
   a) Possible answers are ADCBFEA, ADCEFBA, ADBFCEA, ADBCEFA, ADBCFEA, ADEFCBA, ADECBFA, or ADECFBA.
   b) Possible answers are EADCBFE, EABDCFE, EABFCDE, EADBFCE, EADBCFE, EAFBDCE, EAFCBDE, or EAFBCDE.

5. Possible circuits are ABCDEA, ABCEDA, ABDCEA, and ABECDA.

```
            C < D — E — A
                E — D — A
A — B <     D < C — E — A
                E — C — NO
            E < D — C — NO
                C — D — A
```

7. $(7-1)! = 6! = 720$

9.  K_6

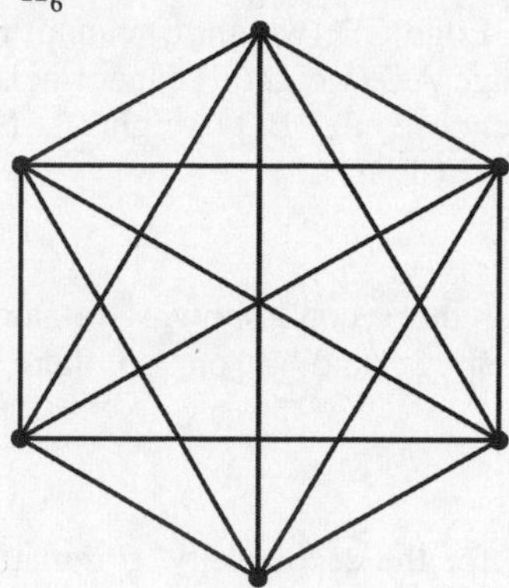

11. a) yes, $\frac{n}{2} = \frac{5}{2} = 2.5$, and each vertex has degree of at least 3.

    b) yes; $n = 5$, and the sum of the degrees of all pairs of non-adjacent vertices is 6.

    c) yes

13. a) no, $\frac{n}{2} = \frac{11}{2} = 5.5$, and each vertex has degree of at most 4.

    b) no; $n = 11$, and the sum of the degrees of all pairs of non-adjacent vertices is at most 8.

    c) no

15. a) The weight of path AGEDCB would be $2 + 4 + 2 + 8 + 4 = 20$.
    b) The weight of path ABCDGEF would be $3 + 4 + 8 + 1 + 4 + 3 = 23$.

17. Since Hamilton circuits occur in pairs, we need only calculate the weight of three Hamilton circuits.

Hamilton Circuit	Weight
ABCDA	3+4+5+3=15
ABDCA	3+6+5+1=15
ACBDA	**1+4+6+3=14**

Since 14 is the smallest number in the right hand column, the circuits ACBDA and ADBCA have minimal weight.

19. Since Hamilton circuits occur in pairs, we need only calculate the weight of three Hamilton circuits.

Hamilton Circuit	Weight
ABCDA	23+14+37+36=110
ABDCA	23+31+37+18=109
ACBDA	**18+14+31+36=99**

Since 99 is the smallest number in the right hand column, the circuits ACBDA and ADBCA have minimal weight.

21. ABDECA

23. ADCFEBA

25. Start by selecting DE which has the smallest weight (weight 2). Next choose edges AB and CD (weight 3). Edge BD (weight 4) cannot next be chosen because that will create a vertex that joins three edges. We can next choose edge AC (weight 5). Edge AE (weight 6) cannot next be chosen because that will create a vertex that joins three edges. Next choose edge BE (weight 6) to complete the circuit of ABEDCA or ACDEBA.

27. Start by selecting CF which has the smallest weight (weight 4). Next choose edge AD (weight 5). Next choose edge BE (weight 6). Next choose edge CE (weight 7). Edge CD (weight 8) cannot next be chosen because that will create a vertex that joins three edges. Edge AC (weight 9) cannot next be chosen because that will create a vertex that joins three edges. Next choose edge BD (weight 9). Next choose edge AF (weight 13) to complete the circuit of ADBECFA or AFCEBDA.

29. 120
Since the tour starts in New York, there are 5 possible choices for the second show, 4 remaining choices for the third show, 3 remaining choices for the fourth show, and so on, so there are $5\times4\times3\times2\times1=120$ possible routes.

31. 40,320
There are 8 possible choices for the first city, 7 remaining choices for the second city, 6 remaining choices for the third city, and so on, so there are $8\times7\times6\times5\times4\times3\times2\times1=40,320$ possible routes.

33. a) This is a K_7 drawing. Since this graph has 7 vertices, there will be 6!=720 Hamilton circuits. We won't count reversals so $\frac{720}{2}=360$ circuits need to be considered.

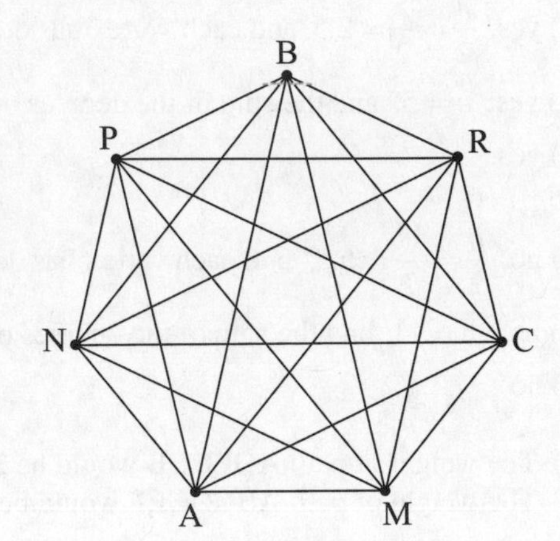

b) If you consider 360 circuits at one per minute, then it would take 360 minutes or 6 hours to do this problem by the brute force algorithm.

c) Using the nearest neighbor algorithm, the path Danielle should go is PNBRAMCP. The cost of this trip would be 210 + 180 + 290 + 90 + 170 + 240 + 420= $1,600.

d) Using the best edge algorithm, the edge with the smallest weight is AR with a weight of 90. The next edge of choice would be MR (weight 120). Edge AM (weight 170) cannot be next chosen because it will create a circuit. Edge BN (weight 180) would be chosen next. Edge NP (weight 210) would be chosen next followed by edge AP (weight 230). Edge PR (weight 240) cannot be chosen because it will create a vertex in which three edges meet. The next choice would be edge CM (weight 240). The remaining choice of BC (weight 380) would be made to complete the circuit. The path would therefore be PNBCMRAP or PARMCBNP. The cost would be 210 + 180 + 380 + 240 + 120 + 90 + 230 = $1,450.

35. Find a table to relate the distance between the stops and Papa John's Pizza.

	A	B	D	E	H	F	M
A	0	2	3	6	4	9	3
B	2	0	3	4	4	7	1
D	3	3	0	5	3	8	2
E	6	4	5	0	4	3	3
H	4	4	3	4	0	5	3
F	9	7	8	3	5	0	6
M	3	1	2	3	3	6	0

a) ABMDHEFA

35. (continued)

b) Using the best edge algorithm, we can first choose edge BM because is has the lowest weight (weight 1). The next edges that can be chosen are AB and DM (weight 2). Of the edges of weight 3, you don't want to choose EM, AM, HM or BD because any of these will cause three edges to meet at one vertex. The next edge of weight 3 to choose would be EF. You now have to make a choice between edges AD and DH. They both have weight 3 and can't both be chosen because the combination of these two edges will cause three edges will meet at a vertex. AD cannot be chosen however because it will close a circuit that does not include all the stops. Therefore, if you next choose DH (weight 3), then the next to be chosen would be EH (weight 4). The next edge to choose would be AF (weight 9) to complete the circuit. The circuits that could be found would be ABMDHEFA or AFEHDMBA

37. a) Find a table to relate the distance between the stops and Fed Ex.

	X	A	B	C	D	E	F	P
X	0	9	8	9	8	5	2	7
A	9	0	3	2	3	4	7	6
B	8	3	0	5	4	3	6	5
C	9	2	5	0	1	6	7	4
D	8	3	4	1	0	5	6	3
E	5	4	3	6	5	0	3	4
F	2	7	6	7	6	3	0	5
P	7	6	5	4	3	4	5	0

Using nearest neighbor algorithm the path followed should be XFEBACDPX

b) Using the best edge algorithm, we can first choose edge CD because it has the lowest weight (weight 1). The next edges that can be chosen are AC and XF (weight 2). Of the edges of weight 3, you can choose EF, DP, AB and BE. If edge AD were chosen then it would cause three vertices to meet. The next edge to choose would be XP (weight 7) to complete the circuit. The circuits that could be found would be XFEBACDPX or XPDCABEFX.

39. A Hamilton circuit goes through every vertex but does not travel over every edge as an Euler circuit does.

41. a) After starting with vertex A, we have 9 choices for the first vertex, 8 choices for the second, 7 for the third, and so on. The total number of ways to choose all the vertices is $9\times8\times7\times6\times5\times4\times3\times2\times1 = 9!$ possible routes.

b) After starting with vertex A, we have $n-1$ choices for the first vertex, $n-2$ choices for the second, $n-3$ for the third, and so on. The total number of ways to choose all the vertices is $(n-1)\times(n-2)\times(n-3)\times\cdots\times3\times2\times1=(n-1)!$ possible routes.

c) 15 is not of the form $n!$.

43. $n!$ grows more rapidly, $2^n < n!$, for $n \geq 4$.

45. A K_{10} figure would have 9!=362,880 circuits. We won't count reversals so $\frac{362,880}{2} = 181,440$ circuits need to be considered. If a computer can examine 1,000 Hamilton circuits per second, then it would take the computer 181.44 seconds. This is just over 3 minutes.

47. Answers will vary. One possible solution is A, C, E, F, B, D.

49. Using the nearest neighbor algorithm starting at A we have the circuit **ABDECA**. The weight of this circuit is 3 + 4 + 2 + 7 + 5 = **21**.
Using the nearest neighbor algorithm starting at B we have the circuit **BACDEB**. The weight of this circuit is 3 + 5 + 3 + 2 + 6=**19**.
Using the nearest neighbor algorithm starting at C we have the circuit **CDEBAC** or **CDEABC**. The weight of these circuits are 3 + 2 + 6 + 3 + 5 = **19** and 3 + 2 + 6 + 3 + 9 = **23**, respectively.
Using the nearest neighbor algorithm starting at D we have the circuit **DEABCD** or **DEBACD**. The weight of these circuits are 2 + 6 + 3 + 9 + 3 = **23** and 2 + 6 + 3 + 5 + 3 = **19**, respectively.
Using the nearest neighbor algorithm starting at E we have the circuit **EDCABE**. The weight of this circuit is 2 + 3 + 5 + 3 + 6 = **19**.
The smallest weight among these is 19. There is essentially one circuit (not counting reversing the order) here that starts and stops at A. It is ABEDCA.

Section 4.3 Directed Graphs

1. a) ABCE or ABCDE
b) ABC
c) ABCE
d) Not possible because you must first go from A to B. From B you cannot go directly to E.
e) Answers may vary. Possible answers include ABCEDA or ABCEBA.

3. a) BCE
b) Answers may vary. Possible answers included ABCFG or ACFG.
c) ACFGCE
d) Not possible because you cannot go to D. It is possible to go from D.
e) ABCFGC

5.

From \ To	A	B	C	D	E
A	1	1	1	0	0
B	1	1	1	1	1
C	1	1	0	2	2
D	1	2	0	1	1
E	2	1	1	1	1

7.

From \ To	A	B	C	D	E	F	G
A	0	1	2	0	1	1	0
B	0	0	1	0	1	1	0
C	0	0	0	0	1	1	1
D	0	0	1	0	2	1	0
E	0	0	0	0	0	0	0
F	0	0	1	0	0	0	1
G	0	0	1	0	1	1	0

9. a) Only Kevin; if anyone else starts the rumor, it would not spread to everyone. (Kevin would be excluded). b) Reverse the direction of the edge KD or RK.

11. The only possible sequence would be chief financial officer, then sales manager, then production manager, then company president, then marketing director. If the secretary starts with the sales manager, then he/she will not be able to obtain the signature of the chief financial officer. If the secretary starts with the production manager, then he/she will not be able to get the signatures of the sales manager and the chief financial officer. If the secretary starts with the company president, then he/she will not be able to get the signatures of the sales manager, chief financial officer, and the production manager. If the secretary starts with the marketing director, then he/she will not be able to get the signatures of the chief financial officer, sales manager, production manager, and the company president.

13.

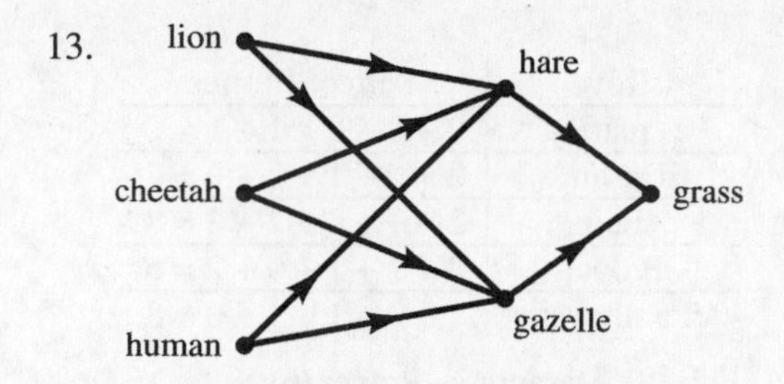

15.

	To					
From	Georgia	Arkansas	LSU	Florida	Mississippi	Alabama
Georgia	0	0	0	0	0	0
Arkansas	1	0	0	0	0	0
LSU	2	1	0	0	0	0
Florida	1	1	2	0	1	1
Mississippi	1	1	1	1	0	1
Alabama	0	0	1	1	1	0

Row	Sum of Entries
Georgia	$0+0+0+0+0+0=0$
Arkansas	$1+0+0+0+0+0=1$
LSU	$2+1+0+0+0+0=3$
Florida	$1+1+2+0+1+1=6$
Mississippi	$1+1+1+1+0+1=5$
Alabama	$0+0+1+1+1+0=3$

Florida (6), Mississippi (5), LSU and Alabama (3), Arkansas (1), and Georgia (3)

17.

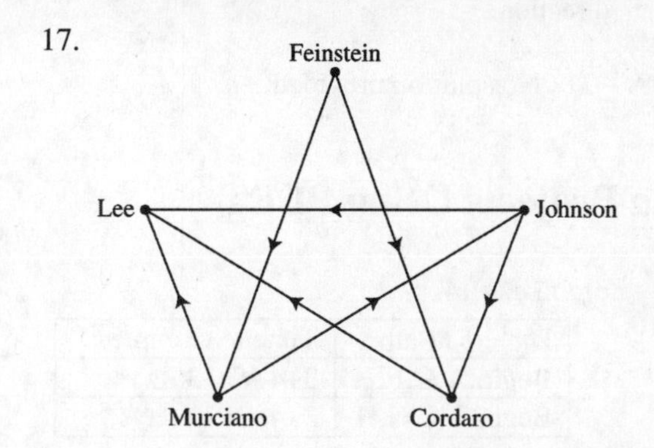

Row	Sum of Entries
Lee	$0+0+0+0+0=0$
Feinstein	$2+0+1+1+1=5$
Johnson	$2+0+0+1+0=3$
Cordaro	$1+0+0+0+0=1$
Murciano	$2+0+1+1+0=4$

Feinstein (5), Murciano (4), Johnson (3), Cordaro (1), and Murciano (0)

	To				
From	Lee	Feinstein	Johnson	Cordaro	Murciano
Lee	0	0	0	0	0
Feinstein	2	0	1	1	1
Johnson	2	0	0	1	0
Cordaro	1	0	0	0	0
Murciano	2	0	1	1	0

19.

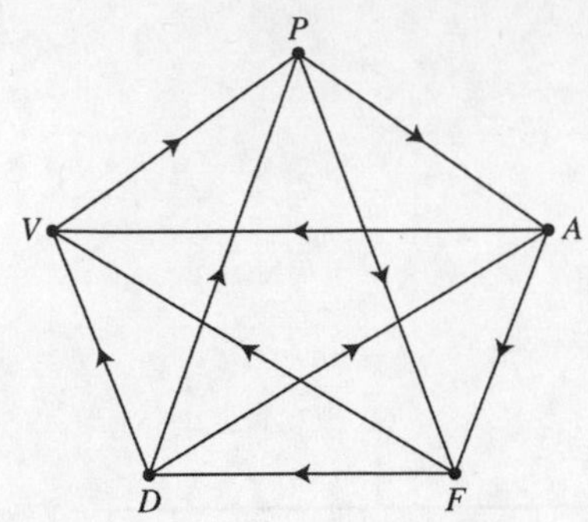

Row	Sum of Entries
Aquafina	0 + 0 + 0 + 1 + 1 = 2
Dasani	2 + 0 + 1 + 2 + 2 = 7
Fuji	2 + 1 + 0 + 2 + 3 = 8
Propel	2 + 1 + 1 + 0 + 2 = 6
Vitaminwater	1 + 0 + 1 + 1 + 0 = 3

Fuji (8), Dasani (7), Propel (6), Vitaminwater (3), and Aquafina (2)

		To				
		Aquafina	Dasani	Fuji	Propel	Vitaminwater
From	Aquafina	0	0	0	1	1
	Dasani	2	0	1	2	2
	Fuji	2	1	0	2	3
	Propel	2	1	1	0	2
	Vitaminwater	1	0	1	1	0

21. $\begin{bmatrix} 0 & 1 & 0 & 0 \\ 0 & 0 & 1 & 0 \\ 1 & 0 & 0 & 1 \\ 1 & 0 & 0 & 0 \end{bmatrix}$

23.

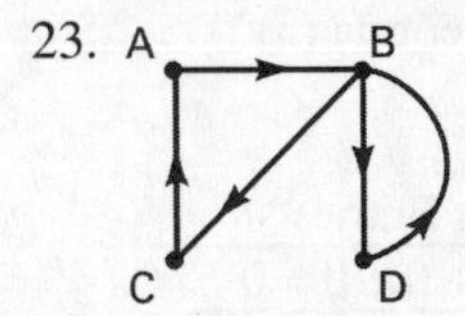

25.

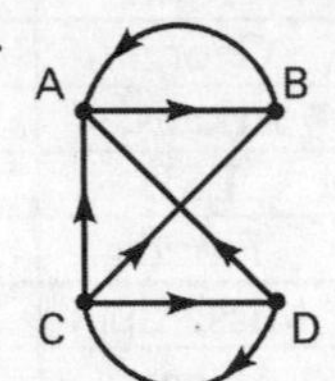

27. When the relationship goes in only one direction.

29. – 31. No solution provided.

Section 4.4 Looking Deeper: Scheduling Projects Using PERT

1. a) Begin,A,B,E,I

Directed Path	Time to Complete
Begin,A,B,D,I	3+6+3+2=14 days
Begin,A,B,E,I	**3+6+5+2=16 days**
Begin,A,C,E,I	3+4+5+2=14 days
Begin,A,C,F,I	3+4+6+2=15 days

b) Begin,A,B,E

Directed Path	Time to Complete
Begin,A,B,E	**3+6+5=14 days**
Begin,A,C,E	3+4+5=12 days

c) On day 14

Directed Path	Time to Complete
Begin,A,C,F,H	**3+4+6=13 days**
Begin,A,C,G,H	3+4+3=10 days

d) After 16 days

Directed Path	Time to Complete
Begin,A,B,D,I	3+6+3+2=14 days
Begin,A,B,E,I	**3+6+5+2=16 days**
Begin,A,C,E,I	3+4+5+2=14 days
Begin,A,C,F,I	3+4+6+2=15 days

1. (continued)

e) 17 days

Directed Path	Time to Complete
Begin,A,B,D,I,J,End	3+6+3+2+1=15 days
Begin,A,B,E,I,J,End	**3+6+5+2+1=17 days**
Begin,A,C,E,I,J,End	3+4+5+2+1=15 days
Begin,A,C,F,I,J,End	3+4+6+2+1=16 days
Begin,A,C,F,H,J,End	**3+4+6+3+1=17 days**
Begin,A,C,G,H,J,End	3+4+3+3+1=14 days

f) Begin,A,B,E,I,J,End or Begin,A,C,F,H,J,End

3. a) Begin,B,E,F,G

Directed Path	Time to Complete
Begin,A,D,G	2+4+2=8 days
Begin,A,E,G	2+2+2=6 days
Begin,B,E,G	3+2+2=7 days
Begin,B,E,F,G	**3+2+3+2=10 days**
Begin,C,F,G	3+3+2=8 days

b) Begin,B,E,F,H

Directed Path	Time to Complete
Begin,A,E,H	2+2+1=5 days
Begin,A,E,F,H	2+2+3+1=8 days
Begin,B,E,H	3+2+1=6 days
Begin,B,E,F,H	**3+2+3+1=9 days**
Begin,C,F,H	3+3+1=7 days

c) Begin,B,E,F,G,I,End

Directed Path	Time to Complete
Begin,A,D,G,I,End	2+4+2+4=12 days
Begin,A,E,G,I,End	2+2+2+4=10 days
Begin,A,E,H,I,End	2+2+1+4=9 days
Begin,A,E,F,G,I,End	2+2+3+2+4=13 days
Begin,A,E,F,H,I,End	2+2+3+1+4=12 days
Begin,B,E,G,I,,End	3+2+2+4=11 days
Begin,B,E,H,I,End	3+2+1+4=10 days
Begin,B,E,F,G,I,End	**3+2+3+2+4=14 days**
Begin,B,E,F,H,I,End	3+2+3+1+4=13 days
Begin,C,F,G,I,End	3+3+2+4=12 days
Begin,C,F,H,I,End	3+3+1+4=11 days

d) Since it will take 3+2=5 days to complete the tasks that precede task F, task F should begin on day 6

Directed Path	Time to Complete
Begin,A,E,F	2+2+3=7 days
Begin,B,E,F	**3+2+3=8 days**
Begin,C,F	3+3=6 days

e) From part a) we see that a critical path for G is Begin, B, E, F, G. Since it will take 3+2+3=8 days to complete the tasks that precede task G, task G should begin on day 9.

f) From part c) we determined that the critical path for "End" was Begin,B,E,F,G,I,End. This implies that it should take 14 days to complete the project.

5. Both tasks A and B can begin on day 1. We must however find the critical paths for tasks C,D,E,F,G,and H.

C: Since it will take 2 days to complete the tasks that precede task C, task C should begin on day 3.

Directed Path	Time to Complete
Begin,A,C	1+3=4 days
Begin,B,C	**2+3=5 days**

D: Since it will take 2 days to complete the tasks that precede task D, task D should begin on day 3.

Directed Path	Time to Complete
Begin,A,D	1+5=6 days
Begin,B,D	**2+5=7 days**

E: Since it will take 2+3=5 days to complete the tasks that precede task E, task E should begin on day 6.

Directed Path	Time to Complete
Begin,A,C,E	1+3+5=9 days
Begin,B,C,E	**2+3+5=10 days**

F: Since it will take 2+5=7 days to complete the tasks that precede task F, task F should begin on day 6.

Directed Path	Time to Complete
Begin,A,C,F	1+3+4=8 days
Begin,A,D,F	1+5+4=10 days
Begin,B,C,F	2+3+4=9 days
Begin,B,D,F	**2+5+4=11 days**

G: Since it will take 2+5=7 days to complete the tasks that precede task G, task G should begin on day 8.

Directed Path	Time to Complete
Begin,A,D,G	1+5+2=8 days
Begin,B,D,G	**2+5+2=9 days**

H: Since it will take 2+5+4=11 days to complete the tasks that precede task H, task H should begin on day 8.

Directed Path	Time to Complete
Begin,A,C,E,H	1+3+5+6=15 days
Begin,A,C,F,H	1+3+4+6=14 days
Begin,A,D,F,H	1+5+4+6=16 days
Begin,A,D,G,H	1+5+2+6=14 days
Begin,B,C,E,H	2+3+5+6=16 days
Begin,B,C,F,H	2+3+4+6=15 days
Begin,B,D,F,H	**2+5+4+6=17 days**
Begin,B,D,G,H	2+5+2+6=15 days

7. Tasks A, B, and C can begin on day 1. We must however find the critical paths for tasks D,E,F,G,H, I, and J.

D: Since it will take 2 days to complete the tasks that precede task D, task D should begin on day 3.

Directed Path	Time to Complete
Begin,A,D	1+2=3 days
Begin,B,D	**2+2=4 days**

E: Since it will take 3 days to complete the tasks that precede task E, task E should begin on day 4.

Directed Path	Time to Complete
Begin,B,E	2+3=5 days
Begin,C,E	**3+3=6 days**

7. (continued)

F: Since it will take 2+2=4 days to complete the tasks that precede task F, task F should begin on day 5.

Directed Path	Time to Complete
Begin,A,D,F	1+2+5=8 days
Begin,B,D,F	**2+2+5=9 days**

G: Since it will take 3+3=6 days to complete the tasks that precede task G, task G should begin on day 7.

Directed Path	Time to Complete
Begin,A,D,G	1+2+4=7 days
Begin,B,D,G	2+2+4=8 days
Begin,B,E,G	2+3+4=9 days
Begin,C,E,G	**3+3+4=10 days**

H: Since it will take 3+3=6 days to complete the tasks that precede task H, task H should begin on day 7.

Directed Path	Time to Complete
Begin,B,E,H	2+3+2=7 days
Begin,C,E,H	**3+3+2=8 days**

I: Since it will take 3+3+4=10 days to complete the tasks that precede task I, task I should begin on day 11.

Directed Path	Time to Complete
Begin,A,D,F,I	1+2+5+1=9 days
Begin,A,D,G,I	1+2+4+1=8 days
Begin,B,D,F,I	2+2+5+1=10 days
Begin,B,E,G,I	2+3+4+1=10 days
Begin,C,E,G,I	**3+3+4+1=11 days**

J: Since it will take 3+3+4=10 days to complete the tasks that precede task J, task J should begin on day 11.

Directed Path	Time to Complete
Begin,A,D,G,J	1+2+4+3=10 days
Begin,B,D,G,J	2+2+4+3=11 days
Begin,B,E,G,J	2+3+4+3=12 days
Begin,B,E,H,J	2+3+2+3=10 days
Begin,C,E,G,J	**3+3+4+3=13 days**
Begin,C,E,H,J	3+3+2+3=11 days

9.

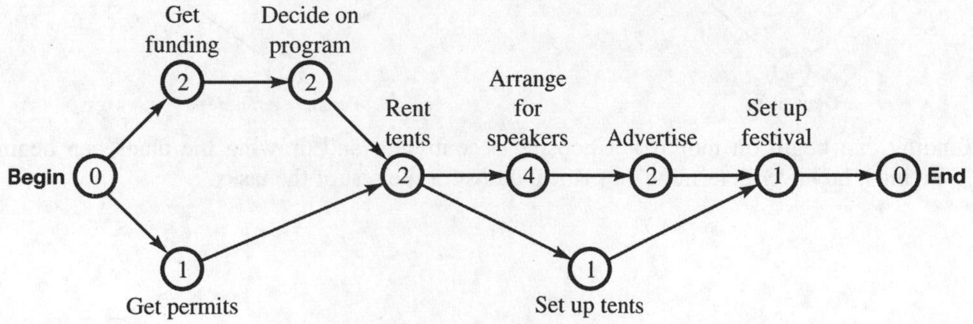

Get funding 2
Decide on program 2
Begin 0
Rent tents 2
Arrange for speakers 4
Advertise 2
Set up festival 1
0 End
Get permits 1
Set up tents 1

Funding and Permits can begin in week 1. Programs can begin in week 3. We must determine the critical paths for the rest of the tasks.

9. (continued)

Rent tents: Since it will take 2+2=4 weeks to complete the tasks that precede renting tents, renting tents should begin on week 5.

Directed Path	Time to Complete
Begin,1,3,4	**2+2+2=6 weeks**
Begin,2,4	1+2=3 weeks

Arrange for speakers, etc.: Since it will take 2+2+2=6 weeks to complete the tasks that precede arranging for speakers, arranging for speakers should begin on week 7.

Directed Path	Time to Complete
Begin,1,3,4,5	**2+2+2+4=10 weeks**
Begin,2,4,5	1+2+4=7 weeks

Advertise: Since it will take 2+2+2+4=10 weeks to complete the tasks that precede arranging for advertising, arranging for advertising should begin on week 11.

Directed Path	Time to Complete
Begin,1,3,4,5,6	**2+2+2+4+2=12 weeks**
Begin,2,4,5,6	1+2+4+2=9 weeks

Set up tents, etc.: Since it will take 2+2+2=6 weeks to complete the tasks that precede setting up tents, setting up tents should begin on week 7.

Directed Path	Time to Complete
Begin,1,3,4,7	**2+2+2+1=7 weeks**
Begin,2,4,7	1+2+1=4 weeks

Set up festival: Since it will take 2+2+2+4+2=12 weeks to complete the tasks that precede setting up the festival, setting up the festival should begin on week 13.

Directed Path	Time to Complete
Begin,1,3,4,5,6,8	**2+2+2+4+2+1=13 weeks**
Begin,2,4,5,6,8	1+2+4+2+1=10 weeks
Begin,1,3,4,7,8	2+2+2+1+1=8 weeks
Begin,2,4,7,8	1+2+1+1=5 weeks

11.

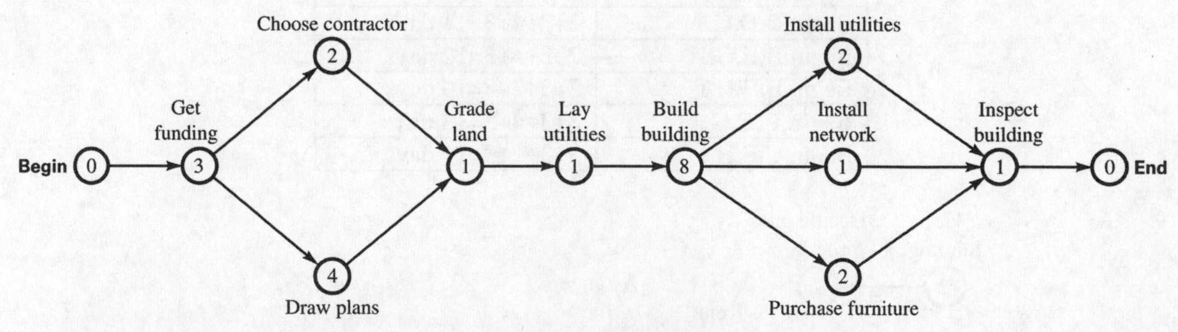

Getting funding can begin on month 1. Choosing a contractor and drawing the plans can begin on month 4. We must however determine the critical paths for the rest of the tasks.

11. (continued)

Grade land: Since it will take 3+4=7 months to complete the tasks that precede grading the land, grading the land should begin on month 8.

Directed Path	Time to Complete
Begin,1,2,4	3+2+1=6 months
Begin,1,3,4	**3+4+1=8 months**

Lay underground utilities: Since it will take 3+4+1=8 months to complete the tasks that precede laying underground utilities, laying underground utilities should begin on month 9.

Directed Path	Time to Complete
Begin,1,2,4,5	3+2+1+1=7 months
Begin,1,3,4,5	**3+4+1+1=9 months**

Build buildings: Since it will take 3+4+1+1=9 months to complete the tasks that precede building buildings, building buildings should begin on month 10.

Directed Path	Time to Complete
Begin,1,2,4,5,6	3+2+1+1+8=15 months
Begin,1,3,4,5,6	**3+4+1+1+8=17 months**

Install utilities in building: Since it will take 3+4+1+1+8=17 months to complete the tasks that precede installing utilities in buildings, installing utilities in buildings should begin on month 12.

Directed Path	Time to Complete
Begin,1,2,4,5,6,7	3+2+1+1+8+2=17 months
Begin,1,3,4,5,6,7	**3+4+1+1+8+2=19 months**

Install computers: Since it will take 3+4+1+1+8=17 months to complete the tasks that precede installing computers, installing computers should begin on month 18.

Directed Path	Time to Complete
Begin,1,2,4,5,6,8	3+2+1+1+8+1=16 months
Begin,1,3,4,5,6,8	**3+4+1+1+8+1=18 months**

Purchase furniture: Since it will take 3+4+1+1+8=17 months to complete the tasks that precede purchase furniture, purchase furniture should begin on month 18.

Directed Path	Time to Complete
Begin,1,2,4,5,6,9	3+2+1+1+8+2=17 months
Begin,1,3,4,5,6,9	**3+4+1+1+8+2=19 months**

Inspect Building: Since it will take 3+4+1+1+8+2=19 months to complete the tasks that precede inspecting the buildings, inspecting the buildings should begin on month 20.

Directed Path	Time to Complete
Begin,1,2,4,5,6,7,10	3+2+1+1+8+2+1=18 months
Begin,1,2,4,5,6,8,10	3+2+1+1+8+1+1=17 months
Begin,1,2,4,5,6,9,10	3+2+1+1+8+2+1=18 months
Begin,1,3,4,5,6,7,10	**3+4+1+1+8+2+1=20 months**
Begin,1,3,4,5,6,8,10	3+4+1+1+8+1+1=19 months
Begin,1,3,4,5,6,9,10	**3+4+1+1+8+2+1=20 months**

13.

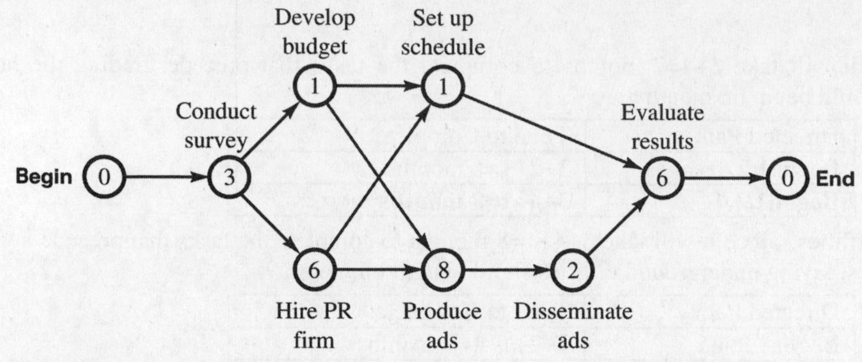

Conducting a survey can begin on month 1. Developing a budget and hiring a PR firm can begin on month 4. We must however determine the critical paths for the rest of the tasks.

Set up production schedule: Since it will take 3+6=9 months to complete the tasks that precede setting up the production schedule, setting up the production schedule should begin on month 10.

Directed Path	Time to Complete
Begin,1,2,4	3+1+1=5 months
Begin,1,3,4	**3+6+1=10 months**

Produce ads: Since it will take 3+6=9 months to complete the tasks that precede producing ads, producing ads should begin on month 10.

Directed Path	Time to Complete
Begin,1,2,5	3+1+8=12 months
Begin,1,3,5	**3+6+8=17 months**

Disseminate ads: Since it will take 3+6+8=17 months to complete the tasks that precede disseminating ads, disseminating ads should begin on month 18.

Directed Path	Time to Complete
Begin,1,2,5,6	3+1+8+2=14 months
Begin,1,3,5,6	**3+6+8+2=19 months**

Evaluate results: Since it will take 3+6+8+2=19 months to complete the tasks that precede evaluating results, evaluating results should begin on month 20.

Directed Path	Time to Complete
Begin,1,2,4,7	3+1+1+6=11 months
Begin,1,3,4,7	3+6+1+6=16 months
Begin,1,2,5,6,7	3+1+8+2+6=20 months
Begin,1,3,5,6,7	**3+6+8+2+6=25 months**

15. – 17. No solution provided.

Chapter Review Exercises

1. a) 12
   b) A, B, C, D, and E are even. F, G, H, and I are odd.
   c) Yes. The graph is connected.
   d) Yes. The bridges are FH, GH, and HI.

2. Answers may vary.
   The objects are represented by vertices. Any two objects that are related are joined by an edge in the graph.

3. Graph (a) can be traced because it has only two odd vertices. Graph (b) cannot be traced because it has more than two odd vertices, namely vertices A, B, D, and E.

4. Answers may vary.
One possible circuit is ABCDB EDFGH FKHIJ KCA.

5. This is not an Eulerian graph. Duplicate edges BC, EI, HL, and NO so that all vertices are even. One possible route would be ABCDH GCBFG KLHLP ONJKO NMIEF JIEA.

6.

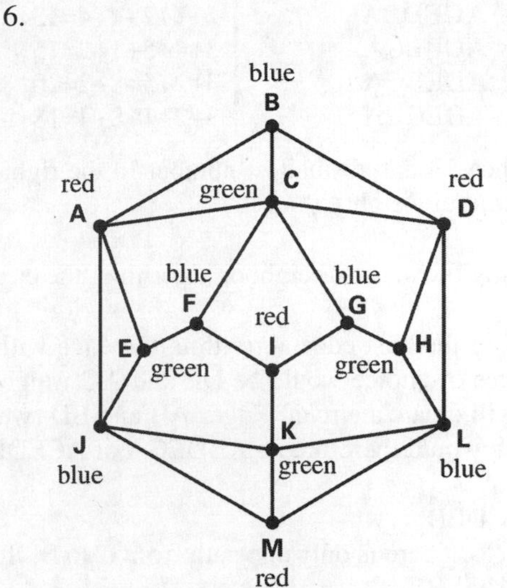

7. The edges in the following graph represent people that cannot be in the same canoe.

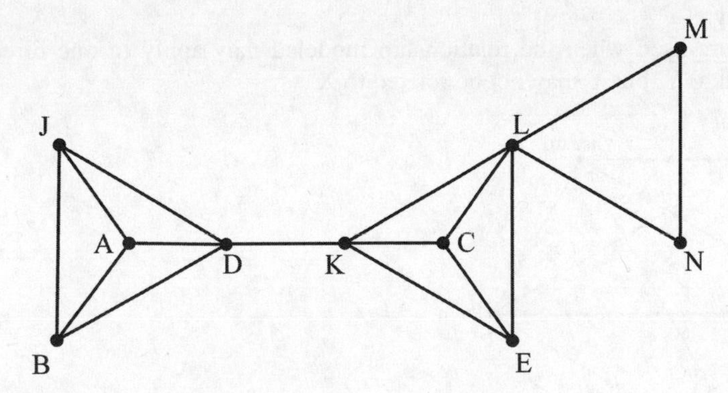

It is possible to color this graph with four colors (indicated by 1, 2, 3, and 4). This tells us that four canoes will be satisfactory. A satisfactory canoe assignment would be:

Canoe 1) Allison, Kami, and Nicole,

Canoe 2) Brandon, Colin, and Marshall,

Canoe 3) Donny and Erica,

Canoe 4) Jim and Lance.

8. ABCDEA, ABCEDA , ABDCEA, ABDECA, ABECDA, and ABEDCA.

9.

Hamilton Circuit	Weight
ABCEDA	4+8+1+2+3=18
ABDECA	**4+4+2+1+2=13**
ACEBDA	2+1+5+4+3=15
ACBEDA	2+8+5+2+3=20
ACEDBA	**2+1+2+4+4=13**
ADBECA	3+4+5+1+2=15
ADEBCA	3+2+5+8+2=20
ADECBA	3+2+1+8+4=18

Since 13 is the smallest number in the right hand column, the circuits ABDECA and ACEDBA have minimal weight.

10. Using the nearest neighbor algorithm, the circuit would be ACEDBA for a weight of 2+1+2+4+4=13.

11. Using the best edge algorithm, the edge with the smallest weight is EC with a weight of 1. The next edges of choice would be DE and AC(weight 2). Edge AD (weight 3) cannot be next chosen because it will create a circuit. Edges AB and BD (weight 4) would be chosen next to complete the circuit. The path would therefore be ABDECA or ACEDBA. The weight would be 4+4+2+1+2=13.

12. a) CDEB
b) No. There is only one path from C to B. It is CDEB.
c) HGEBF
d) FGEB and FHGEB

13. Answers may vary.
Directed graphs are used when the relationship modeled may apply in one direction. For example, object X is related to Y, but Y may not be related to X.

14.

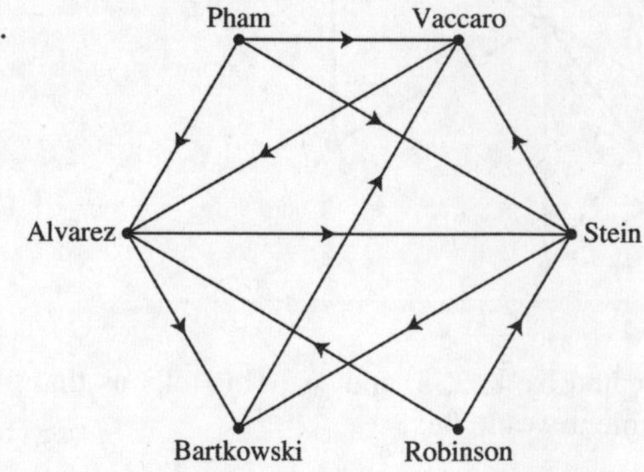

		To					
		Pham	Vaccaro	Robinson	Alvarez	Bartkowski	Stein
	Pham	0	2	0	2	2	2
	Vaccaro	0	0	0	1	1	1
From	Robinson	0	1	0	1	2	2
	Alvarez	0	2	0	0	2	1
	Bartkowski	0	1	0	1	0	0
	Stein	0	2	0	1	1	0

14. (continued)

Committee Member	Sum of Entries
Pham	0 + 2 + 0 + 2 + 2 + 2 = 8
Vaccaro	0 + 0 + 0 + 1 + 1 + 1 = 3
Robinson	0 + 1 + 0 + 1 + 2 + 2 = 6
Alvarez	0 + 2 + 0 + 0 + 2 + 1 = 5
Bartkowski	0 + 1 + 0 + 1 + 0 + 0 = 2
Stein	0 + 2 + 0 + 1 + 1 + 0 = 4

Thus, Pham is most influential.

15. a) Begin, A,C,F,H

Directed Path	Time to Complete
Begin,A,E,H	4+4+2=10 days
Begin,A,C,H	4+5+2=11 days
Begin,A,C,F,H	**4+5+5+2=16 days**
Begin,B,C,H	3+5+2=10 days
Begin,B,C,F,H	3+5+5+2=10 days
Begin,B,D,F,H	3+3+5+2=13 days

b) Begin,A,C,F,I,End

Directed Path	Time to Complete
Begin,A,E,H,End	4+4+2=10 days
Begin,A,C,H,End	4+5+2=11 days
Begin,A,C,F,H,End	4+5+5+2=16 days
Begin,A,C,F,I,End	**4+5+5+4=18 days**
Begin,B,C,H,End	3+5+2=10 days
Begin,B,C,F,H,End	3+5+5+2=16 days
Begin,B,C,F,I,End	3+5+5+4=17 days
Begin,B,D,F,H,End	3+3+5+2=13 days
Begin,B,D,F,I,End	3+3+5+4=15 days
Begin,B,D,G,I,End	3+3+5+4=15 days

c) Since it will take 4+5=10 days to complete the tasks that precede task F, task F should begin on day 10.

Directed Path	Time to Complete
Begin,A,C,F	**4+5+5=14 days**
Begin,B,C,F	3+5+5=13 days
Begin,B,D,F	3+3+5=11 days

d) From part b) we determined that the critical path for "End" was Begin,A,C,F,I,End. This implies that it should take 18 days to complete the project.

16.

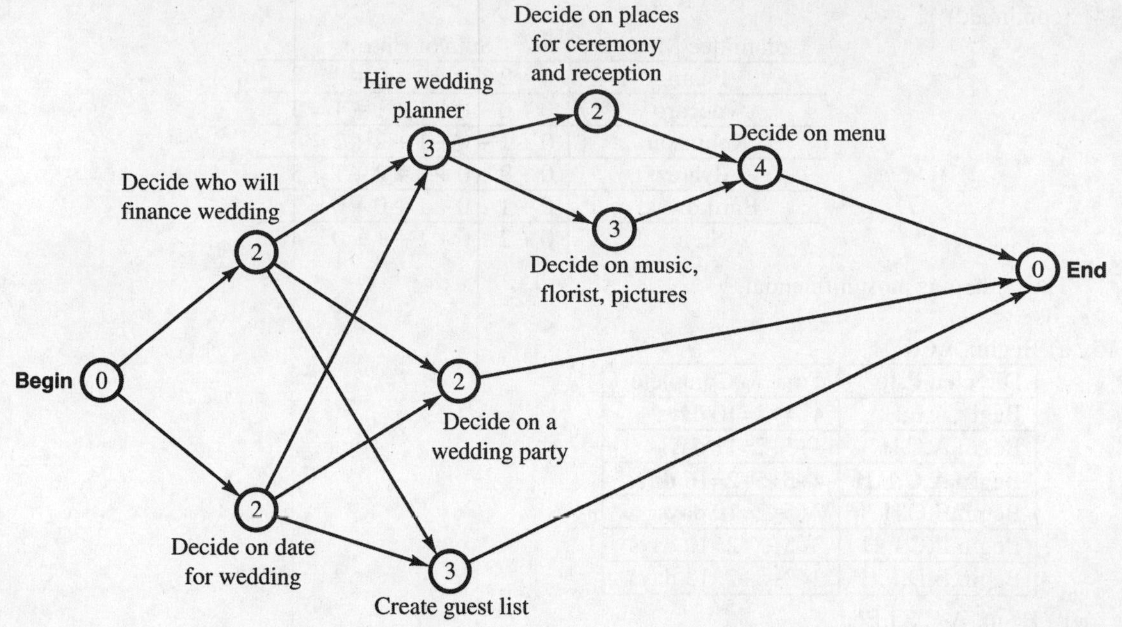

Deciding on a date and who will finance the wedding can begin in week 1. Hiring the wedding planner, deciding on the wedding party, and creating the guest list can begin in week 3. We must determine the critical paths for the rest of the tasks.

Decide on places for ceremony and reception: Since it will take 2+3=5 weeks to complete the tasks that precede deciding on places for ceremony and reception, deciding on places for ceremony and reception should begin on week 6.

Directed Path	Time to Complete
Begin,1,3,6	**2+3+3=8 weeks**
Begin,2,3,6	**2+3+3=8 weeks**

Decide on music, florist, pictures.: Since it will take 2+3=5 weeks to complete the tasks that precede deciding on music, florist, and pictures, Decide on music, florist, and pictures should begin on week 6.

Directed Path	Time to Complete
Begin,1,3,7	**2+3+3=8 weeks**
Begin,2,3,7	**2+3+3=8 weeks**

Decide on menu: Since it will take 2+3+3=8 weeks to complete the tasks that precede deciding on menu, deciding on menu should begin on week 9.

Directed Path	Time to Complete
Begin,1,3,6,8	2+3+2+4=11 weeks
Begin,1,3,7,8	**2+3+3+4=12 weeks**
Begin,1,3,6,8	2+3+2+4=11 weeks
Begin,1,3,7,8	**2+3+3+4=12 weeks**

Chapter Test

1. a) 9
   b) E and F are odd; A, B, C, D, G, and H are even.
   c) Yes, the graph is connected.
   d) Yes, EF is a bridge.

2. a) This graph cannot be traced because it has more than two odd vertices, namely A, B, C, E, G, and H.
   b) can be traced because it has two odd vertices.

3. Answers may vary. One possible circuit is ABDACEGIHGFEDFA.

4. ADBCEA, ADBECA, ADCBEA, ADCEBA, ADEBCA, ADECBA.

5. This is not an Eulerian graph. Duplicate the edges as shown so that all vertices are even. One possible route is numbered on the diagram.

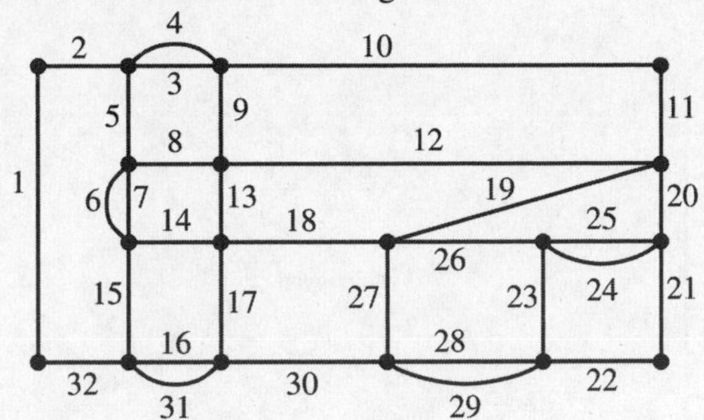

6. Two colors are required.

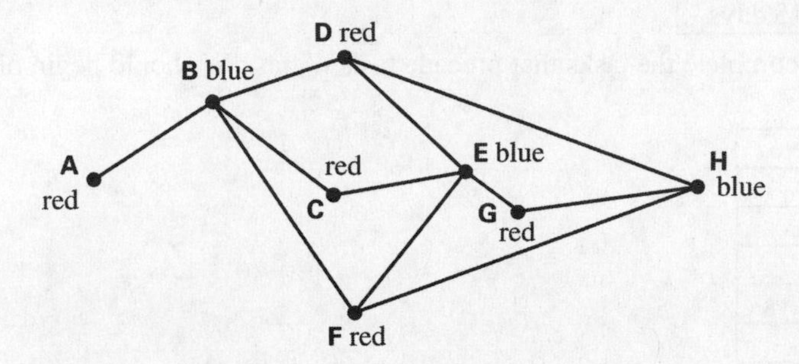

7.

Hamilton Circuit	Weight
ABCEDA	5+5+3+7+8=28
ABDECA	**5+2+7+3+2=19**
ACEBDA	**2+3+4+2+8=19**
ACBEDA	2+5+4+7+8=26
ACEDBA	2+3+7+2+5=19
ADBECA	8+2+4+3+2=19
ADEBCA	8+7+4+5+2=26
ADECBA	8+7+3+5+5=28

Since 19 is the smallest number in the right hand column, the circuits ABDECA and ACEBDA have minimal weight.

8. Using the nearest neighbor algorithm, the circuit would be ACEBDA for a weight of 2+3+4+2+8=19.

9. Using the best edge algorithm, the edges with the smallest weight are AC and BD (weight 2). The next edge of choice would be CE (weight 3). Edge BE (weight 4) cannot be next chosen because it will create a circuit. Edges AB and AD (weight 5) would be chosen next to complete the circuit. The path would therefore be ADBECA or ACEBDA. The weight would be 2+3+4+2+8=19.

10. a) BAEF
b) There is no path from C to G since no edges go towards G.

11. a) Begin,X,P,B,Z

Directed Path	Time to Complete
Begin,P,B,Z	5+7+3=15 days
Begin,X,P,B,Z	**3+5+7+3=18 days**
Begin,X,Y,B,Z	3+4+7+3=17 days
Begin,X,Y,G,Z	3+4+5+3=15 days

b) Begin,X,P,B,Z,End

Directed Path	Time to Complete
Begin,D,E,W,End	6+4+2=12 days
Begin,P,E,W,End	5+4+2=11 days
Begin,P,B,W,End	5+7+2=14 days
Begin,P,B,Z,End	5+7+3=15 days
Begin,X,P,E,W,End	3+5+4+2=14 days
Begin,X,P,B,W,End	3+5+7+2=17 days
Begin,X,P,B,Z,End	**3+5+7+3=18 days**
Begin,X,Y,B,W,End	3+4+7+2=16 days
Begin,X,Y,B,Z,End	3+4+7+3=17 days
Begin,X,Y,G,Z,End	3+4+5+3=15 days

c) Since it will take 3+5+7=15 days to complete the tasks that precede task W, task W should begin on day 16.

Directed Path	Time to Complete
Begin,D,E,W	6+4+2=12 days
Begin,P,E,W	5+4+2=11 days
Begin,P,B,W	5+7+2=14 days
Begin,X,P,B,W	**3+5+7+2=17 days**
Begin,X,Y,B,W	3+4+7+2=days

d) From part b) we determined that the critical path for "End" was Begin,X,P,B,Z,End. This implies that it should take 18 days to complete the project.

12.

		To					
		A	B	C	D	E	F
From	A	0	0	1	0	0	0
	B	1	0	3	1	0	1
	C	0	0	0	0	0	0
	D	1	1	2	0	1	2
	E	0	0	1	1	0	1
	F	0	1	2	1	1	0

Person	Sum of Entries
A	0 + 0 + 1 + 0 + 0 + 0 = 1
B	1 + 0 + 3 + 1 + 0 + 1 = 6
C	0 + 0 + 0 + 0 + 0 + 0 = 0
D	1 + 1 + 2 + 0 + 1 + 2 = 7
E	0 + 0 + 1 + 1 + 0 + 1 = 3
F	0 + 1 + 2 + 1 + 1 + 0 = 5

Thus, D is most influential.

Chapter 5

Numeration Systems: Does It Matter How We Name Numbers?

Section 5.1 The Evolution of Numeration Systems

1. $300+20+4=324$

3. $20,000+1,000+300+20+4=21,324$

5. $2,000,000+100,000+20,000+1,000+200+10=2,121,210$

7. 𓍢𓍢𓎆𓎆𓎆𓎆𓏺𓏺𓏺𓏺𓏺

9. 𓆼𓆼𓆼𓍢𓍢𓎆𓎆𓎆𓎆𓏺𓏺𓏺𓏺𓏺

11. 𓆐𓆐𓂭𓂭𓂭𓂭𓆼𓆼𓆼𓆼𓆼𓍢𓍢𓍢𓎆

13.

𓍢𓍢𓍢𓎆𓎆𓏺𓏺𓏺𓏺𓏺
+𓍢𓍢𓍢𓍢𓍢𓍢𓍢𓍢𓎆𓎆𓏺𓏺𓏺𓏺𓏺𓏺𓏺𓏺
𓍢𓍢𓍢𓍢𓍢𓍢𓍢𓍢𓍢𓍢𓍢𓎆𓎆𓎆𓎆𓏺𓏺𓏺𓏺𓏺𓏺𓏺𓏺𓏺𓏺𓏺𓏺𓏺

ten scrolls equals 1 lotus flower

ten strokes equals 1 heel bone

𓆼𓍢𓎆𓎆𓎆𓎆𓎆𓏺𓏺𓏺

15.

𓁨𓂭𓂭𓂭𓂭𓆼𓍢𓍢𓎆
+𓁨𓆐𓂭𓂭𓂭𓂭𓂭𓂭𓂭𓂭𓍢
𓁨𓁨𓆐𓂭𓂭𓂭𓂭𓂭𓂭𓂭𓂭𓂭𓂭𓂭𓂭𓆼𓍢𓍢𓍢𓎆

ten pointing fingers equals 1 fish

𓁨𓁨𓆐𓆐𓂭𓆼𓍢𓍢𓍢𓎆

17.

one scroll equals ten heel bones

𓍢𓍢𓍢𓍢𓍢𓍢𓎆𓎆𓏺𓏺𓏺𓏺𓏺𓏺𓏺
−𓍢𓍢𓍢𓎆𓎆𓎆𓎆𓎆𓏺𓏺𓏺𓏺

𓍢𓍢𓍢𓍢𓍢𓎆𓎆𓎆𓎆𓎆𓎆𓎆𓎆𓎆𓎆𓎆𓎆𓏺𓏺𓏺𓏺𓏺𓏺𓏺
−𓍢𓍢𓍢𓎆𓎆𓎆𓎆𓎆𓏺𓏺𓏺𓏺
𓍢𓍢𓎆𓎆𓎆𓎆𓎆𓎆𓎆𓏺𓏺𓏺

19.

one pointing finger equals ten lotus flowers

one scroll equals ten heel bones

𓂭𓂭𓂭𓍢𓍢𓍢𓍢𓍢
−𓂭𓂭𓆼𓆼𓆼𓆼𓍢𓍢𓍢𓎆𓎆𓎆𓎆𓎆𓏺𓏺𓏺𓏺

one heel bone equals ten staffs

𓂭𓂭𓆼𓆼𓆼𓆼𓆼𓆼𓆼𓆼𓆼𓆼𓍢𓍢𓍢𓍢𓎆𓎆𓎆𓎆𓎆𓎆𓎆𓎆𓎆𓎆
−𓂭𓂭𓆼𓆼𓆼𓆼𓍢𓍢𓍢𓎆𓎆𓎆𓎆𓎆𓏺𓏺𓏺𓏺

𓂭𓂭𓆼𓆼𓆼𓆼𓆼𓆼𓆼𓆼𓆼𓆼𓍢𓍢𓍢𓍢𓎆𓎆𓎆𓎆𓎆𓎆𓎆𓎆𓎆𓏺𓏺𓏺𓏺𓏺𓏺𓏺𓏺𓏺𓏺
−𓂭𓂭𓆼𓆼𓆼𓆼𓍢𓍢𓍢𓎆𓎆𓎆𓎆𓎆𓏺𓏺𓏺𓏺
𓆼𓆼𓆼𓆼𓆼𓆼𓍢𓎆𓎆𓎆𓎆𓏺𓏺𓏺𓏺𓏺𓏺

21. $14 = 2 + 4 + 8$

Powers of 2	Times 43
1	43
2	43 + 43 = 86
4	86 + 86 = 172
8	172 + 172 = 344

86 + 172 + 344 = 602

23. $21 = 1 + 4 + 16$

Powers of 2	Times 126
1	126
2	126 + 126 = 252
4	252 +252 = 504
8	504 + 504 = 1,008
16	1,008 + 1,008 = 2,016
32	1,936 + 1,936 = 3,872

126 + 504 + 2,016 = 2,646

25. $500 + 50 + 10 + (5 - 1) =$
$500 + 60 + 4 = 564$

27. $1{,}000 + (1{,}000 - 100) + 50 + 10 + 3 = 1{,}000 + 900 + 60 + 3 = 1{,}963$

29. $5 \times 1{,}000 + 2{,}000 + 500 + (50 - 10) + (5 - 1) = 5{,}000 + 2{,}000 + 500 + 40 + 4 =$
$7{,}000 + 500 + 40 + 4 = 7{,}544$

31. $500 \times 100 + 200 + 50 + 10 + 2 = 50{,}000 + 200 + 60 + 2 = 50{,}262$

33. $5 \times 1{,}000 \times 100 + 1{,}000 + (500 - 100) + 20 = 500{,}000 + 1{,}000 + 400 + 20 = 501{,}420$

35. CCLXXVIII

37. CDXLIV

39. $\overline{\text{IV}}$DCCXCV

41. $\overline{\text{LXXXIX}}$CDXXIII

43. $4 \times 100 + 3 \times 10 + 6 = 400 + 30 + 6 = 436$

45. $5 \times 1{,}000 + 6 \times 10 + 7 = 5{,}000 + 60 + 7 = 5{,}067$

47. $9 \times 1{,}000 + 9 \times 100 + 9 \times 10 + 9 = 9{,}000 + 900 + 90 + 9 = 9{,}999$

49. 四百九十五

51. 二千八百五

53. 九千八百四十六

55. $2 \times 1{,}000 + 6 \times 100 + 5 \times 10 = 2{,}000 + 600 + 50 = 2{,}650$ B.C.

57.

𓆐𓆐𓂭𓂭𓂭𓆼𓆼𓆼𓆼𓍢𓍢𓍢𓍢𓍢𓎆𓎆𓎆𓎆
+𓆐𓆼𓆼𓎆𓎆𓎆𓎆𓎆𓎆𓎆𓎆
𓆐𓆐𓆐𓂭𓂭𓂭𓆼𓆼𓆼𓆼𓆼𓆼𓍢𓍢𓍢𓍢𓍢𓎆𓎆𓎆𓎆𓎆𓎆𓎆𓎆𓎆𓎆𓎆𓎆

ten heel bones equals 1 scroll

𓆐𓆐𓆐𓂭𓂭𓂭𓆼𓆼𓆼𓆼𓆼𓆼𓍢𓍢𓍢𓍢𓍢𓍢𓎆𓎆

$$\begin{array}{r} \scriptstyle 1 \\ 234{,}540 \\ +102{,}080 \\ \hline 336{,}620 \end{array}$$

59. 36; 9 𓆼's, 9 𓍢's, 9 𓎆's, and 9 𓏺's

61. 54; 9 𓆐's, 9 𓂭's, 9 𓆼's, 9 𓍢's, 9 𓎆's, and 9 𓏺's

63. MCMXXXIX

65. MCMXCIV

67.

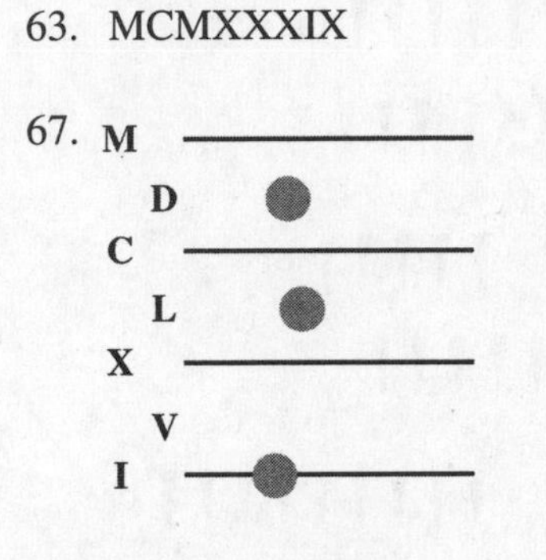

DLI

69. M, D, C, L, X, V, I

DCCIIII

71. 一千五百二十三 ; 一千二十七

73. A number tells "how many." A numeral is a symbol for writing a number.

75. They did not have the concept of place value and therefore had to write each power of ten that they used in their numerals.

77. 9,999,999

79. In the solution to this exercise, the symbol √. is being used instead of √.

a) ∝∝∝ ∇ ⊗⊗⊗⊗√.√. ; ∝∝∝∝∝∝∝ ∇∇√.√.√.

b)

∝∝∝ ∇ ⊗⊗⊗⊗√.√.
\+ ∝∝∝∝∝∝∝ ∇∇√.√.√.
∝∝∝∝∝∝∝∝∝∝ ∇∇∇ ⊗⊗⊗⊗√.√.√.√.√.

regroup ≈ ∇∇∇ ⊗⊗⊗⊗√.√.√.√.√.

c)

∝∝∝∝∝∝∝ ∇∇√.√.√.
− ∝∝∝ ∇ ⊗⊗⊗⊗√.√.

regroup

∝∝∝∝∝∝ ∇ ⊗⊗⊗⊗⊗⊗⊗⊗⊗⊗√.√.√.
− ∝∝∝ ∇ ⊗⊗⊗⊗√.√.
∝∝∝ ⊗⊗⊗⊗⊗⊗ √.

81. Answers may vary.

$\frac{1}{2}+\frac{1}{6}$

83. Answers may vary.

$\frac{1}{4}+\frac{1}{28}$

Section 5.2 Place Value Systems

1. $(10+2)\times 60+(10+1)=12\times 60+11=720+11=731$

3. $2\times 60^2+(20-4)\times 60+(20+7)=2\times 3,600+16\times 60+27=7,200+960+27=8,187$

5. $8,235=2\times 60^2+17\times 60+15$ or
$8,235=2\times 60^2+(20-3)\times 60+15$

7. $18,397=5\times 60^2+6\times 60+37$ or
$18,397=5\times 60^2+6\times 60+(40-3)$

9. $123,485=34\times 60^2+18\times 60+5$ or
$123,485=34\times 60^2+(20-2)\times 60+5$

11. $188,289=52\times 60^2+18\times 60+9$ or
$188,289=52\times 60^2+(20-2)\times 60+9$

13. $2\times 20\times 18\times 20=14,400$
$12\times 20\times 18=4,320$
$0\times 20=0$
$13=13$
18,733

15. $2\times 20\times 18\times 20^2=288,000$
$0\times 20\times 18\times 20=0$
$6\times 20\times 18=2,160$
$0\times 20=0$
$13=13$
290,173

17.

19.

21. five hundred

23. five thousand

25. $2\times 10^4+5\times 10^3+3\times 10^2+8\times 10^1+9\times 10^0$

27. $2\times 10^5+7\times 10^4+8\times 10^3+0\times 10^2+6\times 10^1+3\times 10^0$

29. $1\times 10^6+2\times 10^5+0\times 10^4+0\times 10^3+0\times 10^2+4\times 10^1+5\times 10^0$

31. 5,368

33. 370,082

35. $17\times 10^2=1\times 10^3+7\times 10^2$

37.
$$\begin{array}{rl} 2,863 = & 2\times10^3 + 8\times10^2 + 6\times10^1 + 3\times10^0 \\ \underline{+\ 425} = & \underline{+\qquad\quad 4\times10^2 + 2\times10^1 + 5\times10^0} \\ & 2\times10^3 + 12\times10^2 + 8\times10^1 + 8\times10^0 \end{array}$$

$$\begin{aligned} 2\times10^3 + 12\times10^2 + 8\times10^1 + 8\times10^0 &= 2\times10^3 + (10+2)\times10^2 + 8\times10^1 + 8\times10^0 \\ &= 2\times10^3 + 10\times10^2 + 2\times10^2 + 8\times10^1 + 8\times10^0 \\ &= 2\times10^3 + 10^3 + 2\times10^2 + 8\times10^1 + 8\times10^0 \\ &= (2+1)\times10^3 + 2\times10^2 + 8\times10^1 + 8\times10^0 \\ &= 3\times10^3 + 2\times10^2 + 8\times10^1 + 8\times10^0 \\ &= 3,288 \end{aligned}$$

39.
$$\begin{array}{rl} 3,482 = & 3\times10^3 + 4\times10^2 + 8\times10^1 + 2\times10^0 \\ \underline{+2,756} = & \underline{+\ 2\times10^3 + 7\times10^2 + 5\times10^1 + 6\times10^0} \\ & 5\times10^3 + 11\times10^2 + 13\times10^1 + 8\times10^0 \end{array}$$

$$\begin{aligned} 5\times10^3 + 11\times10^2 + 13\times10^1 + 8\times10^0 &= 5\times10^3 + (10+1)\times10^2 + (10+3)\times10^1 + 8\times10^0 \\ &= 5\times10^3 + 10\times10^2 + 1\times10^2 + 10\times10^1 + 3\times10^1 + 8\times10^0 \\ &= 5\times10^3 + 10^3 + 1\times10^2 + 10^2 + 3\times10^1 + 8\times10^0 \\ &= (5+1)\times10^3 + (1+1)\times10^2 + 3\times10^1 + 8\times10^0 \\ &= 6\times10^3 + 2\times10^2 + 3\times10^1 + 8\times10^0 \\ &= 6,238 \end{aligned}$$

41.
$$\begin{array}{rl} 926 = & 9\times10^2 + 2\times10^1 + 6\times10^0 \\ \underline{-784} = & \underline{-\ 7\times10^2 + 8\times10^1 + 4\times10^0} \end{array}$$

$$\begin{aligned} 9\times10^2 + 2\times10^1 + 6\times10^0 &= (8+1)\times10^2 + 2\times10^1 + 6\times10^0 \\ &= 8\times10^2 + 1\times10^2 + 2\times10^1 + 6\times10^0 \\ &= 8\times10^2 + 10\times10^1 + 2\times10^1 + 6\times10^0 \\ &= 8\times10^2 + (10+2)\times10^1 + 6\times10^0 \\ &= 8\times10^2 + 12\times10^1 + 6\times10^0 \end{aligned}$$

$$\begin{array}{rl} 926 = & 8\times10^2 + 12\times10^1 + 6\times10^0 \\ \underline{-784} = & \underline{-\ 7\times10^2 + 8\times10^1 + 4\times10^0} \\ & 1\times10^2 + 4\times10^1 + 2\times10^0 \end{array}$$

or 142

43.
$$\begin{array}{rl} 5,238 = & 5\times10^3 + 2\times10^2 + 3\times10^1 + 8\times10^0 \\ \underline{-1,583} = & \underline{-\ 1\times10^3 + 5\times10^2 + 8\times10^1 + 3\times10^0} \end{array}$$

$$\begin{aligned} 5\times10^3 + 2\times10^2 + 3\times10^1 + 8\times10^0 &= 5\times10^3 + (1+1)\times10^2 + 3\times10^1 + 8\times10^0 \\ &= 5\times10^3 + 1\times10^2 + 1\times10^2 + 3\times10^1 + 8\times10^0 \\ &= 5\times10^3 + 1\times10^2 + 10\times10^1 + 3\times10^1 + 8\times10^0 \\ &= 5\times10^3 + 1\times10^2 + (10+3)\times10^1 + 8\times10^0 \\ &= 5\times10^3 + 1\times10^2 + 13\times10^1 + 8\times10^0 \end{aligned}$$

43. (continued)

$$
\begin{aligned}
5\times10^3+1\times10^2+13\times10^1+8\times10^0 &= (4+1)\times10^3+1\times10^2+13\times10^1+8\times10^0\\
&= 4\times10^3+1\times10^3+1\times10^2+13\times10^1+8\times10^0\\
&= 4\times10^3+10\times10^2+1\times10^2+13\times10^1+8\times10^0\\
&= 4\times10^3+(10+1)\times10^2+13\times10^1+8\times10^0\\
&= 4\times10^3+11\times10^2+13\times10^1+8\times10^0
\end{aligned}
$$

$$
\begin{array}{rl}
5{,}238 = & 4\times10^3+11\times10^2+13\times10^1+8\times10^0\\
\underline{-1{,}583} = & \underline{-\ \ 1\times10^3+5\times10^2+8\times10^1+3\times10^0}\\
& 3\times10^3+6\times10^2+5\times10^1+5\times10^0
\end{array}
$$

or 3,655

45.

47.

Note: In part b of the solutions to Exercises 49 – 53 the smaller number is placed below the larger in the conventional method.

49. a)

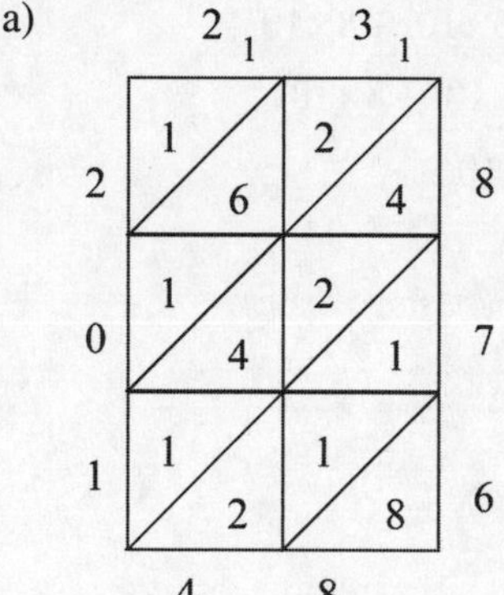

b)

$$
\begin{array}{r}
876\\
\times\ 23\\
\hline
18\\
210\\
2400\\
120\\
1400\\
16000\\
\hline
20{,}148
\end{array}
$$

(18, 210, 2400: 876×3; 120, 1400, 16000: 876×20)

51. a)

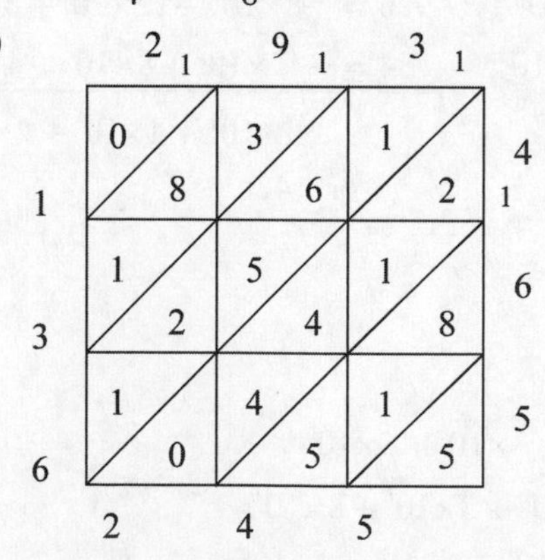

b)

$$
\begin{array}{r}
465\\
\times\ 293\\
\hline
15\\
180\\
1200\\
450\\
5400\\
36000\\
1000\\
12000\\
800000\\
\hline
136{,}245
\end{array}
$$

(15, 180, 1200: 465×3; 450, 5400, 36000: 465×90; 1000, 12000, 800000: 465×200)

53. $ad = 24$, $cd = 20$, $ae = 42$, and $be = 56 \Rightarrow ad = 6 \cdot 4$, $cd = 5 \cdot 4$, $ae = 6 \cdot 7$, and $be = 8 \cdot 7$

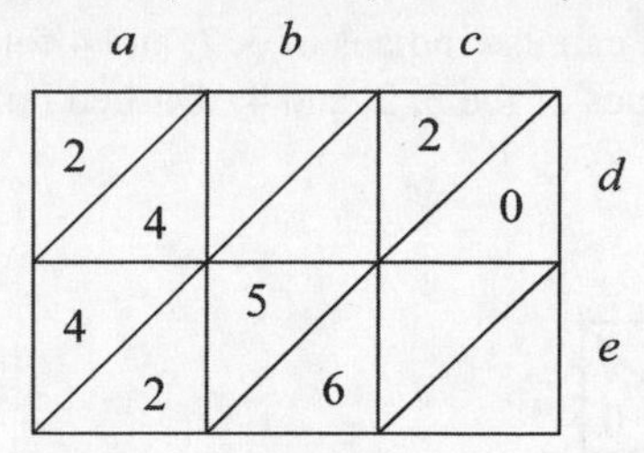

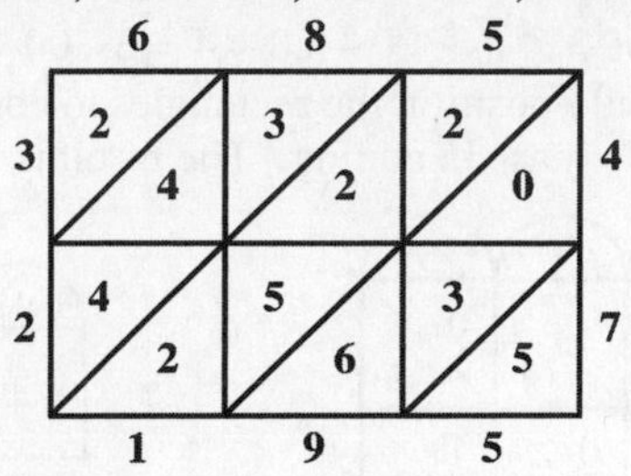

$685 \times 47 = 32,195$

55. 3,936

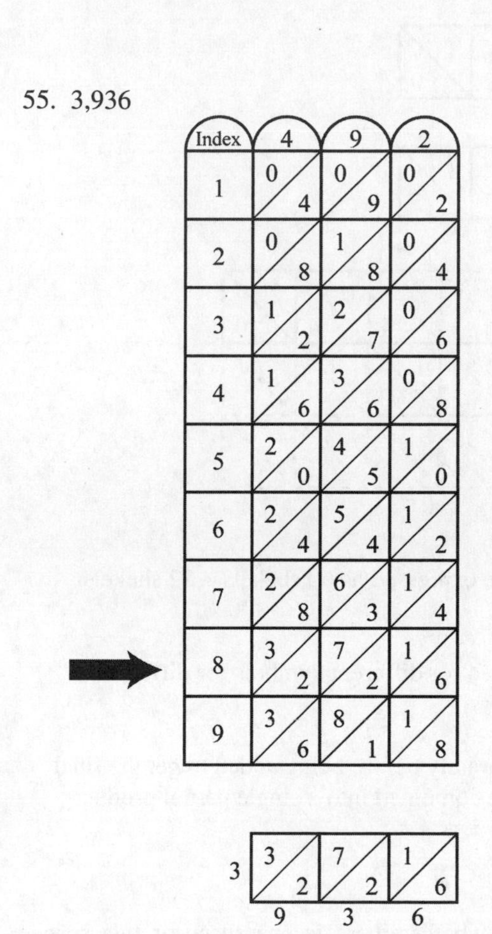

57. 5,544

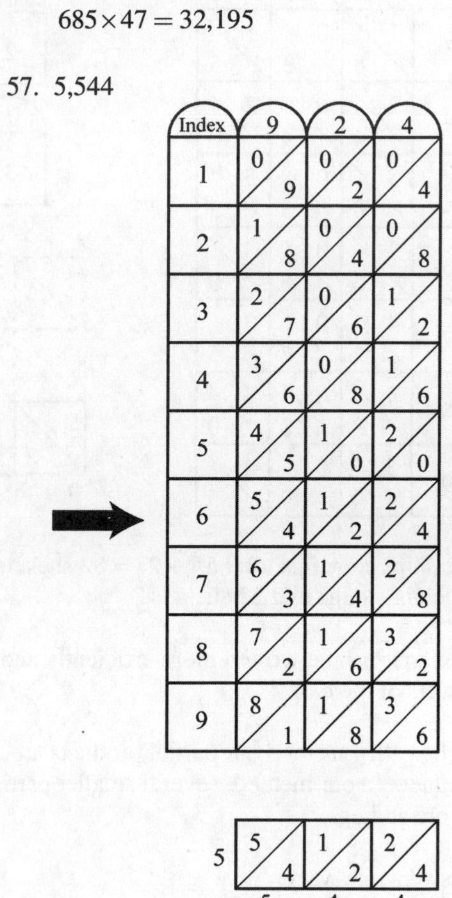

59. Answers may vary.

If you think of 324 as $300+20+4=3\cdot 100+2\cdot 10+4$ then you can use indices of 3, 2, and 4 for 615. You would position the rectangles to consider the place values of the 3, 2, and 4. Dashed boxes are included for ease in adding. The result is 199,260.

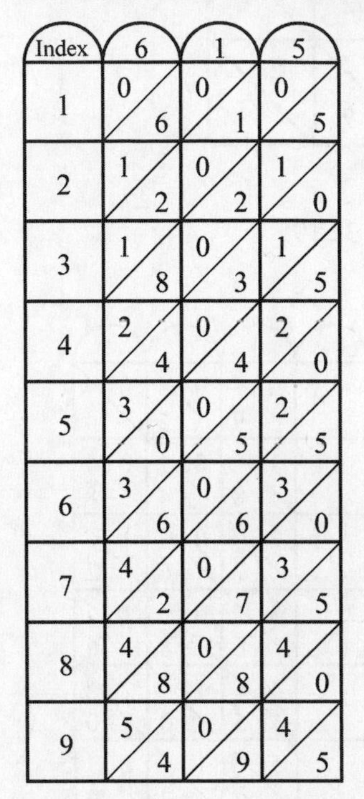

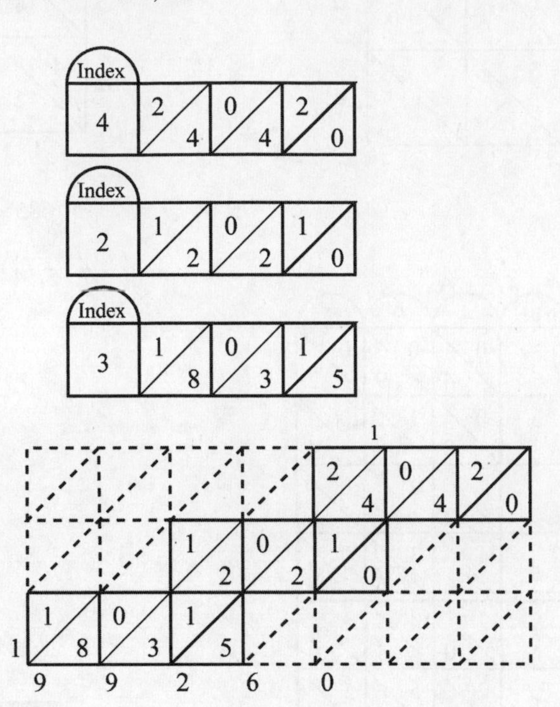

61. The ships contain a total 55 + 27 = 82 shekels, which can be expressed as 60 shekels + 22 shekels, which is 1 mina and 22 shekels.

63. Numerals can be written more efficiently and there is no need for different symbols for different powers of the base.

65. In the galley method, all partial products are computed separately before being added to get the final product. In our method, several smaller partial products are combined into a single partial product before adding.

67. Answers may vary.

In writing ▼ ▼▼▼▼▼, it could be difficult to tell whether there is one space or two spaces between the two groups of symbols. If there is one space, the number is $1\times 60+5=60+5=65$. If there are two spaces, the number is $1\times 60^2+5=3,600+5=3,605$.

69. Answers may vary.

The ancient Chinese system requires a separate symbol for each power of ten, e.g. one billion, ten billion, and so on. The Hindu–Arabic system uses different powers of 10, 10^9, 10^{10}, and so on. The positioning of symbols in the numeral indicates these powers of ten.

71. 四三六

73. 五〇六七

75. ħOOħ

77. $0.3=3\times 10^{-1}$; 0.07×10^{-2}; 0.005×10^{-3}

79. $2\times 10^2+5\times 10^0+6\times 10^{-1}+3\times 10^{-4}$

Section 5.3 Calculating in Other Bases

1. 23_5 and 30_5

3. 1010_2 and 1100_2

5. EE_{16} and $F0_{16}$

7. $$\begin{aligned} 432_5 &= 4\times5^2+3\times5^1+2\times5^0 \\ &= 4\times25+3\times5+2\times1 \\ &= 100+15+2 \\ &= 117 \end{aligned}$$

9. $$\begin{aligned} 504_6 &= 5\times6^2+0\times6^1+4\times6^0 \\ &= 5\times36+0\times6+4\times1 \\ &= 180+0+4 \\ &= 184 \end{aligned}$$

11. $$\begin{aligned} 100111_2 &= 1\times2^5+0\times2^4+0\times2^3+1\times2^2+1\times2^1+1\times2^0 \\ &= 1\times32+0\times16+0\times8+1\times4+1\times2+1\times1 \\ &= 32+0+0+4+2+1 \\ &= 39 \end{aligned}$$

13. $$\begin{aligned} 1110101_2 &= 1\times2^6+1\times2^5+1\times2^4+0\times2^3+1\times2^2+0\times2^1+1\times2^0 \\ &= 1\times64+1\times32+1\times16+0\times8+1\times4+0\times2+1\times1 \\ &= 64+32+16+0+4+0+1 \\ &= 117 \end{aligned}$$

15. $$\begin{aligned} 267_8 &= 2\times8^2+6\times8^1+7\times8^0 \\ &= 2\times64+6\times8+7\times1 \\ &= 128+48+7 \\ &= 183 \end{aligned}$$

17. $$\begin{aligned} 704_8 &= 7\times8^2+0\times8^1+4\times8^0 \\ &= 7\times64+0\times8+4\times1 \\ &= 448+0+4 \\ &= 452 \end{aligned}$$

19. $$\begin{aligned} 2F4_{16} &= 2\times16^2+15\times16^1+4\times16^0 \\ &= 2\times256+15\times16+4\times1 \\ &= 512+240+4 \\ &= 756 \end{aligned}$$

21. $$\begin{aligned} D08_{16} &= 13\times16^2+0\times16^1+8\times16^0 \\ &= 13\times256+0\times16+8\times1 \\ &= 3328+0+8 \\ &= 3{,}336 \end{aligned}$$

23. 2314_5

$$\begin{array}{l|ll} 5 & 334 & 4 \\ 5 & 66 & 1 \\ 5 & 13 & 3 \\ 5 & 2 & 2 \\ & 0 & \end{array}$$

25. 12302_6

$$\begin{array}{l|ll} 6 & 1838 & 2 \\ 6 & 306 & 0 \\ 6 & 51 & 3 \\ 6 & 8 & 2 \\ 6 & 1 & 1 \\ & 0 & \end{array}$$

27. 1100111_2

$$\begin{array}{r|l r} 2 & 103 & 1 \\ 2 & 51 & 1 \\ 2 & 25 & 1 \\ 2 & 12 & 0 \\ 2 & 6 & 0 \\ 2 & 3 & 1 \\ 2 & 1 & 1 \\ & 0 & \end{array}$$

29. 1011110_2

$$\begin{array}{r|l r} 2 & 94 & 0 \\ 2 & 47 & 1 \\ 2 & 23 & 1 \\ 2 & 11 & 1 \\ 2 & 5 & 1 \\ 2 & 2 & 0 \\ 2 & 1 & 1 \\ & 0 & \end{array}$$

31. 6513_8

$$\begin{array}{r|l r} 8 & 3403 & 3 \\ 8 & 425 & 1 \\ 8 & 53 & 5 \\ 8 & 6 & 6 \\ & 0 & \end{array}$$

33. $AE8_{16}$

$$\begin{array}{r|l l} 16 & 2792 & 8 \\ 16 & 174 & E\,(14) \\ 16 & 10 & A\,(10) \\ & 0 & \end{array}$$

35. DEA_{16}

$$\begin{array}{r|l l} 16 & 3562 & A\,(10) \\ 16 & 222 & E\,(14) \\ 16 & 13 & D\,(13) \\ & 0 & \end{array}$$

37.
$$\begin{array}{r} \overset{1}{3}412_5 \\ +\ 231_5 \\ \hline 4143_5 \end{array}$$

39.
$$\begin{array}{r} \overset{1}{2}\overset{1}{7}35_9 \\ +\,3246_9 \\ \hline 6082_9 \end{array}$$

41.
$$\begin{array}{r} \overset{1}{5}\overset{1}{4}\overset{1}{1}5_7 \\ +\,2436_7 \\ \hline 11154_7 \end{array}$$

43.
$$\begin{array}{r} 2A\overset{1}{1}8_{16} \\ +\ 43B_{16} \\ \hline 2E53_{16} \end{array}$$

45.
$$\begin{array}{r} 11011_2 \\ +\,10101_2 \\ \hline 110000_2 \end{array}$$
(carries: 1 1 1 1 1)

47.
$$\begin{array}{r} 2\overset{3}{\not4}\overset{1}{1}2_5 \\ -\ 321_5 \\ \hline 2041_5 \end{array}$$

49.
$$\begin{array}{r} \overset{3}{\not4}\,\overset{5}{\not2}\,\overset{1}{6}3_7 \\ -\,2436_7 \\ \hline 1524_7 \end{array}$$

51.
$$\begin{array}{r} A\overset{7}{\not8}\overset{1}{3}_{16} \\ -\ 43B_{16} \\ \hline 648_{16} \end{array}$$

53.
$$\begin{array}{r} 11\overset{0}{\not1}\overset{1}{0}11_2 \\ -\ 10101_2 \\ \hline 100110_2 \end{array}$$

55.
$$\begin{array}{r} 41_5 \\ \times\ 23_5 \\ \hline 223_5 \\ {}_1132 \\ \hline 2043_5 \end{array}$$

57.
$$\begin{array}{r} \overset{1}{\overset{1}{3}}02_5 \\ \times\ 43_5 \\ \hline \overset{1}{1}411_5 \\ 2213 \\ \hline 24041_5 \end{array}$$

59. $114_5\ R\,11_5$

$$\begin{array}{r} 114 \\ 24_5\,\overline{)\,3412_5} \\ \underline{24} \\ 101 \\ \underline{24} \\ 222 \\ \underline{211} \\ 11 \end{array}$$

61. $44_5\ R\,24_5$

$$\begin{array}{r} 44 \\ 42_5\,\overline{)\,4132_5} \\ \underline{323} \\ 402 \\ \underline{323} \\ 24 \end{array}$$

63. 1355_8 $2ED_{16}$

1 011 101 101 10 1110 1101
1 3 5 5 2 E D

67. 10100110_2

2 4 6
010 100 110

65. 1751_8 $3E9_{16}$

1 111 101 001 11 1110 1001
1 7 5 1 3 E 9

69. 101000111110_2

A 3 E
1010 0011 1110

71. 754_{16}

3 5 2 4
011 101 010 100
0111 0101 0100
7 5 4

73. 1654_8

3 *A* *C*
0011 1010 1100
001 110 101 100
1 6 5 4

For Exercises 75 – 77 refer to the following table.

65	66	67	68	69	70	71	72	73	74	75	76	77
A	B	C	D	E	F	G	H	I	J	K	L	M

78	79	80	81	82	83	84	85	86	87	88	89	90
N	O	P	Q	R	S	T	U	V	W	X	Y	Z

75. CANDY

1000011 1000001 1001110 1000100 1011001

$2^6+2^1+2^0=$ $2^6+2^0=$ $2^6+2^3+2^2+2^1=$ $2^6+2^2=$ $2^6+2^4+2^3+2^0=$

$64+2+1=67$ $64+1=65$ $64+8+4+2=78$ $64+4=68$ $64+16+8+1=89$

77. LOVE

1001100 1001111 1010110 1000101

$2^6+2^3+2^2=$ $2^6+2^3+2^2+2^1+2^0=$ $2^6+2^4+2^2+2^1=$ $2^6+2^2+2^0=$

$64+8+4=76$ $64+8+4+2+1=79$ $64+16+4+2=86$ $64+4+1=69$

79. 21 quarters, 1 dime, 1 nickel, and three pennies

81. *b*; Since it is the base on what appears to the smaller value.

83. 6; Since $1_b + 4_b = 5_b$ and $5_b + 3_b = 12_b$.

85. Base 8: 64 facts; The eight symbols 0, 1, ... 7 are combined with themselves.
Base 16: 256 facts; The sixteen symbols 0, 1, ...9, *A*, *B*, ... *F* are combined with themselves.

87. One; Three binary places can represent 000_2 to 111_2 (decimal 0 to 7), which is the same as 0_8 to 7_8, which is one octal place.

89. 4123_5

$$\begin{aligned}201221_3 &= 2\times3^5 + 0\times3^4 + 1\times3^3 + 2\times3^2 + 2\times3^1 + 1\times3^0\\ &= 2\times243 + 0\times81 + 1\times27 + 2\times9 + 2\times3 + 1\times1\\ &= 486 + 27 + 18 + 6 + 1\\ &= 538\end{aligned}$$

$$\begin{array}{r|rl} 5 & 538 & 3\\ 5 & 107 & 2\\ 5 & 21 & 1\\ 5 & 4 & 4\\ & 0 & \end{array}$$

91. 3774_9

$$\begin{aligned}B05_{16} &= 11\times16^2 + 0\times16^1 + 5\times16^0\\ &= 11\times256 + 0\times16 + 5\times1\\ &= 3072 + 5\\ &= 2821\end{aligned}$$

$$\begin{array}{r|rl} 9 & 2821 & 4\\ 9 & 313 & 7\\ 9 & 34 & 7\\ 9 & 3 & 3\\ & 0 & \end{array}$$

93. We will use the letters P, G, B, and Y to represent the pink, green, blue, and yellow faces, respectively.
G, B, Y, GP, GG, GB, GY, BP, BG, BB, BY, YP, YG, YB, YY, GPP, GPG, GPB

95. GYP
$28 = 1\times4^2 + 3\times4^1 + 0\times4^0$

97. $x = 4, y = 5$

99. $x = A, y = 5$

Section 5.4 Looking Deeper: Modular Systems

1. 7 since $43 = 3\times12 + 7$

3. 4 since $39 = 5\cdot7 + 4$

5. True since $59 - 35 = 24$ is evenly divisible by 12.

7. True since $43 - 11 = 32$ is evenly divisible by 8.

9. False since $78 - 46 = 32$ is not evenly divisible by 10.

11. 2 since $3 + 4 \equiv 2 \pmod 5$

13. 7 since $4 - 9 \equiv 7 \pmod{12}$

15. 0 since $4\times9 \equiv 0 \pmod{12}$

17. 6 since $2\times5 + 4 \equiv 2 \pmod 8$

19. 7 since $6 - 3\times5 \equiv 7 \pmod 8$

21. 6

$6+0 \equiv 4 \pmod 8$; false

$6+1 \equiv 4 \pmod 8$; false

$6+2 \equiv 4 \pmod 8$; false

$6+3 \equiv 4 \pmod 8$; false

$6+4 \equiv 4 \pmod 8$; false

$6+5 \equiv 4 \pmod 8$; false

$6+6 \equiv 4 \pmod 8$; **true**

$6+7 \equiv 4 \pmod 8$; false

23. 6

$3-0 \equiv 4 \pmod 7$; false

$3-1 \equiv 4 \pmod 7$; false

$3-2 \equiv 4 \pmod 7$; false

$3-3 \equiv 4 \pmod 7$; false

$3-4 \equiv 4 \pmod 7$; false

$3-5 \equiv 4 \pmod 7$; false

$3-6 \equiv 4 \pmod 7$; **true**

25. 3

$2-0 \equiv 8 \pmod 9$; false

$2-1 \equiv 8 \pmod 9$; false

$2-2 \equiv 8 \pmod 9$; false

$2-3 \equiv 8 \pmod 9$; **true**

$2-4 \equiv 8 \pmod 9$; false

$2-5 \equiv 8 \pmod 9$; false

$2-6 \equiv 8 \pmod 9$; false

$2-7 \equiv 8 \pmod 9$; false

$2-8 \equiv 8 \pmod 9$; false

27. 4

$3\times0 \equiv 5 \pmod 7$; false

$3\times1 \equiv 5 \pmod 7$; false

$3\times2 \equiv 5 \pmod 7$; false

$3\times3 \equiv 5 \pmod 7$; false

$3\times4 \equiv 5 \pmod 7$; **true**

$3\times5 \equiv 5 \pmod 7$; false

$3\times6 \equiv 5 \pmod 7$; false

29. 3, 8

$4\times0 \equiv 2 \pmod{10}$; false

$4\times1 \equiv 2 \pmod{10}$; false

$4\times2 \equiv 2 \pmod{10}$; false

$4\times3 \equiv 2 \pmod{10}$; **true**

$4\times4 \equiv 2 \pmod{10}$; false

$4\times5 \equiv 2 \pmod{10}$; false

$4\times6 \equiv 2 \pmod{10}$; false

$4\times7 \equiv 2 \pmod{10}$; false

$4\times8 \equiv 2 \pmod{10}$; **true**

$4\times9 \equiv 2 \pmod{10}$; false

31. For $x \equiv 3 \pmod 7$, x is in the congruence class containing 3 modulo 7 which is

3, 10, 17, **24**, 31, 38, …

For $x \equiv 4 \pmod 5$, x is in the congruence class containing 4 modulo 5 which is

4, 9, 14, 19, **24**, 29, …

The smallest number that is found in both of these classes is 24.

33. For $x \equiv 2 \pmod{10}$, x is in the congruence class containing 2 modulo 10 which is

2, 12, 22, 32, **42**, 52, 62, 72...

For $x \equiv 2 \pmod{8}$, x is in the congruence class containing 2 modulo 8 which is

2, 10, 18, 26, 34, **42**, 50, 58 ...

For $x \equiv 0 \pmod{6}$, x is in the congruence class containing 0 modulo 6 which is

0, 6, 12, 18, 24, 30, 36, **42**, ...

The smallest number that is found in these three classes is 42.

35. 3 since $80 = 11 \cdot 7 + 3$

37. This problem is the same as finding the smallest positive integer that solves both $x \equiv 4 \pmod{6}$ and $x \equiv 3 \pmod{5}$.

For $x \equiv 4 \pmod{6}$, x is in the congruence class containing 4 modulo 6 which is

4, 10, 16, 22, **28**, 34, ...

For $x \equiv 3 \pmod{5}$, x is in the congruence class containing 3 modulo 5 which is

3, 8, 13, 18, 23, **28**, ...

The smallest number that is found in both of these classes is 28.

39. 23;
$23 \equiv 2 \pmod{3}$
$23 \equiv 3 \pmod{5}$
$23 \equiv 2 \pmod{7}$

41. We are looking for years that are congruent to 2 modulo 12. Since this book was written in 2010 and Chris appears to be less than 30, we will start in the year $2010 - 29 = 1981$.

$1981 \equiv 2 \pmod{12}$; false
$1982 \equiv 2 \pmod{12}$; **true**
$1983 \equiv 2 \pmod{12}$; false
$1984 \equiv 2 \pmod{12}$; false
$1985 \equiv 2 \pmod{12}$; false
$1986 \equiv 2 \pmod{12}$; false
$1987 \equiv 2 \pmod{12}$; false
$1988 \equiv 2 \pmod{12}$; false
$1989 \equiv 2 \pmod{12}$; false
$1990 \equiv 2 \pmod{12}$; false
$1991 \equiv 2 \pmod{12}$; false
$1992 \equiv 2 \pmod{12}$; false

Chris was born in 1982.

Note: If we use 2011 for the starting year, the answer will still be 1982. If a starting year after 2011 is used, the answer will be 1982 + 12 = 1994.

43. 9

10×0	0
9×3	27
8×0	0
7×7	49
6×5	30
$5\times d$	$5d$
4×2	8
3×8	24
2×3	6
Total	$144+5d$

Since 9 is the check digit, we need to find the smallest integer that satisfies $11-9=2\equiv(144+5d)(\text{mod}\,11)$.

$2\equiv(144+5\times0)(\text{mod}\,11)$; false
$2\equiv(144+5\times1)(\text{mod}\,11)$; false
$2\equiv(144+5\times2)(\text{mod}\,11)$; false
$2\equiv(144+5\times3)(\text{mod}\,11)$; false
$2\equiv(144+5\times4)(\text{mod}\,11)$; false
$2\equiv(144+5\times5)(\text{mod}\,11)$; false
$2\equiv(144+5\times6)(\text{mod}\,11)$; false
$2\equiv(144+5\times7)(\text{mod}\,11)$; false
$2\equiv(144+5\times8)(\text{mod}\,11)$; false
$2\equiv(144+5\times9)(\text{mod}\,11)$; **true**
$2\equiv(144+5\times10)(\text{mod}\,11)$; false

45. 6

10×0	0
9×0	0
8×6	48
7×1	7
6×3	18
5×4	20
4×9	36
$3\times d$	$3d$
2×0	0
Total	$129+3d$

Since 7 is the check digit, we need to find the smallest integer that satisfies $11-7=4\equiv(129+3d)(\text{mod}\,11)$.

$4\equiv(129+3\times0)(\text{mod}\,11)$; false
$4\equiv(129+3\times1)(\text{mod}\,11)$; false
$4\equiv(129+3\times2)(\text{mod}\,11)$; false
$4\equiv(129+3\times3)(\text{mod}\,11)$; false
$4\equiv(129+3\times4)(\text{mod}\,11)$; false
$4\equiv(129+3\times5)(\text{mod}\,11)$; false
$4\equiv(129+3\times6)(\text{mod}\,11)$; **true**
$4\equiv(129+3\times7)(\text{mod}\,11)$; false
$4\equiv(129+3\times8)(\text{mod}\,11)$; false
$4\equiv(129+3\times9)(\text{mod}\,11)$; false
$4\equiv(129+3\times10)(\text{mod}\,11)$; false

47. 6

Digits	4	5	6	3	2	6	2	5	2	1	0	4	3	5	3	c
Doubled Digits	8		12		4		4		4		0		6		6	?
Digit sums	8	5	3	3	4	6	4	5	4	1	0	4	6	5	6	

Sum of digits is 64, so $c=10-(64(\text{mod}\,10))=10-4=6$

49. 3

Digits	3	1	6	2	4	4	2	5	3	4	8	2	2	9	1	c
Doubled Digits	6		12		8		4		6		16		4		2	?
Digit sums	6	1	3	2	8	4	4	5	6	4	7	2	4	9	2	

Sum of digits is 67, so $c=10-(67(\text{mod}\,10))=10-7=3$

51. Answers may vary. One possible answer is to perform the operations as usual and then count off the result on an m-hour clock.

53. a) Check digits help insure that numbers have no errors.

    b) Modular arithmetic is often used in computing check digits.

55. $15^5 \pmod{35} = 759375 \pmod{35} = 15$

57. $5^5 \pmod{35} = 3125 \pmod{35} = 10$

59. If $2 / x \equiv 5 \pmod{6}$ then $2 \equiv 5x \pmod{6}$

    $2 \equiv (5\times 0) \pmod{6}$; false

    $2 \equiv (5\times 1) \pmod{6}$; false

    $2 \equiv (5\times 2) \pmod{6}$; false

    $2 \equiv (5\times 3) \pmod{6}$; false

    $2 \equiv (5\times 4) \pmod{6}$; **true**

    $2 \equiv (5\times 5) \pmod{6}$; false   Answer: 4

Chapter Review Exercises

1. $1{,}000{,}000 + 200{,}000 + 30{,}000 + 2{,}000 + 200 + 10 = 1{,}232{,}210$

2. one scroll equals ten heel bones   one heel bone equals ten staffs

3. $37 = 1 + 4 + 32$

Powers of 2	Times 53
1	53
2	53 + 53 = 106
4	106 + 106 = 212
8	212 + 212 = 424
16	424 + 424 = 848
32	848 + 848 = 1,696

1,696 + 212 +53 = 1,961

4. MMMMDCCXCV

5. $7\times 1{,}000 + 5\times 100 + 9\times 10 + 3 = 7{,}000 + 500 + 90 + 3 = 7{,}593$

6. Answers may vary.
   No. A number tells us "how many," whereas a numeral is a symbol or symbols that we use to represent the number.

7. Answers may vary.
   The Roman system had the subtraction principle, which allowed them to write certain numbers more efficiently. They also had a multiplication principle.

8. One numeral could represent several objects; numerals could be changed to cheat customers.

9. $11,292 = 3\times 60^2 + 8\times 60 + 12$ or

$11,292 = 3\times 60^2 + (10-2)\times 60 + 12$

10.
$$\begin{array}{rl} 4,237 = & 4\times10^3 + 2\times10^2 + 3\times10^1 + 7\times10^0 \\ \underline{-2,673} = - & \underline{2\times10^3 + 6\times10^2 + 7\times10^1 + 3\times10^0} \end{array}$$

$$\begin{aligned} 4\times10^3 + 2\times10^2 + 3\times10^1 + 7\times10^0 &= 4\times10^3 + (1+1)\times10^2 + 3\times10^1 + 7\times10^0 \\ &= 4\times10^3 + 1\times10^2 + 1\times10^2 + 3\times10^1 + 7\times10^0 \\ &= 4\times10^3 + 1\times10^2 + 10\times10^1 + 3\times10^1 + 7\times10^0 \\ &= 4\times10^3 + 1\times10^2 + (10+3)\times10^1 + 7\times10^0 \\ &= 4\times10^3 + 1\times10^2 + 13\times10^1 + 7\times10^0 \end{aligned}$$

$$\begin{aligned} 4\times10^3 + 1\times10^2 + 13\times10^1 + 7\times10^0 &= (3+1)\times10^3 + 1\times10^2 + 13\times10^1 + 7\times10^0 \\ &= 3\times10^3 + 1\times10^3 + 1\times10^2 + 13\times10^1 + 7\times10^0 \\ &= 3\times10^3 + 10\times10^2 + 1\times10^2 + 13\times10^1 + 7\times10^0 \\ &= 3\times10^3 + (10+1)\times10^2 + 13\times10^1 + 7\times10^0 \\ &= 3\times10^3 + 11\times10^2 + 13\times10^1 + 7\times10^0 \end{aligned}$$

$$\begin{array}{rl} 4,237 = & 3\times10^3 + 11\times10^2 + 13\times10^1 + 7\times10^0 \\ \underline{-2,673} = - & \underline{2\times10^3 + 6\times10^2 + 7\times10^1 + 3\times10^0} \\ & 1\times10^3 + 5\times10^2 + 6\times10^1 + 4\times10^0 \end{array}$$

or 1,564

11. 4,738

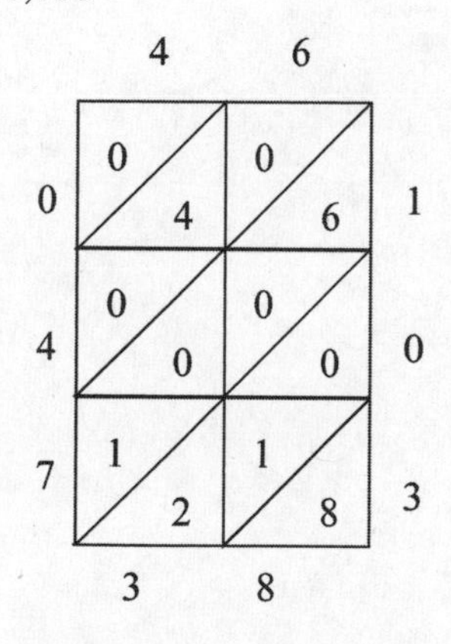

12. Answers may vary.
We need to be able to indicate which powers of the base (say ten) that are missing from the numeral.

13. a) $1\times 20\times 18 + 15 = 360 + 15 = 375$

b) $3\times 20\times 18 + 18\times 20 + 0 = 1080 + 360 = 1440$

14. $342_5 = 3\times5^2 + 4\times5^1 + 2\times5^0$
$= 3\times25 + 4\times5 + 2\times1$
$= 75 + 20 + 2$
$= 97$

and

$B3D_{16} = 11\times16^2 + 3\times16^1 + 13\times16^0$
$= 11\times256 + 3\times16 + 13\times1$
$= 2816 + 48 + 13$
$= 2,877$

15. 6513_8

$$\begin{array}{r|l l} 8 & 3403 & 3 \\ 8 & 425 & 1 \\ 8 & 53 & 5 \\ 8 & 6 & 6 \\ & 0 & \end{array}$$

16.
$$\begin{array}{r} {\scriptstyle 11111} \\ 10111_2 \\ +\ 11001_2 \\ \hline 110000_2 \end{array}$$

17. 134_5 R 20_5

$$\begin{array}{r} 134 \\ 23_5 \overline{)4312_5} \\ \underline{23} \\ 201 \\ \underline{124} \\ 222 \\ \underline{202} \\ 20 \end{array}$$

18. 1342_8 $2E2_{16}$

1 011 100 010 10 1110 0010
1 3 4 2 2 E 2

19. 1431_5

$463_7 = 4\times7^2 + 6\times7^1 + 3\times7^0$
$= 4\times49 + 6\times7 + 3\times1$
$= 196 + 42 + 3$
$= 241$

$$\begin{array}{r|l l} 5 & 241 & 1 \\ 5 & 48 & 3 \\ 5 & 9 & 4 \\ 5 & 1 & 1 \\ & 0 & \end{array}$$

20. 4 since $76 = 9\times12 + 4$

21. a) True since $72 - 54 = 18$ is evenly divisible by 6.
b) False since $75 - 29 = 46$ is not evenly divisible by 11.

22. a) 7 since $8 + 11 \equiv 7 \pmod{12}$

b) 9 since $7 - 9 \equiv 9 \pmod{11}$

c) 5 since $4\times8 \equiv 5 \pmod 9$

23. a) 3, 7, 11

$3\times 0\equiv 9\pmod{12}$; false

$3\times 1\equiv 9\pmod{12}$; false

$3\times 2\equiv 9\pmod{12}$; false

$3\times 3\equiv 9\pmod{12}$; **true**

$3\times 4\equiv 9\pmod{12}$; false

$3\times 5\equiv 9\pmod{12}$; false

$3\times 6\equiv 9\pmod{12}$; false

$3\times 7\equiv 9\pmod{12}$; **true**

$3\times 8\equiv 9\pmod{12}$; false

$3\times 9\equiv 9\pmod{12}$; false

$3\times 10\equiv 9\pmod{12}$; false

$3\times 11\equiv 9\pmod{12}$; **true**

b) $\varnothing$

$4\times 0\equiv 3\pmod{8}$; false

$4\times 1\equiv 3\pmod{8}$; false

$4\times 2\equiv 3\pmod{8}$; false

$4\times 3\equiv 3\pmod{8}$; false

$4\times 4\equiv 3\pmod{8}$; false

$4\times 5\equiv 3\pmod{8}$; false

$4\times 6\equiv 3\pmod{8}$; false

$4\times 7\equiv 3\pmod{8}$; false

Chapter Test

1. MMMDCLXXXV

2. Hindu–Arabic numeration has zero and place value.

3.
$$\begin{aligned}264_7 &= 2\times 7^2+6\times 7^1+4\times 7^0\\ &= 2\times 49+6\times 7+4\times 1\\ &= 98+42+4\\ &= 144\end{aligned}$$
and
$$\begin{aligned}A3E_{16} &= 10\times 16^2+3\times 16^1+14\times 16^0\\ &= 10\times 256+3\times 16+14\times 1\\ &= 2560+48+14\\ &= 2{,}622\end{aligned}$$

4. $2\times 1{,}000{,}000+1\times 100{,}000+2\times 10{,}000+3\times 1{,}000+4\times 100+2\times 10=$
$2{,}000{,}000+100{,}000+20{,}000+3{,}000+400+20=2{,}123{,}420$

5. a) $1\times 20\times 18+19=360+19=379$

b) $3\times 20\times 18+12\times 20=1{,}080+240=1{,}320$

6. a) False since $57-43=14$ is not evenly divisible by 8.

b) True since $52-16=36$ is evenly divisible by 6.

7.

one astonished man equals ten tadpoles

one tadpole equals ten pointing fingers

one scroll equals ten heel bones

8. $5\times1{,}000+3\times100+4\times10+8=5{,}000+300+40+8=5{,}348$

9. 3 since $57=9\times6+3$

10. 10, X, ∩

11.
$$\begin{array}{r} 101101_2 \\ +110101_2 \\ \hline 1100010_2 \end{array}$$

12. $10{,}937=3\times60^2+2\times60+17$ or

$10{,}937=3\times60^2+2\times60+(20-3)$

13. 15,946

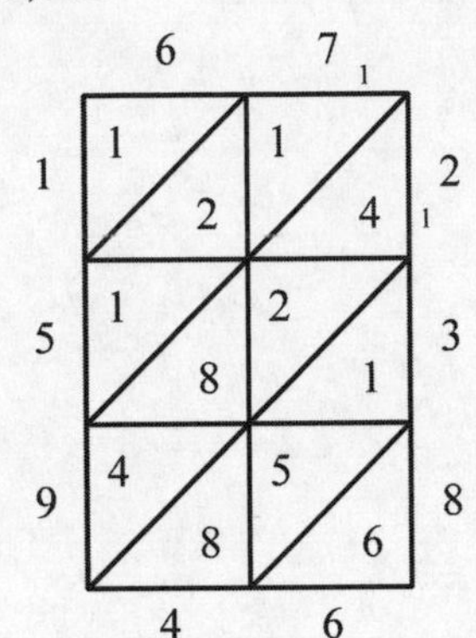

14. 4401_8

$$\begin{array}{r|rl} 8 & 2305 & 1 \\ 8 & 288 & 0 \\ 8 & 36 & 4 \\ 8 & 4 & 4 \\ & 0 & \end{array}$$

15. a) 8 since $9+12\equiv8\pmod{13}$

b) 8 since $6-10\equiv8\pmod{12}$

c) 2 since $5\times4\equiv2\pmod{6}$

16.
$$\begin{array}{rl} 1{,}738= & 1\times10^3+7\times10^2+3\times10^1+8\times10^0 \\ 526=+ & \qquad\quad 5\times10^2+2\times10^1+6\times10^0 \\ \hline & 1\times10^3+12\times10^2+5\times10^1+14\times10^0 \end{array}$$

$$\begin{aligned} 1\times10^3+12\times10^2+5\times10^1+14\times10^0 &=1\times10^3+(10+2)\times10^2+5\times10^1+(10+4)\times10^0 \\ &=1\times10^3+10\times10^2+2\times10^2+5\times10^1+10\times10^0+4\times10^0 \\ &=1\times10^3+1\times10^3+2\times10^2+5\times10^1+1\times10^1+4\times10^0 \\ &=(1+1)\times10^3+2\times10^2+(5+1)\times10^1+4\times10^0 \\ &=2\times10^3+2\times10^2+6\times10^1+4\times10^0 \\ &=2{,}264 \end{aligned}$$

17. Answers may vary. Possible answers include: More symbols were required to write their numerals.

18. a) 2, 5, 8, 11

$4\times 0\equiv 8\pmod{12}$; false
$4\times 1\equiv 8\pmod{12}$; false
$4\times 2\equiv 8\pmod{12}$; **true**
$4\times 3\equiv 8\pmod{12}$; false
$4\times 4\equiv 8\pmod{12}$; false
$4\times 5\equiv 8\pmod{12}$; **true**
$4\times 6\equiv 8\pmod{12}$; false
$4\times 7\equiv 8\pmod{12}$; false
$4\times 8\equiv 8\pmod{12}$; **true**
$4\times 9\equiv 8\pmod{12}$; false
$4\times 10\equiv 8\pmod{12}$; false
$4\times 11\equiv 8\pmod{12}$; **true**

b) $\varnothing$

$4\times 0\equiv 7\pmod{10}$; false
$4\times 1\equiv 7\pmod{10}$; false
$4\times 2\equiv 7\pmod{10}$; false
$4\times 3\equiv 7\pmod{10}$; false
$4\times 4\equiv 7\pmod{10}$; false
$4\times 5\equiv 7\pmod{10}$; false
$4\times 6\equiv 7\pmod{10}$; false
$4\times 7\equiv 7\pmod{10}$; false
$4\times 8\equiv 7\pmod{10}$; false
$4\times 9\equiv 7\pmod{10}$; false

19. $110_5\ R2_5$

$$\begin{array}{r} 110_5 \\ 24_5\overline{)3142_5} \\ \underline{24} \\ 24 \\ \underline{24} \\ \underline{2} \\ 2 \end{array}$$

20. $59=1+2+8+16+32$

Powers of 2	Times 26
1	26
2	26+ 26= 52
4	52+ 52= 104
8	104+ 104= 208
16	208 + 208 = 416
32	416+ 416= 832

832 + 416 + 208 + 52 + 26 = 1,534

21. 6536_8 $\quad$ D5E_{16}

110 101 011 110
6 5 3 6

1101 0101 1110
D 5 *E*

22. 3113_5

$$\begin{array}{r} 42_5 \\ \times 34_5 \\ \hline 323 \\ \underline{231} \\ 3133_5 \end{array}$$

Chapter 6

Number Theory and the Real Number System: Understanding the Numbers All Around Us

Section 6.1 Number Theory

1. true

3. false

5. false

7. true

9. 141,270 is divisible by **2** because the last digit, 0, is divisible by 2.
141,270 is divisible by **3** because the sum of the digits, 15, is divisible by 3.
141,270 is not divisible by 4 because the number formed by the last two digits, 70, is not divisible by 4.
141,270 is divisible by **5** because the last digit, 0, is a 0 or a 5.
141,270 is divisible by **6** because the number is divisible by both 2 and 3.
141,270 is not divisible by 8 because the number formed by the last three digits, 270, is not divisible by 8.
141,270 is not divisible by 9 because the sum of the digits, 15, is not divisible by 9.
141,270 is divisible by **10** because the last digit, 0, is 0.

11. 47,385 is not divisible by 2 because the last digit, 5, is not divisible by 2.
47,385 is divisible by **3** because the sum of the digits, 27, is divisible by 3.
47,385 is not divisible by 4 because the number formed by the last two digits, 85, is not divisible by 4.
47,385 is divisible by **5** because the last digit, 5, is a 0 or a 5.
47,385 is not divisible by 6 because the number is not divisible by both 2 and 3.
47,385 is not divisible by 8 because the number formed by the last three digits, 385, is not divisible by 8.
47,385 is divisible by **9** because the sum of the digits, 27, is divisible by 9.
47,385 is not divisible by 10 because the last digit, 5, is not 0.

13. yes

15. no

17. Answers may vary. Possible values for a are 4, 12, 20, 28, 36, etc.

19. Answers may vary. Possible values for a are 20, 60, 100, 140, 180, etc.

21. Since $81 < 95 < 100$, $9 < \sqrt{95} < 10$. Thus $a = 9$.

23. Since $144 < 153 < 169$, $12 < \sqrt{153} < 13$. Thus $a = 12$.

For Exercise 25 refer to the diagram on the next page.

25. 53, 59, 61, 67, 71, 73, 79, 83, 89, 97

27. 231 is composite. It can be factored as $3 \times 7 \times 11$.

29. 113 is prime.

31. 227 is prime

33. 119 is composite. It can be factored as 7×17.

35. $980 = 2^2 \times 5 \times 7^2$

37. $621 = 3^3 \times 23$

39. $319 = 11 \times 29$

For Exercise 25 refer to the following diagram.

~~1~~	2	3	~~4~~	5	~~6~~	7	~~8~~	~~9~~	~~10~~
11	~~12~~	13	~~14~~	~~15~~	~~16~~	17	~~18~~	19	~~20~~
~~21~~	~~22~~	23	~~24~~	~~25~~	~~26~~	~~27~~	~~28~~	29	~~30~~
31	~~32~~	~~33~~	~~34~~	~~35~~	~~36~~	37	~~38~~	~~39~~	~~40~~
41	~~42~~	43	~~44~~	~~45~~	~~46~~	47	~~48~~	~~49~~	~~50~~
~~51~~	~~52~~	53	~~54~~	~~55~~	~~56~~	~~57~~	~~58~~	59	~~60~~
61	~~62~~	~~63~~	~~64~~	~~65~~	~~66~~	67	~~68~~	~~69~~	~~70~~
71	~~72~~	73	~~74~~	~~75~~	~~76~~	~~77~~	~~78~~	79	~~80~~
~~81~~	~~82~~	83	~~84~~	~~85~~	~~86~~	~~87~~	~~88~~	89	~~90~~
~~91~~	~~92~~	~~93~~	~~94~~	~~95~~	~~96~~	97	~~98~~	~~99~~	~~100~~
101	~~102~~	103	~~104~~	~~105~~	~~106~~	107	~~108~~	109	~~110~~
~~111~~	~~112~~	113	~~114~~	~~115~~	~~116~~	~~117~~	~~118~~	~~119~~	~~120~~

41. $20 = 2^2 \times 5$ and $24 = 2^3 \times 3$
$GCD(20,24) = 2^2 = 4$
$LCM(20,24) = 2^3 \times 3 \times 5 = 120$

43. $56 = 2^3 \times 7$ and $70 = 2 \times 5 \times 7$
$GCD(56,70) = 2 \times 7 = 14$
$LCM(56,70) = 2^3 \times 5 \times 7 = 280$

45. $216 = 2^3 \times 3^3$ and $288 = 2^5 \times 3^2$
$GCD(216,288) = 2^3 \times 3^2 = 72$
$LCM(216,288) = 2^5 \times 3^3 = 864$

47. $147 = 3 \times 7^2$ and $567 = 3^4 \times 7$
$GCD(147,567) = 3 \times 7 = 21$
$LCM(147,567) = 3^4 \times 7^2 = 3969$

49. 28

$$280\overline{\smash{)}588}\ \text{quotient } 2,\quad 588 - 560 = 28 \qquad 28\overline{\smash{)}280}\ \text{quotient } 10,\quad 280 - 280 = 0$$

51. 45

$$495\overline{\smash{)}1575}\ \text{quotient } 3,\ 1575-1485 = 90 \qquad 90\overline{\smash{)}495}\ \text{quotient } 5,\ 495-450 = 45 \qquad 45\overline{\smash{)}90}\ \text{quotient } 2,\ 90 - 24\ \ 0$$

53. $GCD(12,27) = 3$

$$12\overline{\smash{)}27}\ \text{quotient } 2,\ 27-24 = 3 \qquad 3\overline{\smash{)}12}\ \text{quotient } 4,\ 12-12 = 0$$

$$LCM(12,27) = \frac{12 \cdot 27}{3} = 4 \cdot 27 = 108$$

55. $GCD(90,120) = 30$

$$90\overline{\smash{)}120}\ \text{quotient } 1,\ 120-90 = 30 \qquad 30\overline{\smash{)}90}\ \text{quotient } 3,\ 90-90 = 0$$

$$LCM(90,120) = \frac{90 \cdot 120}{30} = 3 \cdot 120 = 360$$

57. $LCM(120,90,84) = 2520$; $LCM(120,90) = 360$, $LCM(360,84) = 2520$

$GCD(120,90,84) = 6$; $GCD(120,90) = 30$, $GCD(30,84) = 6$

59. $LCM(64,56,100) = 11{,}200$; $LCM(64,56) = 448$; $LCD(448,100) = 11{,}200$

$GCD(64,56,100) = 4$; $GCD(64,56) = 8$, $GCD(8,100) = 4$

61. $GCD(24,16) = 8$, so she could use stacks of 2, 4, or 8 books.

63. $LCM(2500,3000) = 15{,}000$ so both services will be due at the same time in 15,000 miles.

65. $LCM(13,17) = 221$, so both types of cicadas will emerge at the same time in 2010 + 221 = 2231.

67. In this exercise we need to find $LCM(20,35)$ which is 140.; 140 minutes

69. In this exercise we need to find $GCD(30,36)$ which is 6.; 6 packages

71. In this exercise we need to find $GCD(21,33)$ which is 3.; 3 feet by 3 feet

73. In this exercise we need to find $LCM(48,54)$ which is 432.; 432 hours

75. 17

77. Three will always divide a multiple of nine, that is, $5\cdot 999+7\cdot 99+1\cdot 9$, so all we have to check is to see if three divides the sum of the digits, $5+7+1+2=15$, which it does.

79. not prime

81. prime

83. a) $2^2-1=3$; $2^3-1=7$; $2^5-1=31$; $2^7-1=127$; $2^{13}-1=8191$

    b) 11; $2^{11}-1=2407=23\times 89$

85. If you refer to the table for Exercise 25, you can easily determine that the next three twin primes are: 59 and 61; 71 and 73; 101 and 103.

87. No; Counterexamples may vary.
    Both 4 and 6 divide 12, but $4\times 6=24$, which does not divide 12.

89. Fifteen will divide a number if both three and five divide the number.

Section 6.2 The Integers

1. $(+8)+(-5)=+3$

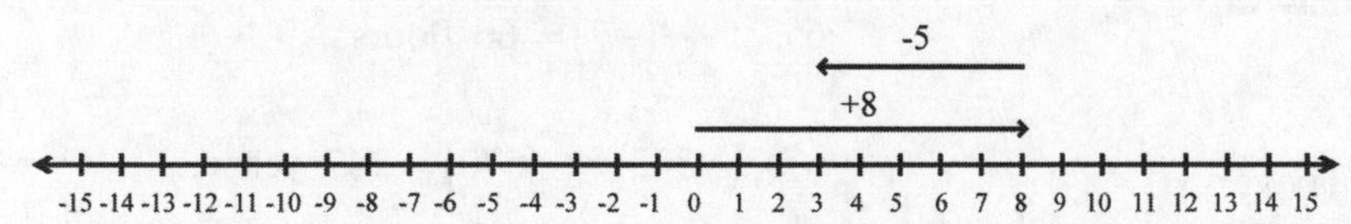

3. It is not practical to draw a number line when calculating with such large numbers; however, you can still imagine movements along the number line. If you visualize moving 128 units to the left and then 137 units to the right, you can see that the net effect is that you moved $137-128=9$ units to the right, so $(-128)+(+137)=+9$.

5. It is not practical to draw a number line when calculating with such large numbers; however, you can still imagine movements along the number line. If you visualize moving 57 units to the right and then 38 units to the left, you can see that the net effect is that you moved $57-38=19$ units to the right, so $(+57)+(-38)=+19$.

7. $(+18)-(-5)=(+18)+(+5)=+23$

9. $(+6)-(+19)=(+6)+(-19)=-13$

11. $(-28)-(+37)=(-28)+(-37)=-65$

13. $(+32)-(-18)=(+32)+(+18)=+50$

15. $(-5)(-7)=+35$

17. $(-7)(+8)=-56$

19. $(+8)(+6)=+48$

21. $(+19)(-2)=-38$

23. $(-8)(-9)=+72$

25. -3

27. $+15$

29. $+5$

31. -6

33. $+6$

35. $(-2)(-3)+(+4)(-8)=$ $(+6)+(-32)=-26$

37. $(-4)(-2)+(+4)(-5)=$ $(+8)+(-20)=-12$

39. $(-6-(-14))\cdot\left(\frac{-12+(-6)}{7-4}\right)=(+8)\cdot\left(\frac{-18}{3}\right)=(+8)\cdot(-6)=-48$

41. $\left(\frac{-1-(-5)}{4-6}\right)\cdot\left(\frac{-4+(-2)}{-5+3}\right)=\left(\frac{4}{-2}\right)\cdot\left(\frac{-6}{-2}\right)=(-2)\cdot(3)=-6$

43. There are an infinite number of integers that are not whole numbers. You can choose from –1, –2, –3, –4,…

45. 0

47. not possible; The whole numbers are 0, 1, 2, 3, … All natural numbers appear in this list.

49. true

51. false; Counterexamples may vary; $(-5)+(+8)=+3$

53. false; counterexamples may vary; $\frac{-6}{-3}=+2$

55. true

57. $29,028-(-28,232)=57,260$ feet

59. $(-28,232)-(-35,433)=7,201$ feet

61. $(134)-(-129)=263$ degrees

63. $(-90)-(-129)=39$ degrees

65. $59-(-7)=66$ floors

67. $(-323)-(-570)=247$ years

69. $(1170)-(-323)-1=1492$ years; We subtract 1 to compensate for the missing year zero.

71. $65+4(-7)=65+(-28)=37$ degrees

73. $12,354+134+27-81+45-129=12,350$

75. Tuesday, since at that time the market will be at its highest value of $12,354+134+27=\$12,515$.

77. The notion of adding signed numbers was used to define subtraction.

79. Yes; when multiplying or dividing signed numbers, if both numbers have the same sign, the result is positive; if the numbers have different signs, the result is negative.

81. If $\frac{8}{0} = x$, then $8 = 0 \cdot x$, which is not possible.

83. For the next four days you will spend \$6, so in four days, your net change in finances will be $-\$24$. Thus, $(+4)(-6) = -24$.

85. For the past three days you received \$7, so three days ago, you had \$21 less. Thus $(-3)(+7) = -21$.

87. Since $a - 8 = -6$, we have $a = 2$.; Since $c + 1 = -6$, we have $c = -7$.; Since $e + 0 = -6$, we have $e = -6$.; Since $b + e + 3 = -6$ and $e = -6$, we have $b = -3$.; Since $b + d + 5$ and $b = -3$, we have $d = -8$.

	-9	5	4	e	$e+0$
	a	-4	b	-1	$a+b+(-5)$
	-2	0	1	-5	-6
	3	c	d	6	$c+d+9$
$b+e+3$	$a-8$	$c+1$	$b+d+5$	$e+0$	-6

Since $a + b + (-5) = (2) + (-3) + (-5) = -6$, $c + d + 9 = (-7) + (-8) + (9) = -6$, and $b + e + 3 = (-3) + (-6) + 3 = -6$, the values satisfy all rows, columns, and diagonals.

89. The sum of the entries, 333, is nine times 37, the center number.

91. Put a point, c, in the middle of the 4-by-4 square and notice that any two numbers that are symmetric with respect to c have the same total. There are eight such pairs in a 4-by-4 box; therefore add two numbers in the box that are symmetric with respect to the center, c, and multiply this total by eight to get the total of all sixteen numbers.

Section 6.3 The Rational Numbers

1. equal; $\frac{2}{3} = \frac{8}{12}$ since $2 \cdot 12 = 3 \cdot 8$ $(24 = 24)$

3. not equal; $\frac{12}{14} \neq \frac{14}{16}$ since $12 \cdot 16 \neq 14 \cdot 14$ $(192 \neq 196)$

5. not equal; $\frac{22}{14} \neq \frac{30}{21}$ since $22 \cdot 21 \neq 14 \cdot 30$ $(462 \neq 420)$

7. equal; $\frac{5}{14} = \frac{15}{42}$ since $5 \cdot 42 = 14 \cdot 15$ $(210 = 210)$

9. $\frac{15}{35} = \frac{\cancel{5} \cdot 3}{\cancel{5} \cdot 7} = \frac{3}{7}$

11. $-\frac{24}{72} = -\frac{\cancel{24} \cdot 1}{\cancel{24} \cdot 3} = -\frac{1}{3}$

13. $\frac{225}{350}=\frac{\cancel{25}\cdot 9}{\cancel{25}\cdot 14}=\frac{9}{14}$

15. $\frac{143}{154}=\frac{\cancel{11}\cdot 13}{\cancel{11}\cdot 14}=\frac{13}{14}$

17. $\frac{2}{3}+\frac{1}{2}=\frac{2\cdot 2+3\cdot 1}{3\cdot 2}=\frac{4+3}{6}=\frac{7}{6}$

19. $\frac{1}{6}-\frac{1}{2}=\frac{1}{6}-\frac{1\cdot 3}{2\cdot 3}=\frac{1}{6}-\frac{3}{6}=-\frac{2}{6}=-\frac{1}{3}$

21. $\frac{7}{16}-\frac{1}{3}=\frac{7\cdot 3-16\cdot 1}{16\cdot 3}=\frac{21-16}{48}=\frac{5}{48}$

23. $\frac{3}{4}+\frac{5}{6}+\frac{7}{8}=\frac{3\cdot 3}{4\cdot 3}+\frac{5\cdot 2}{6\cdot 2}+\frac{7}{8}=\frac{9}{12}+\frac{10}{12}+\frac{7}{8}=\frac{19}{12}+\frac{7}{8}=\frac{19\cdot 2}{12\cdot 2}+\frac{7\cdot 3}{8\cdot 3}=\frac{38}{24}+\frac{21}{24}=\frac{59}{24}$

25. $\frac{1}{8}-\frac{2}{3}+\frac{1}{2}=\frac{1\cdot 3}{8\cdot 3}-\frac{2\cdot 8}{3\cdot 8}+\frac{1}{2}=\frac{3}{24}-\frac{16}{24}+\frac{1}{2}=\frac{-13}{24}+\frac{1}{2}=\frac{-13}{24}+\frac{1\cdot 12}{2\cdot 12}=\frac{-13}{24}+\frac{12}{24}=\frac{-1}{24}=-\frac{1}{24}$

27. $\frac{\overset{1}{\cancel{2}}}{3}\cdot\frac{1}{\underset{1}{\cancel{2}}}=\frac{1}{3}$

29. $\frac{1}{6}\div\frac{1}{2}=\frac{1}{\underset{3}{\cancel{6}}}\cdot\frac{\overset{1}{\cancel{2}}}{1}=\frac{1}{3}$

31. $\frac{7}{8}\div\left(-\frac{5}{24}\right)=\frac{7}{\underset{1}{\cancel{8}}}\cdot\left(-\frac{\overset{3}{\cancel{24}}}{5}\right)=-\frac{21}{5}$

33. $\left(\frac{7}{\underset{6}{\cancel{18}}}\cdot\left(-\frac{\overset{1}{\cancel{3}}}{4}\right)\right)\div\left(\frac{7}{9}\right)=-\frac{7}{24}\div\frac{7}{9}=-\frac{\overset{1}{\cancel{7}}}{\underset{8}{\cancel{24}}}\cdot\frac{\overset{3}{\cancel{9}}}{\underset{1}{\cancel{7}}}=-\frac{3}{8}$

35. $\left(\frac{11}{30}\div\left(-\frac{1}{6}\right)\right)\cdot\left(\frac{15}{4}\right)=\left(\frac{11}{\underset{5}{\cancel{30}}}\cdot\left(-\frac{\overset{1}{\cancel{6}}}{1}\right)\right)\cdot\left(\frac{15}{4}\right)=-\frac{11}{\underset{1}{\cancel{5}}}\frac{\overset{3}{\cancel{15}}}{4}=-\frac{33}{4}$

37. $\frac{5}{3}\cdot\left(\frac{8}{15}+\frac{2}{3}\right)=\frac{5}{3}\cdot\left(\frac{8}{15}+\frac{2\cdot 5}{3\cdot 5}\right)=\frac{5}{3}\cdot\left(\frac{8}{15}+\frac{10}{15}\right)=\frac{\overset{1}{\cancel{5}}}{\underset{1}{\cancel{3}}}\cdot\left(\frac{\overset{\overset{2}{\cancel{6}}}{\cancel{18}}}{\underset{\underset{1}{\cancel{3}}}{\cancel{15}}}\right)=\frac{2}{1}=2$

39. $\frac{22}{27}\cdot\left(\frac{3}{11}+\frac{2}{3}\right)=\frac{22}{27}\cdot\left(\frac{3\cdot 3+11\cdot 2}{11\cdot 3}\right)=\frac{22}{27}\cdot\left(\frac{9+22}{33}\right)=\frac{\overset{2}{\cancel{22}}}{27}\cdot\left(\frac{31}{\underset{3}{\cancel{33}}}\right)=\frac{62}{81}$

41. $\frac{7}{30}\div\left(\frac{1}{6}-\frac{3}{14}\right)=\frac{7}{30}\div\left(\frac{1\cdot 7}{6\cdot 7}-\frac{3\cdot 3}{14\cdot 3}\right)=\frac{7}{30}\div\left(\frac{7}{42}-\frac{9}{42}\right)=\frac{7}{30}\div\left(-\frac{2}{42}\right)=\frac{7}{\underset{5}{\cancel{30}}}\cdot\left(-\frac{\overset{7}{\cancel{42}}}{2}\right)=-\frac{49}{10}$

43. $\frac{79}{5}=15\frac{4}{5}$

45. $\frac{29}{3}=9\frac{2}{3}$

47. $$\begin{array}{r} 6 \\ 4\overline{)27} \\ \underline{24} \\ 3 \end{array} \; ; \; 6\frac{3}{4}$$

49. $$\begin{array}{r} 8 \\ 15\overline{)121} \\ \underline{120} \\ 1 \end{array} \; ; \; 8\frac{1}{15}$$

51. $\dfrac{2\cdot 4+3}{4}=\dfrac{8+3}{4}=\dfrac{11}{4}$

53. $\dfrac{9\cdot 6+1}{6}=\dfrac{54+1}{6}=\dfrac{55}{6}$

55. 0.75

$$\begin{array}{r} 0.75 \\ 4\overline{)3.00} \\ \underline{28} \\ 20 \\ \underline{20} \\ 0 \end{array}$$

57. 0.1875

$$\begin{array}{r} 0.1875 \\ 16\overline{)3.0000} \\ \underline{16} \\ 140 \\ \underline{128} \\ 120 \\ \underline{112} \\ 80 \\ \underline{80} \\ 0 \end{array}$$

59. $0.\overline{81}$

$$\begin{array}{r} 0.81 \\ 11\overline{)9.00} \\ \underline{88} \\ 20 \\ \underline{11} \\ 90 \end{array}$$

61. $0.\overline{307692}$

$$\begin{array}{r} 0.307692 \\ 13\overline{)4.000000} \\ \underline{39} \\ 10 \\ \underline{0} \\ 100 \\ \underline{91} \\ 90 \\ \underline{78} \\ 120 \\ \underline{117} \\ 30 \\ \underline{26} \\ 40 \end{array}$$

63. $0.64=\dfrac{64}{100}=\dfrac{16\cdot 4}{25\cdot 4}=\dfrac{16}{25}$

65. $0.836=\dfrac{836}{1000}=\dfrac{209\cdot 4}{250\cdot 4}=\dfrac{209}{250}$

67. $12.2=\dfrac{122}{10}=\dfrac{61\cdot 2}{5\cdot 2}=\dfrac{61}{5}$

69. $$\begin{array}{rl} 10\cdot x = & 4.444444\ldots \\ \underline{-\quad x =} & \underline{-0.444444\ldots} \end{array}$$

$$9\cdot x=4\Rightarrow x=\frac{4}{9}$$

71. $$\begin{array}{rl} 1000\cdot x = & 189.189189\ldots. \\ \underline{-\quad x =} & \underline{-\ 0.189189\ldots} \end{array}$$

$$999\cdot x=189\Rightarrow x=\frac{189}{999}=\frac{7\cdot 27}{37\cdot 27}=\frac{7}{37}$$

73. $\begin{array}{r} 100\cdot x = \ 31.81818... \\ - \quad x = -\ 0.31818... \\ \hline \end{array}$

$$99\cdot x = 31.5 \Rightarrow x = \frac{31.5}{99} = \frac{31.5\cdot 10}{99\cdot 10} = \frac{315}{990} = \frac{7\cdot 45}{22\cdot 45} = \frac{7}{22}$$

or $\begin{array}{r} 1000\cdot x = \ 318.1818... \\ -\ 10\cdot x = -\ \ 3.1818... \\ \hline \end{array}$

$$990\cdot x = 315 \Rightarrow x = \frac{315}{990} = \frac{7\cdot 45}{22\cdot 45} = \frac{7}{22}$$

75. Andre spends $\frac{1}{3}+\frac{1}{4}+\frac{1}{6}=\frac{1\cdot 4+3\cdot 1}{3\cdot 4}+\frac{1}{6}=\frac{4+3}{12}+\frac{1}{6}=\frac{7}{12}+\frac{1}{6}=$ $\frac{7}{12}+\frac{1\cdot 2}{6\cdot 2}=\frac{7}{12}+\frac{2}{12}=\frac{9}{12}=\frac{3\cdot 3}{4\cdot 3}=\frac{3}{4}$ of his paycheck on rent, food, and utilities. That leaves $1-\frac{3}{4}=\frac{4}{4}-\frac{3}{4}=\frac{1}{4}$ of his paycheck for other expenses.

77. $\frac{1}{3}+\frac{1}{4}=\frac{1\cdot 4+3\cdot 1}{3\cdot 4}=\frac{4+3}{12}=\frac{7}{12}$ of the purse goes to the winner and the person in second place. That leaves $1-\frac{7}{12}=\frac{12}{12}-\frac{7}{12}=\frac{5}{12}$ for the last four players. Each would get $\frac{1}{4}\cdot\frac{5}{12}=\frac{5}{48}$ of the purse.

79. $\frac{4}{11}+\frac{1}{4}=\frac{4\cdot 4}{11\cdot 4}+\frac{1\cdot 11}{4\cdot 11}=\frac{16}{44}+\frac{11}{44}=\frac{16+11}{44}=\frac{27}{44}$

81. The height of the room is $17\cdot 12=204$ inches. We need to find $204\div 8\frac{1}{2}$ and determine if it represents a natural number. You could perform long division or simplify $\frac{204}{8\frac{1}{2}}$. Using the second approach we have $\frac{204}{8\frac{1}{2}}=\frac{204}{\frac{17}{2}}=\frac{\overset{12}{\cancel{204}}}{1}\cdot\frac{2}{\underset{1}{\cancel{17}}}=\frac{24}{1}=24$. Thus we will have 24 rows and the tiles do not need to be cut.

83. $2\frac{1}{3}+4\frac{3}{8}+3\frac{1}{4}=\frac{7}{3}+\frac{35}{8}+\frac{13}{4}=\frac{7\cdot 8}{3\cdot 8}+\frac{35\cdot 3}{8\cdot 3}+\frac{13\cdot 6}{4\cdot 6}=\frac{56}{24}+\frac{105}{24}+\frac{78}{24}=\frac{239}{24}=9\frac{23}{24}$ miles

85. $2\left(32\frac{2}{3}+18\frac{1}{2}\right)=2\left(\frac{98}{3}+\frac{37}{2}\right)=2\left(\frac{98\cdot 2}{3\cdot 2}+\frac{37\cdot 3}{2\cdot 3}\right)=2\left(\frac{196}{6}+\frac{11}{6}\right)=\overset{1}{\cancel{2}}\left(\frac{307}{\underset{3}{\cancel{6}}}\right)=\frac{307}{3}=102\frac{1}{3}$ feet

87. In order to increase the recipe that serves 4 to a recipe that serves 10, you must increase the ingredients by 2.5 or $2\frac{1}{2}$ times. You should use $2\frac{1}{2}\cdot 1\frac{1}{2}=\frac{5}{2}\cdot\frac{3}{2}=\frac{15}{4}=3\frac{3}{4}$ tablespoons of lemon juice.

89. $5\frac{1}{4}=5.25$ and $7\frac{1}{8}=7.125$

The unit cost of the $5\frac{1}{4}$-ounce tube would be about 49.5 cents per ounce (rounded).

$$\begin{array}{r} 0.4952 \\ 5.25\overline{)2.600000} \\ \underline{2100} \\ 5000 \\ \underline{4725} \\ 2750 \\ \underline{2625} \\ 1250 \\ \underline{1050} \\ 200 \end{array}$$

The unit cost of the $7\frac{1}{8}$-ounce tube would be about 50.5 cents per ounce (rounded).

$$\begin{array}{r} 0.5052 \\ 7.125\overline{)3.6000000} \\ \underline{35625} \\ 3750 \\ \underline{0} \\ 37500 \\ \underline{35625} \\ 18750 \\ \underline{14250} \\ 4500 \end{array}$$

The smaller tube is a better buy.

91. $1-\left(\frac{3}{5}+\frac{7}{40}\right)=1-\left(\frac{3\cdot 8}{5\cdot 8}+\frac{7}{40}\right)=1-\left(\frac{24}{40}+\frac{7}{40}\right)=\frac{40}{40}-\frac{31}{40}=\frac{9}{40}$

93. He can get $12\div 3\frac{3}{8}=12\div\frac{27}{8}=\frac{\overset{4}{\cancel{12}}}{1}\cdot\frac{8}{\underset{9}{\cancel{27}}}=\frac{32}{9}=3\frac{5}{9}$ strips from a twelve-foot strip and $15\div 3\frac{3}{8}=15\div\frac{27}{8}=\frac{\overset{5}{\cancel{15}}}{1}\cdot\frac{8}{\underset{9}{\cancel{27}}}=\frac{40}{9}=4\frac{4}{9}$ strips from a fifteen-foot strip. Since $\frac{5}{9}>\frac{4}{9}$, there is more waste from the twelve-foot strips.

95. Answers may vary. Possible examples include:
$\frac{2}{7}+\frac{3}{7}=\frac{5}{7}$ and $\frac{2}{7}+\frac{3}{7}=\frac{2+3}{7+7}=\frac{5}{14}$, but $\frac{5}{7}\neq\frac{5}{14}$, so the claim is not true.

97. Examples will vary.

99. No solution provided.

101. $\dfrac{\frac{5}{6}}{\frac{2}{3}}=\dfrac{\frac{5}{6}\cdot\frac{3}{2}}{\frac{2}{3}\cdot\frac{3}{2}}=\dfrac{\frac{5}{6}\cdot\frac{3}{2}}{1}=\frac{5}{\underset{2}{\cancel{6}}}\cdot\frac{\overset{1}{\cancel{3}}}{2}=\frac{5}{4}$

103. Each piece would be $10\frac{7}{8}\div 4=\frac{87}{8}\div\frac{4}{1}=\frac{87}{8}\cdot\frac{1}{4}=\frac{87}{32}=2\frac{23}{32}$ feet

105. There are many approaches to this exercise. One approach is to first determine the distance from the edge of the mirror to the edge of the wall with the mirror centered on the wall. We can do this by finding the difference between the length of the wall and the mirror and then dividing this difference by 2. We can then add to this measurement one-third of the length of the mirror. This will yield the desired measurement.

105. (continued)

Distance from edge of mirror to edge of wall:

$$\left(72-40\tfrac{1}{2}\right)\div 2=\left(\tfrac{72}{1}-\tfrac{81}{2}\right)\div 2=\left(\tfrac{72\cdot 2}{1\cdot 2}-\tfrac{81}{2}\right)\div 2=\left(\tfrac{144}{2}-\tfrac{81}{2}\right)\div 2=\left(\tfrac{63}{2}\right)\div 2=\tfrac{63}{2}\cdot\tfrac{1}{2}=\tfrac{63}{4}=15\tfrac{3}{4} \text{ inches}$$

One third of the length the mirror: $\frac{1}{3}\cdot\left(40\frac{1}{2}\right)=\frac{1}{\cancel{3}_{1}}\cdot\frac{\cancel{81}^{27}}{2}=\frac{27}{2}=13\frac{1}{2}$ inches

Distance from the edge of the wall to a nail would therefore be:

$$15\frac{3}{4}+13\frac{1}{2}=15+\frac{3}{4}+13+\frac{1}{2}=(15+13)+\left(\frac{3}{4}+\frac{1}{2}\right)=28+\left(\frac{3}{4}+\frac{1\cdot 2}{2\cdot 2}\right)=$$

$$=28+\left(\frac{3}{4}+\frac{2}{4}\right)=28+\frac{5}{4}=28+1\frac{1}{4}=29\frac{1}{4} \text{ inches}$$

107. The 10 inch wide boards would have the least waste.

Since $8\div 3\frac{1}{4}=8\div\frac{13}{4}=\frac{8}{1}\cdot\frac{4}{13}=\frac{32}{13}=2\frac{6}{13}$, you can get 2 strips from the 8 inch wide pieces. The waste per board would be $8-2\times 3\frac{1}{4}=8-\frac{\cancel{2}^{1}}{1}\cdot\frac{13}{\cancel{4}_{2}}=\frac{8\cdot 2}{1\cdot 2}-\frac{13}{2}=\frac{16}{2}-\frac{13}{2}=\frac{3}{2}=1.5$ inches.

The total waste would be $6\times 1.5=9$ inches.

Since $10\div 3\frac{1}{4}=10\div\frac{13}{4}=\frac{10}{1}\cdot\frac{4}{13}=\frac{40}{13}=3\frac{1}{13}$, you can get 3 strips from the 10 inch wide pieces.

The waste per board would be $10-3\times 3\frac{1}{4}=10-\frac{3}{1}\cdot\frac{13}{4}=\frac{10\cdot 4}{1\cdot 4}-\frac{39}{4}=\frac{40}{4}-\frac{39}{4}=\frac{1}{4}=0.25$ inches.

The total waste would be $4\times 0.25=1$ inch.

Since $12\div 3\frac{1}{4}=12\div\frac{13}{4}=\frac{12}{1}\cdot\frac{4}{13}=\frac{48}{13}=3\frac{9}{13}$, you can get 3 strips from the 10 inch wide pieces.

The waste per board would be $12-3\times 3\frac{1}{4}=12-\frac{3}{1}\cdot\frac{13}{4}=\frac{12\cdot 4}{1\cdot 4}-\frac{39}{4}=\frac{48}{4}-\frac{39}{4}=\frac{9}{4}=2.25$ inches.

The total waste would be $4\times 2.25=9$ inches.

Since $15\div 3\frac{1}{4}=15\div\frac{13}{4}=\frac{15}{1}\cdot\frac{4}{13}=\frac{60}{13}=4\frac{8}{13}$, you can get 3 strips from the 10 inch wide pieces.

The waste per board would be $15-4\times 3\frac{1}{4}=15-\frac{\cancel{4}^{1}}{1}\cdot\frac{13}{\cancel{4}_{1}}=15-13=2$ inches.

The total waste would be $4\times 2=8$ inches.

Section 6.4 The Real Number System

1. rational

3. irrational

5. rational

7. irrational

9. Examples may vary; $\sqrt{9}=3$

11. $\sqrt{18}=\sqrt{9\cdot 2}=\sqrt{9}\sqrt{2}=3\sqrt{2}$

13. $\sqrt{75}=\sqrt{25\cdot 3}=\sqrt{25}\sqrt{3}=5\sqrt{3}$

15. $\sqrt{189}=\sqrt{9\cdot 21}=\sqrt{9}\sqrt{21}=3\sqrt{21}$

17. not possible

19. $\sqrt{20}+6\sqrt{5}=\sqrt{4\cdot 5}+6\sqrt{5}=\sqrt{4}\sqrt{5}+6\sqrt{5}=2\sqrt{5}+6\sqrt{5}=8\sqrt{5}$

21. not possible

$\sqrt{50}+2\sqrt{75}=\sqrt{25\cdot 2}+2\sqrt{25\cdot 3}=\sqrt{25}\sqrt{2}+2\sqrt{25}\sqrt{3}=5\sqrt{2}+2\cdot 5\sqrt{3}=5\sqrt{2}+10\sqrt{3}$

23. $\sqrt{18}\sqrt{2}=\sqrt{18\cdot 2}=\sqrt{9\cdot 2\cdot 2}=\sqrt{9}\sqrt{2\cdot 2}=3\sqrt{4}=3\cdot 2=6$

25. $\sqrt{12}\sqrt{15}=\sqrt{12\cdot 15}=\sqrt{4\cdot 3\cdot 3\cdot 5}=\sqrt{4}\sqrt{3\cdot 3\cdot 5}=2\sqrt{9\cdot 5}=2\sqrt{9}\sqrt{5}=2\cdot 3\sqrt{5}=6\sqrt{5}$

27. $\sqrt{28}\sqrt{21}=\sqrt{28\cdot 21}=\sqrt{4\cdot 7\cdot 7\cdot 3}=\sqrt{4}\sqrt{7\cdot 7\cdot 3}=2\sqrt{49\cdot 3}=2\sqrt{49}\sqrt{3}=2\cdot 7\sqrt{3}=14\sqrt{3}$

29. $\dfrac{\sqrt{24}}{\sqrt{6}}=\sqrt{\dfrac{24}{6}}=\sqrt{4}=2$

31. $\dfrac{\sqrt{32}}{\sqrt{18}}=\sqrt{\dfrac{32}{18}}=\sqrt{\dfrac{16}{9}}=\dfrac{\sqrt{16}}{\sqrt{9}}=\dfrac{4}{3}$

33. $\dfrac{\sqrt{96}}{\sqrt{72}}=\dfrac{\sqrt{16\cdot 6}}{\sqrt{36\cdot 2}}=\dfrac{4\sqrt{6}}{6\sqrt{2}}=\dfrac{4}{6}\cdot\dfrac{\sqrt{6}}{\sqrt{2}}=\dfrac{2}{3}\sqrt{\dfrac{6}{2}}=\dfrac{2}{3}\sqrt{3}=\dfrac{2\sqrt{3}}{3}$

35. $\dfrac{3}{\sqrt{5}}=\dfrac{3\cdot\sqrt{5}}{\sqrt{5}\cdot\sqrt{5}}=\dfrac{3\sqrt{5}}{\sqrt{5\cdot 5}}=\dfrac{3\sqrt{5}}{\sqrt{25}}=\dfrac{3\sqrt{5}}{5}$

37. $\dfrac{12}{\sqrt{6}}=\dfrac{12\cdot\sqrt{6}}{\sqrt{6}\cdot\sqrt{6}}=\dfrac{12\sqrt{6}}{\sqrt{6\cdot 6}}=\dfrac{12\sqrt{6}}{\sqrt{36}}=\dfrac{12\sqrt{6}}{6}=2\sqrt{6}$

39. $\dfrac{10}{\sqrt{22}}=\dfrac{10\sqrt{22}}{\sqrt{22}\cdot\sqrt{22}}=\dfrac{10\sqrt{22}}{\sqrt{22\cdot 22}}=\dfrac{10\sqrt{22}}{\sqrt{484}}=\dfrac{10\sqrt{22}}{22}=\dfrac{5\sqrt{22}}{11}$

41. $\dfrac{3\sqrt{10}}{\sqrt{72}}=\dfrac{3\cdot\sqrt{10}\sqrt{2}}{\sqrt{36\cdot 2}\cdot\sqrt{2}}=\dfrac{3\sqrt{20}}{\sqrt{144}}=\dfrac{3\sqrt{4\cdot 5}}{12}=\dfrac{3\sqrt{4}\cdot\sqrt{5}}{12}=\dfrac{3\cdot 2\sqrt{5}}{12}=\dfrac{6\sqrt{5}}{12}=\dfrac{\sqrt{5}}{2}$

An easier approach for this exercise however is to simplify as much as possible before rationalizing the denominator.

$\dfrac{3\sqrt{10}}{\sqrt{72}}=3\sqrt{\dfrac{10}{72}}=3\sqrt{\dfrac{5}{36}}=3\dfrac{\sqrt{5}}{\sqrt{36}}=3\dfrac{\sqrt{5}}{6}=\dfrac{\sqrt{5}}{2}$

43. Answers may vary.

0.435; 0.43512112111211112...

45. Answers may vary.

$0.\overline{4578}=0.457845784578...$

$0.45\overline{78}=0.457878787878...$

0.45785; 0.45785121121112111112...

47. Answers may vary.

$\dfrac{4}{7}=0.\overline{571428}$

$\dfrac{5}{7}=0.\overline{714285}$

0.6; 0.612112111211112...

49. a, d, b, c
   a) 0.345345000000
   b) 0.345345345345...
   c) 0.345454545454...
   d) 0.345345343434...

51. a, c, d, b
   a) 0.4444444...
   b) 0.5555555...
   c) 0.4545454...
   d) 0.5545454...

53. false; $\sqrt{2}\times\sqrt{2}=2,$ which is rational.

55. true

57. false; We cannot divide by the rational number 0.

59. distributive

61. commutative (addition)

63. associative (addition)

65. identity element for addition

67. commutative (addition)

69. $D=\sqrt{2(6+42)}=\sqrt{2\cdot 48}=\sqrt{2\cdot 3\cdot 16}=\sqrt{6\cdot 16}=\sqrt{16}\cdot\sqrt{6}=4\sqrt{6}\approx 9.8$ miles.

71. $v=2\sqrt{5\cdot 120}=2\sqrt{600}=2\sqrt{100\cdot 6}=2\sqrt{100}\cdot\sqrt{6}=2\cdot 10\cdot\sqrt{6}=20\sqrt{6}\approx 49.0$ mph.

73. $T=2\pi\sqrt{\dfrac{20}{32}}=2\pi\sqrt{\dfrac{5}{8}}=2\pi\sqrt{\dfrac{5\cdot 2}{8\cdot 2}}=2\pi\sqrt{\dfrac{10}{16}}=2\pi\dfrac{\sqrt{10}}{4}=\pi\dfrac{\sqrt{10}}{2}\approx 4.97$ seconds

75. $c=\sqrt{9^2+12^2}=\sqrt{81+144}=\sqrt{225}=15$ feet

77. $F=\sqrt{(120)^2+(160)^2}=\sqrt{40000}=200$ pounds

79. $t=\dfrac{\sqrt{1600}}{4}=\dfrac{40}{4}=10$ seconds

81. a) $t=\dfrac{\sqrt{2640}}{4}=\approx 12.85$ seconds

   b) $v=32\left(\dfrac{\sqrt{2640}}{4}\right)\approx 411$ feet per second

   c) $411\dfrac{ft}{s}\left(\dfrac{3600\ s}{1\ hr}\right)=1{,}479{,}600\dfrac{ft}{hr};\ 1{,}479{,}600\dfrac{ft}{hr}\left(\dfrac{1\ mi}{5280\ ft}\right)=280.23\dfrac{mi}{hr}$

83. a) 1.414213562 is a terminating decimal and hence a rational number, however, $\sqrt{2}$ is irrational.

   b) π is an irrational number and thus cannot be equal to the rational number $\dfrac{22}{7}$.

85. No, because a number that is rational and irrational would have to have both a repeating and a nonrepeating decimal expansion.

87. "I like green eggs and ham."

89. – 91. No solution provided.

93. a) We can cancel common factors from the numerator and denominator.
b) Square both sides of a).
c) Multiply through equation by b^2.
d) 2 divides the left side of c), so 2 divides the right side of c).
e) If a were odd then a^2 would be odd, but it is not.
f) a is even.
g) Substitute $2k$ for a in c).
h) Simplify $(2k)^2 = (2k)(2k) = 4k^2$.
i) Divide both sides of h) by 2.
j) 2 divides b^2, since 2 divides $2k^2$.
k) If b were odd, then b^2 would be odd.
l) We assumed that $\frac{a}{b}$ was reduced, so a and b cannot both be even.

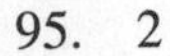

95. 2

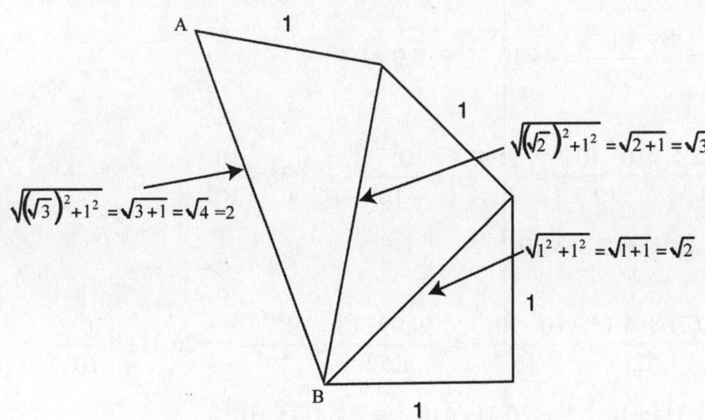

97. Answers may vary. 5, 12, and 13 is one possible solution.

Section 6.5 Exponents and Scientific Notation

1. 32

3. $-2^4 = -(2^4) = -(16) = -16$

5. $-3^2 = -(3^2) = -(9) = -9$

7. 9

9. $3^0 = 1$

11. $3^{2+4} = 3^6 = 729$

13. $(7^2)^3 = 7^{2 \cdot 3} = 7^6 = 117{,}649$

15. $5^4 \cdot 5^{-6} = 5^{4+(-6)} = 5^{-2} = \frac{1}{5^2} = \frac{1}{25}$

17. $(3^2)^{-3} = 3^{2(-3)} = 3^{-6} = \frac{1}{3^6} = \frac{1}{729}$

19. $(-3)^{-2}(-3)^3 = (-3)^{-2+3} = (-3)^1 = -3$

21. $\frac{5^9}{5^7} = 5^{9-7} = 5^2 = 25$

23. $\frac{6^{-2}}{6^{-4}} = 6^{-2-(-4)} = 6^2 = 36$

25. $\left(3^{-4}\right)^0 = 3^{-4(0)} = 3^0 = 1$

27. $\frac{2^9}{4^3} = \frac{2^9}{\left(2^2\right)^3} = \frac{2^9}{2^6} = 2^{9-6} = 2^3 = 8$

29. 4.356×10^6

31. 7.83×10^2

33. 2.4×10^{-3}

35. 8.0×10^{-3}

37. 32,500

39. 0.00178

41. 63

43. 0.00000045

45. 2.381×10^7

47. 8.4×10^2

49. $\left(3\times10^6\right)\left(2\times10^5\right) = (3\cdot2)\times\left(10^6\cdot10^5\right) = 6.0\times10^{6+5} = 6.0\times10^{11}$

51. $\left(1.2\times10^{-3}\right)\left(3\times10^5\right) = (1.2\cdot3)\times\left(10^{-3}\cdot10^5\right) = 3.6\times10^{-3+5} = 3.6\times10^2$

53. $\left(8\times10^{-2}\right)\div\left(2\times10^3\right) = \frac{8\times10^{-2}}{2\times10^3} = \frac{8}{2}\times\frac{10^{-2}}{10^3} = 4\times10^{-2-3} = 4.0\times10^{-5}$

55. $$\frac{\left(5.44\times10^8\right)\left(2.1\times10^{-3}\right)}{\left(3.4\times10^6\right)} = \frac{5.44\cdot2.1}{3.4}\times\frac{10^8\cdot10^{-3}}{10^6} = \frac{11.424}{3.4}\times\frac{10^{8+(-3)}}{10^6} = 3.36\times\frac{10^5}{10^6}$$
$$= 3.36\times10^{5-6} = 3.36\times10^{-1}$$

57. $$\frac{\left(9.6368\times10^3\right)\left(4.15\times10^{-6}\right)}{\left(1.52\times10^4\right)} = \frac{9.6368\cdot4.15}{1.52}\times\frac{10^3\cdot10^{-6}}{10^4} = \frac{39.99272}{1.52}\times\frac{10^{3+(-6)}}{10^4} = 26.311\times\frac{10^{-3}}{10^4}$$
$$= 26.311\times10^{-3-4} = 26.311\times10^{-7} = 2.6311\times10^{-6}$$

59. $$67{,}300{,}000\times1{,}200 = \left(6.73\times10^7\right)\left(1.2\times10^3\right) = (6.73\cdot1.2)\times\left(10^7\cdot10^3\right)$$
$$= 8.076\times10^{7+3} = 8.076\times10^{10}$$

61. $$6{,}800{,}000\times2{,}300{,}000 = \left(6.8\times10^6\right)\left(2.3\times10^6\right) = (6.8\cdot2.3)\times\left(10^6\cdot10^6\right)$$
$$= 15.64\times10^{6+6} = 15.64\times10^{12} = 1.564\times10^{13}$$

63. $$0.00016\times0.0025 = \left(1.6\times10^{-4}\right)\left(2.5\times10^{-3}\right) = (1.6\cdot2.5)\times\left(10^{-4}\cdot10^{-3}\right)$$
$$= 4.0\times10^{-4+(-3)} = 4.0\times10^{-7}$$

65. $$10{,}000{,}000{,}000{,}000{,}000{,}000 = 10\times1{,}000{,}000\times1{,}000{,}000{,}000{,}000 = 10^1\times10^6\times10^{12}$$
$$= 10^{1+6+12} = 10^{19} = 1.0\times10^{19}$$

67. $3{,}720{,}000{,}000{,}000 = 3.720\times1{,}000{,}000{,}000{,}000 = 3.720\times10^{12}$

69. $0.0000001 = \frac{1}{10,000,000} = \frac{1}{10 \times 1,000,000} = \left(\frac{1}{10^1}\right)\left(\frac{1}{10^6}\right)$
$= 10^{-1} \times 10^{-6} = 10^{(-1)+(-6)} = 10^{-7} = 1.0 \times 10^{-7}$

71. $0.000000000000014 = \frac{14}{1,000,000,000,000,000} = \frac{14}{1,000,000 \times 1,000,000,000} = \frac{14}{10^6 \times 10^9}$
$= 14 \times 10^{-6} \times 10^{-9} = 1.4 \times 10^1 \times 10^{-6} \times 10^{-9} = 1.4 \times 10^{1+(-6)+(-9)} = 1.4 \times 10^{-14}$

73. $\frac{3233 \times 10^9}{249 \times 10^6} = \frac{3.233 \times 10^{12}}{2.49 \times 10^8} = \frac{3.233}{2.49} \times \frac{10^{12}}{10^8} \approx 1.3 \times 10^{12-8} = 1.3 \times 10^4 = \$13,000$ per person.

75. $308 \times 10^6 - 203 \times 10^6 = (308 - 203) \times 10^6 = 105 \times 10^6 = 1.05 \times 10^2 \times 10^6$
$= 1.05 \times 10^{2+6} = 1.05 \times 10^8 = 105$ million people

77. $\frac{685 \times 10^9}{308 \times 10^6} = \frac{6.85 \times 10^9}{3.08 \times 10^8} = \frac{6.85}{3.08} \times \frac{10^9}{10^8} \approx 2.224 \times 10^{9-8} = 2.224 \times 10^1$ times larger.

79. (a) $454 \cdot 130 = 59020 = 5.902 \times 10^4$

(b) $\frac{5.902 \times 10^4}{1 \times 10^{-3}} = \frac{5.902}{1} \times \frac{10^4}{10^{-3}} = 5.902 \times 10^{4-(-3)} = 5.902 \times 10^7$

(c) The 130 pound person weighs as much as 59 million mosquitoes.

81. $\frac{296 \times 10^6}{3.5 \times 10^6} = \frac{2.96 \times 10^8}{3.5 \times 10^6} = \frac{2.96}{3.5} \times \frac{10^8}{10^6} \approx 0.846 \times 10^{8-6} = 0.846 \times 10^2$
$= 8.46 \times 10^1 = 84.6$ people per square mile.

83. $\frac{439.3 \times 10^9}{301 \times 10^6} = \frac{4.393 \times 10^{11}}{3.01 \times 10^8} = \frac{4.393}{3.01} \times \frac{10^{11}}{10^8} \approx 1.459 \times 10^{11-8} = 1.459 \times 10^3 = \$1,459$ per person

85. Assuming there are 365 days in a year, we have $60 \times 60 \times 24 \times 365$ seconds in one year.
$\frac{1.0 \times 10^9}{60 \times 60 \times 24 \times 365} = \frac{1.0 \times 10^9}{31,536,000} = \frac{1.0 \times 10^9}{3.1536 \times 10^7} = \frac{1.0}{3.1536} \times \frac{10^9}{10^7} \approx 0.317 \times 10^{9-7}$
$= 0.317 \times 10^2 = 3.17 \times 10^1 = 31.7$ years

87. $\frac{4.1 \times 10^{12}}{12,800} = \frac{4.1 \times 10^{12}}{1.28 \times 10^4} = \frac{4.1}{1.28} \times \frac{10^{12}}{10^4} \approx 3.2 \times 10^{12-4} = 3.2 \times 10^8 = 320 \times 10^6 = 320$ million

89. $\frac{2.8 \times 10^9}{25,000} = \frac{2.8 \times 10^9}{2.5 \times 10^4} = \frac{2.8}{2.5} \times \frac{10^9}{10^4} = 1.12 \times 10^{9-4} = 1.12 \times 10^5$ hours

Since there are $24 \times 365 = 8,760 = 8.76 \times 10^3$ hours in a year, this corresponds to $\frac{1.12 \times 10^5}{8.76 \times 10^3} \approx$ $0.13 \times 10^{5-3} = 0.13 \times 10^2 = 13$ years.

91. $\dfrac{8.424\times10^{11}}{(2.6\times10^{6})(10,000)}=\dfrac{8.424\times10^{11}}{(2.6\times10^{6})(1\times10^{4})}=\dfrac{8.424\times10^{11}}{2.6\times10^{6+4}}=\dfrac{8.424}{2.6}\times\dfrac{10^{11}}{10^{10}}$

$=3.24\times10^{11-10}=3.24\times10^{1}=32.4$ seconds

93. There are $1\,yr\left(\dfrac{365\,d}{1\,yr}\right)\left(\dfrac{24\,hr}{1\,d}\right)\left(\dfrac{60\,min}{1\,hr}\right)\left(\dfrac{60\,s}{1\,min}\right)=3.1536\times10^{7}$ seconds in a year, so light will travel

$(1.86\times10^{5})(3.1536\times10^{7})=(1.86)(3.1536)\times(10^{5})(10^{7})\approx5.87\times10^{12}$ miles in a year.

95. a) 125 is too large, a must be between 1 and 10; 1.25436×10^{5}
b) 0.537 is too small, b must be between 1 and 10; 5.37×10^{7}

97. Counterexamples may vary.
On the left side, we add first and then square; on the right, we square and then add

$(2+3)^{2}\neq2^{2}+3^{2}$ since

$5^{2}\neq4+9$ since

$25\neq13$

99. $451.6\times10^{9}-6.2\times10^{9}=(451.6-6.2)\times10^{9}=445.5\times10^{9}=4.445\times10^{2}\times10^{9}$

$=4.445\times10^{2+9}=4.445\times10^{11}=\$444.5$ billion

101. no; Examples may vary.

$(2^{3})^{4}\neq2^{(3^{4})}$

$8^{4}\neq2^{81}$

$4,096\neq2,417,851,639,229,258,349,412,352$

103. $(6.14\,in.)(1.0\times10^{9})\left(\dfrac{1\,ft}{12\,in.}\right)\left(\dfrac{1\,mi}{5280\,ft}\right)\approx9.690657\times10^{4}\,mi=96,906.57\,mi$

105. 2×10^{6} times as much

1970; $\$1.25\times10^{-2}$ / byte

$\dfrac{600}{48,000}=\dfrac{6.0\times10^{2}}{4.8\times10^{4}}=1.25\times10^{-2}$

2011; $\$6.25\times10^{-9}$ / byte

$\dfrac{400}{64\times10^{9}}=\dfrac{4.0\times10^{2}}{6.4\times10^{10}}=6.25\times10^{-9}$

So, $\dfrac{1.25\times10^{-2}}{6.25\times10^{-9}}=2\times10^{6}$

Section 6.6 Looking Deeper: Sequences

1. This is an arithmetic sequence. There is a common difference of 3 between consecutive terms. The next two terms are 17 and 20.

3. This is a geometric sequence. There is a common ratio of 3 between consecutive terms. The next two terms are 648 and 1,944.

5. This is a geometric sequence. There is a common ratio of $\frac{1}{2}$ between consecutive terms. The next two terms are $\frac{1}{16}$ and $\frac{1}{32}$.

7. This is an arithmetic sequence. There is a common difference of -5 between consecutive terms. The next two terms are -10 and -15.

9. This is a geometric sequence. There is a common ratio of 2 between consecutive terms. The next two terms are 32 and 64.

11. This is an arithmetic sequence. There is a common difference of 0.5 between consecutive terms. The next two terms are 3.5 and 4.0.

13. a) Since $a_1 = 5$, $d = 3$, and $n = 11$, we have $a_{11} = 5 + (11-1)\cdot 3 = 5 + 10\cdot 3 = 5 + 30 = 35$.

b) Since $a_1 = 5$, $a_{11} = 35$, and $n = 11$, the sum of the first 11 terms is $\frac{11(5+35)}{2} = \frac{11\cdot 40}{2} = 11\cdot 20 = 220$.

15. a) Since $a_1 = 2$, $d = 6$, and $n = 15$, we have $a_{15} = 2 + (15-1)\cdot 6 = 2 + 14\cdot 6 = 2 + 84 = 86$.

b) Since $a_1 = 2$, $a_{15} = 86$, and $n = 15$, the sum of the first 15 terms is $\frac{15(2+86)}{2} = \frac{15\cdot 88}{2} = 15\cdot 44 = 660$.

17. a) Since $a_1 = 1$, $d = 0.5$, and $n = 20$, we have $a_{20} = 1 + (20-1)\cdot 0.5 = 1 + 19\cdot 0.5 = 1 + 9.5 = 10.5$.

b) Since $a_1 = 1$, $a_{20} = 10.5$, and $n = 20$, the sum of the first 20 terms is $\frac{20(1+10.5)}{2} = \frac{20\cdot 11.5}{2} = 10\cdot 11.5 = 115$.

19. We need to find the sum of the first 21 terms of this sequence and subtract the sum of the first 13 terms. Since $a_1 = 5$, $d = 3$, and $n = 21$, we have $a_{21} = 5 + (21-1)\cdot 3 = 5 + 20\cdot 3 = 5 + 60 = 65$. Thus, the sum of the first 21 terms is $\frac{21(5+65)}{2} = \frac{21\cdot 70}{2} = 21\cdot 35 = 735$. Since $a_1 = 5$, $d = 3$, and $n = 13$, we have $a_{13} = 5 + (13-1)\cdot 3 = 5 + 12\cdot 3 = 5 + 36 = 41$. Thus the sum of the first 13 terms is $\frac{13(5+41)}{2} = \frac{13\cdot 46}{2} = 13\cdot 23 = 299$. The desired sum is therefore $735 - 299 = 436$.

21. Since $a_1 = 1$, $r = 3$, and $n = 11$, we have $a_{11} = 1\cdot 3^{11-1} = 1\cdot 3^{10} = 1\cdot 59{,}049 = 59{,}049$.

23. Since $a_1 = 1$, $r = \frac{1}{2}$, and $n = 7$, we have $a_7 = 1\cdot\left(\frac{1}{2}\right)^{7-1} = 1\cdot\left(\frac{1}{2}\right)^6 = 1\cdot\frac{1}{64} = \frac{1}{64}$.

25. Since $a_1 = 2$, $r = 0.1$, and $n = 6$, we have $a_6 = 2\cdot 0.1^{6-1} = 2\cdot 0.1^5 = 2\cdot 0.00001 = 0.00002$.

27. 144

$F_{13} = F_{11} + F_{12}$

$233 = F_{12} + 89$

$144 = F_{12}$

29. 377

$F_{15} = F_{13} + F_{14}$

$610 = 233 + F_{14}$

$377 = F_{14}$

31. $a_1 = 2{,}875$, $d = 35$, and $n = 12$, thus $a_{12} = 2{,}875 + (12-1)\cdot 35 = 2{,}875 + 11\cdot 35 = 2{,}875 + 385 = 3{,}260$.

    Therefore, $\frac{12(2{,}875 + 3{,}260)}{2} = \frac{12\cdot 6{,}135}{2} = 6\cdot 6{,}135 = \$36{,}810$ is the amount earned in one year.

33. In this exercise, you need to find the seventh term of this geometric sequence. The reason for the seventh term is because you are initially making a deposit. The amount in the account in six years is the amount of money you have starting the seventh year. The common ratio would be 1.035 because you retain the amount plus receive a percentage of the amount from one year to the next. Thus the amount you will receive will be $1{,}200(1.035)^{7-1} = 1{,}200(1.035)^6 \approx \$1{,}475.11$.

35. In this geometric sequence, the height after the fifth bounce would be the sixth term where $a_1 = 8$ and $r = \frac{7}{8}$. $a_6 = 8\cdot\left(\frac{7}{8}\right)^{6-1} = 8\cdot\left(\frac{7}{8}\right)^5 = 8\cdot\frac{16{,}807}{32{,}768} = \frac{16{,}807}{4{,}096} \approx 4.10$ feet

37. This is a geometric sequence where $a_1 = 195$, $r = 1.01134$, and $n = 10$. The common ratio is 1.01134 because you retain the population from one year to the next and then add a percentage of the population. Thus the population in 2020 would be $195\cdot 1.01134^{10-1} = 195\cdot 1.01134^9 \approx 216$ million.

39. – 41. Answers will vary.

43. Answers may vary.

    Consider the sum $a_k + a_{k+1} + \ldots + a_{n-1} + a_n$ and the same sum written in reverse order as $a_n + a_{n-1} + \ldots + a_{k+1} + a_k$. If we place these sums under one another and add, we get

$$\begin{array}{ccccccccc} a_k & + & a_{k+1} & +\ldots+ & a_{n-1} & + & a_n \\ a_n & + & a_{n-1} & +\ldots+ & a_{k+1} & + & a_k \\ \hline \multicolumn{7}{c}{(a_k + a_n) + (a_{k+1} + a_{n-1}) + \ldots + (a_{n-1} + a_{k+1}) + (a_n + a_k).} \end{array}$$

    It can be shown that each term represents the same quantity, namely $a_k + a_n$. For example $a_{k+1} + a_{n-1} = (a_k + d) + (a_n - d) = a_k + a_n$, where d is the common difference. This term is being added $n - k + 1$ times, so $(n-k+1)(a_k + a_n)$ is twice the desired sum, so the formula would be $\frac{(n-k+1)(a_k + a_n)}{2}$.

    An alternate, lengthier approach would be to find the sum of the first n terms of the original sequence and subtract from that the sum of the first $k-1$ terms.

$$\frac{n(a_1 + a_n)}{2} - \frac{(k-1)(a_1 + a_{k-1})}{2}$$

    Since $a_n = a_1 + (n-1)\cdot d = a_1 + nd - d$ and $a_{k-1} = a_1 + ((k-1)-1)\cdot d = a_1 + (k-2)\cdot d = a_1 + kd - 2d$, we have

43. (continued)

$$\frac{n(a_1+a_n)}{2}-\frac{(k-1)(a_1+a_{k-1})}{2}=\frac{n(a_1+a_1+nd-d)}{2}-\frac{(k-1)(a_1+a_1+kd-2d)}{2}$$

$$=\frac{n(2a_1+nd-d)}{2}-\frac{(k-1)(2a_1+kd-2d)}{2}$$

$$=\frac{2na_1+n^2d-nd}{2}-\frac{2ka_1+k^2d-2kd-2a_1-kd+2d}{2}$$

$$=\frac{na_1+na_1+n^2d-nd-2ka_1-k^2d+2kd+2a_1+kd-2d}{2}$$

$$=\frac{n(a_1+nd-d)+na_1-ka_1+nkd-nkd-ka_1-k^2d+kd+kd+a_1+(a_1+kd-d)-d+nd-nd}{2}$$

$$=\frac{n(a_1+nd-d)-ka_1-ka_1+n(a_1+kd-d)-nkd-k^2d+kd+kd+(a_1+nd-d)+(a_1+kd-d)}{2}$$

$$=\frac{n(a_1+(n-1)d)-ka_1-ka_1+n(a_1+(k-1)d)-nkd-k^2d+kd+kd+(a_1+(n-1)d)+(a_1+(k-1)d)}{2}$$

$$=\frac{na_n-ka_1-ka_1+na_k-nkd-k^2d+kd+kd+a_n+a_k}{2}$$

$$=\frac{na_n+a_n-k(a_1+nd-d)-k(a_1+kd-d)+na_k+a_k}{2}$$

$$=\frac{na_n+a_n-k(a_1+(n-1)d)-k(a_1+(k-1)d)+na_k+a_k}{2}$$

$$=\frac{na_n+a_n-ka_n-ka_k+na_k+a_k}{2}=\frac{na_n-ka_n+a_n+na_k-ka_k+a_k}{2}=\frac{(n-k+1)a_n+(n-k+1)a_k}{2}$$

$$=\frac{(n-k+1)(a_n+a_k)}{2}$$

45. $F_n{}^2+F_{n+1}{}^2=F_{2n+1}$

n	F_n	$F_n{}^2+F_{n+1}{}^2$
1	1	$1^2+1^2=1+1=2=F_3=F_{2+1}$
2	1	$1^2+2^2=1+4=5=F_5=F_{4+1}$
3	2	$2^2+3^2=4+9=13=F_7=F_{6+1}$
4	3	$3^2+5^2=9+25=34=F_9=F_{8+1}$
5	5	$5^2+8^2=25+64=89=F_{11}=F_{10+1}$
6	8	$8^2+13^2=64+169=233=F_{13}=F_{12+1}$
7	13	$13^2+21^2=169+441=610=F_{15}=F_{14+1}$
8	21	
9	34	
10	55	
11	89	
12	144	
13	233	
14	377	
15	610	

47. $335 = 21\times3+34\times8$

n	F_n	$F_{n-1}+F_{n-2}$
1	3	
2	8	
3	11	$3+8=1\times3+1\times8$
4	19	$8+11=1\times8+(1\times3+1\times8)=1\times3+2\times8$
5	30	$11+19=(1\times3+1\times8)+(1\times3+2\times8)=2\times3+3\times8$
6	49	$19+30=(1\times3+2\times8)+(2\times3+3\times8)=3\times3+5\times8$
7	79	$30+49=(2\times3+3\times8)+(3\times3+5\times8)=5\times3+8\times8$
8	128	$49+79=(3\times3+5\times8)+(5\times3+8\times8)=8\times3+13\times8$
9	207	$79+128=(5\times3+8\times8)+(8\times3+13\times8)=13\times3+21\times8$
10	335	$128+207=(8\times3+13\times8)+(13\times3+21\times8)=21\times3+34\times8$

49. $F_5 = \dfrac{\left(\frac{1+\sqrt{5}}{2}\right)^5+\left(\frac{1-\sqrt{5}}{2}\right)^5}{\sqrt{5}} = \dfrac{\frac{80\sqrt{5}+176}{32}+\frac{176-80\sqrt{5}}{32}}{\sqrt{5}} = \dfrac{\frac{352}{32}}{\sqrt{5}} = \dfrac{11}{\sqrt{5}} \approx 4.9193$, rounded to the nearest whole number would be 5.

51. $F_9 = \dfrac{\left(\frac{1+\sqrt{5}}{2}\right)^9+\left(\frac{1-\sqrt{5}}{2}\right)^9}{\sqrt{5}} = \dfrac{\frac{8,704\sqrt{5}+19,456}{512}+\frac{19,456-8,704\sqrt{5}}{512}}{\sqrt{5}} = \dfrac{\frac{38,912}{512}}{\sqrt{5}} = \dfrac{76}{\sqrt{5}} \approx 33.9882$, rounded to the nearest whole number would be 34.

Chapter Review Exercises

1. Refer to the table for Exercise 25 in the solutions for Section 6.1.
   71, 73, 79, 83, 89

2. Since $169 < 180 < 196$, $13 < \sqrt{180} < 14$. Thus $a = 13$.

3. 191 is prime. You need to check for divisibility by 2, 3, 5, 7, 11, and 13.
   441 is composite. $441 = 3^2\times7^2$

4. 1,080,036 is divisible by **3** because the sum of the digits, 18, is divisible by 3.
   1,080,036 is divisible by **4** because the number formed by the last two digits, 36, is divisible by 4.
   1,080,036 is not divisible by 5 because the last digit, 6, is not a 0 or a 5.
   1,080,036 is divisible by **6** because the number is divisible by both 2 and 3. (You know the number is divisible by 2 because the last digit, 6, is divisible by 2.)
   1,080,036 is not divisible by 8 because the number formed by the last three digits, 36, is not divisible by 8.
   1,080,036 is divisible by **9** because the sum of the digits, 18, is divisible by 9.
   1,080,036 is not divisible by 10 because the last digit, 6, is not 0.

5. $1,584 = 2^4\times3^2\times11$ and $1,320 = 2^3\times3\times5\times11$; $GCD(396,330) = 2^3\times3\times11 = 264$
   $LCM\,(1584,1320) = 2^4\times3^2\times5\times11 = 7,920$

6. To calculate the GCD, multiply the *smallest* powers of any primes that are common to both numbers. To calculate the LCM, multiply the *largest* powers that occur in either number.

7. 2,772 is divisible by **9** because the sum of the digits, 18, is divisible by 9.
2,772 is divisible by **4** because the number formed by the last two digits, 72, is divisible by 4.
2,772 is not divisible by 8 because the number formed by the last three digits, 270, is not divisible by 8.

8. No digits follow the decimal point when one number divides another.

9. a) 9
b) 11
c) 72
d) -16

10. For the past four days you received $3, so four days ago, you had $12 less. Thus $(-4)(+3)=-12$.

11. If $\frac{-24}{-8}=c$, then $-24=(-8)\cdot c$. Thus $c=+3$.

12. $17-(-3)=20$ degrees

13. If $\frac{5}{0}=c$, then $5=0\cdot c$, which is not possible.

14. $\left(\frac{-3+9}{-4-2}\right)\cdot\left(\frac{12-(-4)}{-7-1}\right)=\left(\frac{6}{-6}\right)\cdot\left(\frac{16}{-8}\right)=(-1)\cdot(-2)=2$

15. $67; For the past four days you lose $6, which can be thought of as $(-4)(-6)$. Therefore, four days ago you had $24 more. Thus, $(-4)(-6)=24$..

16. not equal; $\frac{28}{65}\neq\frac{14}{35}$ since $28\cdot35\neq65\cdot14$ $(980\neq910)$

17. a) $\frac{4}{9}\cdot\left(\frac{3}{4}-\frac{1}{3}\right)=\frac{4}{9}\cdot\left(\frac{3\cdot3-4\cdot1}{4\cdot3}\right)=\frac{4}{9}\cdot\left(\frac{9-4}{12}\right)=\frac{\overset{1}{\cancel{4}}}{9}\cdot\left(\frac{5}{\underset{3}{\cancel{12}}}\right)=\frac{5}{27}$

b) $\frac{3}{7}\div\left(\frac{2}{3}+\frac{3}{14}\right)=\frac{3}{7}\div\left(\frac{2\cdot14+3\cdot3}{3\cdot14}\right)=\frac{3}{7}\div\left(\frac{28+9}{42}\right)=\frac{3}{7}\div\left(\frac{37}{42}\right)=\frac{3}{\underset{1}{\cancel{7}}}\cdot\frac{\overset{6}{\cancel{42}}}{37}=\frac{18}{37}$

18. $8\overline{)53}$ quotient 6; 48; remainder 5; $6\frac{5}{8}$

19. $\left(3\frac{1}{2}\right)\left(4\frac{1}{7}\right)=\frac{7}{2}\cdot\frac{29}{7}=\frac{\overset{1}{\cancel{7}}}{2}\cdot\frac{29}{\underset{1}{\cancel{7}}}=\frac{29}{2}=14\frac{1}{2}$

20. a) $0.375=\frac{375}{1000}=\frac{3\cdot125}{8\cdot125}=\frac{3}{8}$

b) $100\cdot x=63.6363\ldots$
$-\quad x=-\ 0.6363\ldots$
$99\cdot x=63\Rightarrow x=\frac{63}{99}=\frac{7\cdot9}{11\cdot9}=\frac{7}{11}$

21. $1\frac{1}{2}\div 12=\left(\frac{3}{2}\right)\cdot\left(\frac{1}{12}\right)=\left(\frac{\overset{1}{\cancel{3}}}{2}\right)\cdot\left(\frac{1}{\underset{4}{\cancel{12}}}\right)=\left(\frac{1}{2}\right)\cdot\left(\frac{1}{4}\right)=\frac{1}{8}$ pounds

22. In order to increase the recipe which serves 16 to a recipe that serves 24, you must increase the ingredients by $1\frac{1}{2}$ times. We should use $1\frac{1}{2}\cdot 2\frac{1}{2}=\frac{3}{2}\cdot\frac{5}{2}=\frac{15}{4}=3\frac{3}{4}$ cups of hot peppers and $1\frac{1}{2}\cdot 3\frac{1}{4}=\frac{3}{2}\cdot\frac{13}{4}=\frac{39}{8}=4\frac{7}{8}$ cups of tomatoes.

23. Answers will vary.

24. Irrational numbers have nonrepeating expansions, and rational numbers have repeating expansions.

25. a) $\sqrt{108}=\sqrt{36\cdot 3}=\sqrt{36}\sqrt{3}=6\sqrt{3}$

b) $\frac{\sqrt{15}}{\sqrt{5}}=\sqrt{\frac{15}{5}}=\sqrt{3}$

26. No, zero does not.

27. Examples may vary.

$$8\div(4\div 2)\neq(8\div 4)\div 2$$
$$8\div 2\neq 2\div 2$$
$$4\neq 1$$

28. a) $3^6\cdot 3^{-2}=3^{6+(-2)}=3^4=81$

b) $\left(2^4\right)^{-2}=2^{4(-2)}=2^{-8}=\frac{1}{2^8}=\frac{1}{256}$

c) $\frac{6^8}{6^5}=6^{8-5}=6^3=216$

d) $\frac{8^{-6}}{8^{-4}}=8^{-6-(-4)}=8^{-2}=\frac{1}{8^2}=\frac{1}{64}$

29. In evaluating -2^4, we raise 2 to the fourth power to get 16 and then negate the result to get –16. In evaluating $(-2)^4$, we first negate 2 and then raise that negative number to the fourth power to get positive 16.

30. $0.000456=4.56\times 10^{-4}$ and $1{,}230{,}000=1.23\times 10^6$

31. $1.325\times 10^6=1{,}325{,}000$ and $8.63\times 10^{-5}=0.0000863$

32. $$\frac{\left(3.6\times 10^3\right)\left(2.8\times 10^{-5}\right)}{\left(4.2\times 10^4\right)}=\frac{3.6\cdot 2.8}{4.2}\times\frac{10^3\cdot 10^{-5}}{10^4}=\frac{10.08}{4.2}\times\frac{10^{3+(-5)}}{10^4}=2.4\times\frac{10^{-2}}{10^4}$$
$$=2.4\times 10^{-2-4}=2.4\times 10^{-6}$$

33. $\frac{6.6\times 10^9}{109\times 10^6}=\frac{6.6\times 10^9}{1.09\times 10^8}=\frac{6.6}{1.09}\times\frac{10^9}{10^8}\approx 6.06\times 10^{9-8}=6.06\times 10^1=60.6$ times

34. a) This is a geometric sequence. There is a common ratio of -2 between consecutive terms. The next two terms are -192 and 384.

b) This is an arithmetic sequence. There is a common difference of 5 between consecutive terms. The next two terms are 36 and 41.

35. a) Since $a_1 = 4$, $d = 3$, and $n = 30$, we have $a_{30} = 4 + (30-1) \cdot 3 = 4 + 29 \cdot 3 = 4 + 87 = 91$.

b) Since $a_1 = 4$, $a_{11} = 91$, and $n = 30$, the sum of the first 30 terms is $\frac{30(4+91)}{2} = \frac{30 \cdot 95}{2} =$

$15 \cdot 95 = 1,425$.

36. Since $a_1 = 4$, $r = 3$, and $n = 10$, we have $a_{10} = 4 \cdot 3^{10-1} = 4 \cdot 3^9 = 4 \cdot 19,683 = 78,732$.

37. $F_{10} = 55$; $F_{15} = 610$

$F_{13} = F_{11} + F_{12}$
$233 = F_{11} + 144$
$89 = F_{11}$
$F_{12} = F_{10} + F_{11}$
$144 = F_{10} + 89$
$55 = F_{10}$

$F_{14} = F_{12} + F_{13} = 144 + 233 = 377$
$F_{15} = F_{13} + F_{14} = 233 + 377 = 610$

Chapter Test

1. Refer to the table for Exercise 25 in the solutions for Section 6.1.
101,103,107,109,113

2. Since $196 < 200 < 225$, $14 < \sqrt{200} < 15$. Thus $a = 14$.

3. 241 is prime. You need to check for divisibility by 2, 3, 5, 7, 11, and 13.
539 is composite. $539 = 7^2 \times 11$

4. 2,542,128 is divisible by **3** because the sum of the digits, 24, is divisible by 3.
2,542,128 is divisible by **4** because the number formed by the last two digits, 28, is divisible by 4.
2,542,128 is not divisible by 5 because the last digit, 8, is not a 0 or a 5.
2,542,128 is divisible by **6** because the number is divisible by both 2 and 3. (You know the number is divisible by 2 because the last digit, 8, is divisible by 2.)
2,542,128 is divisible by **8** because the number formed by the last three digits, 128, is divisible by 8.
2,542,128 is not divisible by 9 because the sum of the digits, 24, is not divisible by 9.
2,542,128 is not divisible by 10 because the last digit, 8, is not 0.

5. $1716 = 2^2 \times 3 \times 11 \times 13$ and $936 = 2^3 \times 3^2 \times 13$; $GCD(1716, 936) = 2^2 \times 3 \times 13 = 156$
$LCM(1716, 936) = 2^3 \times 3^2 \times 11 \times 13 = 10,296$

6. To calculate the GCD, multiply the *smallest* powers of any primes that are common to both numbers. To calculate the LCM, multiply the *largest* powers that occur in either number.

$GCD(a,b) = 2^3 \times 3^6 \times 5^2 \times 7^2$
$LCM(a,b) = 2^8 \times 3^9 \times 5^3 \times 7^9$

7. a) –6
b) +3
c) –54
d) +8

8. a) This is an arithmetic sequence. There is a common difference of 3 between consecutive terms. The next two terms are 27 and 30.

   b) This is a geometric sequence. There is a common ratio of 3 between consecutive terms. The next two terms are 486 and 1,458.

9. If you spend $5 for the next seven days, you will have $35 less. Thus $(+7)(-5)=-35$.

10. If $\frac{-21}{3}=c$, then $-21=(3)\cdot c$. Thus $c=-7$.

11. $50-(-19)=69$ degrees.

12. If $\frac{8}{0}=c$, then $8=0\cdot c$, which is not possible.

13. $F_{14}=377;\ F_{20}=6,765$

$$F_{17}=F_{15}+F_{16}$$
$$1597=F_{15}+987$$
$$610=F_{15}$$
$$F_{16}=F_{14}+F_{15}$$
$$987=F_{14}+610$$
$$377=F_{14}$$

$$F_{18}=F_{16}+F_{17}987+1597=2584$$
$$F_{19}=F_{17}+F_{18}=1597+2584=4181$$
$$F_{20}=F_{18}+F_{19}=2584+4181=6765$$

14. not equal; $\frac{198}{213}\neq\frac{68}{72}$ since $198\cdot 72\neq 68\cdot 213$ $(14,256\neq 14,484)$

15. a) $\frac{3}{5}\cdot\left(\frac{7}{9}-\frac{3}{4}\right)=\frac{3}{5}\cdot\left(\left(\frac{7\cdot 4-3\cdot 9}{9\cdot 4}\right)\right)=\frac{3}{5}\cdot\left(\frac{28-27}{36}\right)=\frac{\overset{1}{\cancel{3}}}{5}\cdot\left(\frac{1}{\underset{12}{\cancel{36}}}\right)=\frac{1}{60}$

    b) $\frac{4}{5}\div\left(\frac{2}{5}-\frac{1}{4}\right)=\frac{4}{5}\div\left(\frac{2\cdot 4-1\cdot 5}{5\cdot 4}\right)=\frac{4}{5}\div\left(\frac{8-5}{20}\right)=\frac{4}{5}\div\left(\frac{3}{20}\right)=\frac{4}{\underset{1}{\cancel{5}}}\cdot\frac{\overset{4}{\cancel{20}}}{3}=\frac{16}{3}$

16.
$$\begin{array}{r} 15 \\ 3\overline{)47} \\ \underline{30} \\ 17 \\ \underline{15} \\ 2 \end{array}\ ;\ 15\frac{2}{3}$$

17. $4\frac{1}{2}\div 3\frac{3}{8}=\frac{9}{2}\div\frac{27}{8}=\frac{\overset{1}{\cancel{9}}}{\underset{1}{\cancel{2}}}\cdot\frac{\overset{4}{\cancel{8}}}{\underset{3}{\cancel{27}}}=\frac{4}{3}$

18. a) $0.573=\frac{573}{1000}$

    b)
$$\begin{array}{rl} 100\cdot x= & 57.5757... \\ -\quad x= & -\ 0.5757... \\ \hline \end{array}$$
$$99\cdot x=57\Rightarrow x=\frac{57}{99}=\frac{19}{33}$$

19. a) Since $a_1 = 11$, $d = 9$, and $n = 20$, we have $a_{20} = 11 + (20-1)\cdot 9 = 11 + 19\cdot 9 = 11 + 171 = 182$.

b) Since $a_1 = 11$, $a_{20} = 182$, and $n = 20$, the sum of the first 20 terms is $\dfrac{20(11+182)}{2} = \dfrac{20\cdot 193}{2} =$

$10\cdot 193 = 1{,}930$.

20. The photo is being enlarged by $5\frac{1}{4}$. The dimensions should be $23\frac{5}{8}$ inches, by $33\frac{1}{4}$ inches.

$5\frac{1}{4}\cdot 4\frac{1}{2} = \frac{21}{4}\cdot\frac{9}{2} = \frac{189}{8} = 23\frac{5}{8}$ $\qquad$ $5\frac{1}{4}\cdot 6\frac{1}{3} = \frac{21}{4}\cdot\frac{19}{3} = \frac{399}{12} = 33\frac{1}{4}$

21. Answers will vary.

22. Rational numbers have repeating or terminating expansions and irrational numbers have nonrepeating expansions.

23. a) $\sqrt{180} = \sqrt{36\cdot 5} = \sqrt{36}\sqrt{5} = 6\sqrt{5}$

b) $\dfrac{3}{\sqrt{15}} = \dfrac{3\cdot\sqrt{15}}{\sqrt{15}\cdot\sqrt{15}} = \dfrac{\overset{1}{\cancel{3}}\cdot\sqrt{15}}{\underset{5}{\cancel{15}}} = \dfrac{\sqrt{15}}{5}$

24. Answers may vary. Possible example is:

$$\frac{5}{2}\div\left(\frac{3}{2}\div\frac{1}{2}\right) \neq \left(\frac{5}{2}\div\frac{3}{2}\right)\div\frac{1}{2}$$
$$\frac{5}{2}\div\left(\frac{3}{2}\cdot\frac{2}{1}\right) \neq \left(\frac{5}{2}\cdot\frac{2}{3}\right)\div\frac{1}{2}$$
$$\frac{5}{2}\div\frac{3}{1} \neq \frac{5}{3}\div\frac{1}{2}$$
$$\frac{5}{2}\cdot\frac{1}{3} \neq \frac{5}{3}\cdot\frac{2}{1}$$
$$\frac{5}{6} \neq \frac{10}{3}$$

25. a) $2^7\cdot 2^{-4} = 2^{7+(-4)} = 2^3 = 8$

b) $\left(3^2\right)^{-2} = 3^{2(-2)} = 3^{-4} = \dfrac{1}{3^4} = \dfrac{1}{81}$

c) $\dfrac{5^7}{5^4} = 5^{7-4} = 5^3 = 125$

d) $\dfrac{3^{-2}}{3^{-5}} = 3^{-2-(-5)} = 3^3 = 27$

26. Since $a_1 = -2$, $r = -3$, and $n = 10$, we have $a_{10} = -2\cdot(-3)^{10-1} = -2\cdot(-3)^9 =$

$-2\cdot(-19{,}683) = 39{,}366$.

27. In evaluating -3^2, 3 is raised to the second power to get 9 before being negated to get –9. In evaluating $(-3)^2$, 3 is negated first and then that negative number is raised to the second power to get 9.

28. $15{,}460{,}000 = 1.546\times 10^7$ and $0.00000000623 = 6.23\times 10^{-9}$

29. $\dfrac{1.322\times 10^9}{3.04\times 10^8} = \dfrac{1.322}{3.04}\times\dfrac{10^9}{10^8} \approx 0.435\times 10^{9-8} = 0.435\times 10^1 = 4.35\times 10^0 = 4.35$ times larger

Chapter 7

Algebraic Models: How Do We Approximate Reality?

Section 7.1 Linear Equations

1. $$\begin{aligned} 3x+4&=5x-6\\ -2x+4&=-6\\ -2x&=-10\\ x&=5 \end{aligned}$$

3. $$\begin{aligned} 4-2y&=8+3y\\ 4&=8+5y\\ -4&=5y\\ \frac{-4}{5}&=y\\ y&=-\frac{4}{5} \end{aligned}$$

5. $$\begin{aligned} \frac{1}{2}x-6&=\frac{1}{5}x+3\\ 10\cdot\left[\frac{1}{2}x-6\right]&=\left[\frac{1}{5}x+3\right]\cdot 10\\ 5x-60&=2x+30\\ 3x-60&=30\\ 3x&=90\\ x&=30 \end{aligned}$$

7. $$\begin{aligned} 0.3y+2&=0.5y-3\\ 10\cdot[0.3y+2]&=[0.5y-3]\cdot 10\\ 3y+20&=5y-30\\ 20&=2y-30\\ 50&=2y\\ 25&=y\\ y&=25 \end{aligned}$$

9. $$\begin{aligned} P&=2l+2w\\ P-2l&=2w\\ \frac{P-2l}{2}&=w\\ w&=\frac{P-2l}{2} \end{aligned}$$

11. $$\begin{aligned} z&=\frac{x-\mu}{\sigma}\\ \sigma\cdot z&=\left[\frac{x-\mu}{\sigma}\right]\cdot\sigma\\ \sigma z&=x-\mu\\ \sigma z-x&=-\mu\\ -(\sigma z-x)&=\mu\\ x-\sigma\ z&=\mu\\ \mu&=x-\sigma z \end{aligned}$$

13. $$\begin{aligned} A&=P(1+rt)\\ A&=P+Prt\\ A-P&=Prt\\ \frac{A-P}{Pt}&=\frac{Prt}{Pt}=r\\ r&=\frac{A-P}{Pt} \end{aligned}$$

15. $$\begin{aligned} 2x+3y&=6\\ 2x&=-3y+6\\ x&=\frac{-3y+6}{2} \end{aligned}$$

17. $$\begin{aligned} V&=lwh\\ \frac{V}{wh}&=\frac{lwh}{wh}=l\\ l&=\frac{V}{wh} \end{aligned}$$

19. $$\begin{aligned} S&=2\pi rh+2\pi r^2\\ S-2\pi r^2&=2\pi rh\\ \frac{S-2\pi r^2}{2\pi r}&=\frac{2\pi rh}{2\pi r}=h\\ h&=\frac{S-2\pi r^2}{2\pi r} \end{aligned}$$

21. x-intercept (when $y = 0$)

$$3x + 2 \cdot 0 = 12$$
$$3x + 0 = 12$$
$$3x = 12$$
$$x = 4 \Rightarrow (4,0)$$

y-intercept (when $x = 0$)

$$3 \cdot 0 + 2y = 12$$
$$0 + 2y = 12$$
$$2y = 12$$
$$y = 6 \Rightarrow (0,6)$$

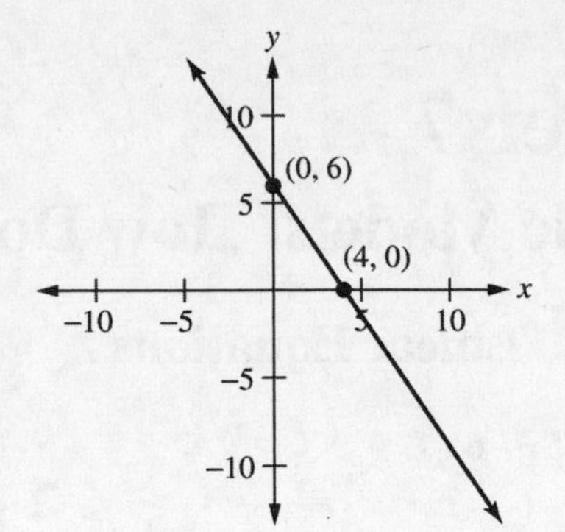

23. x-intercept (when $y = 0$)

$$4x - 3 \cdot 0 = 16$$
$$4x - 0 = 16$$
$$4x = 16$$
$$x = 4 \Rightarrow (4,0)$$

y-intercept (when $x = 0$)

$$4 \cdot 0 - 3y = 16$$
$$0 - 3y = 16$$
$$-3y = 16$$
$$y = -\frac{16}{3} \Rightarrow \left(0, -\tfrac{16}{3}\right)$$

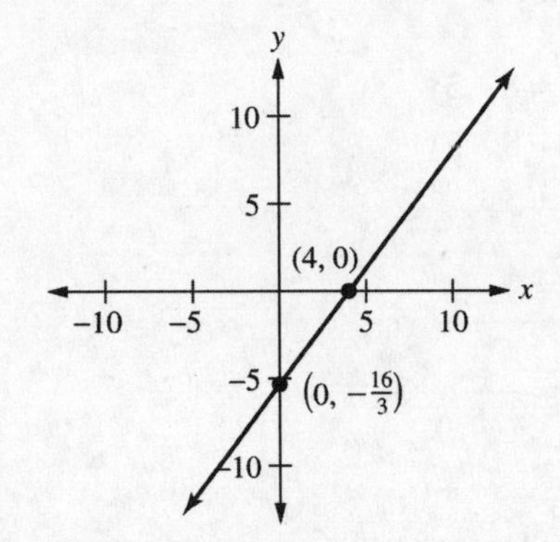

25. x-intercept (when $y = 0$)

$$\frac{1}{3}x + \frac{1}{2} \cdot 0 = 3$$
$$\frac{1}{3}x + 0 = 3$$
$$\frac{1}{3}x = 3$$
$$x = 9 \Rightarrow (9,0)$$

y-intercept (when $x = 0$)

$$\frac{1}{3} \cdot 0 + \frac{1}{2}y = 3$$
$$0 + \frac{1}{2}y = 3$$
$$\frac{1}{2}y = 3$$
$$y = 6 \Rightarrow (0,6)$$

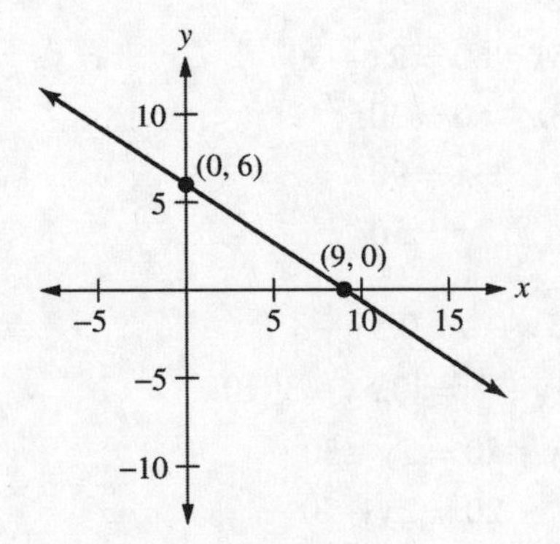

27. x-intercept (when $y = 0$)

$$\frac{1}{6}x - 2\cdot 0 = \frac{3}{4} \Rightarrow \frac{1}{6}x - 0 = \frac{3}{4}$$

$$\frac{1}{6}x = \frac{3}{4} \Rightarrow 12\cdot\left[\frac{1}{6}x\right] = \left[\frac{3}{4}\right]\cdot 12$$

$$2x = 9 \Rightarrow x = \frac{9}{2} \Rightarrow \left(\tfrac{9}{2}, 0\right)$$

y-intercept (when $x = 0$)

$$\frac{1}{6}\cdot 0 - 2y = \frac{3}{4} \Rightarrow \frac{1}{6}x - 0 = \frac{3}{4}$$

$$0 - 2y = \frac{3}{4}$$

$$-2y = \frac{3}{4} \Rightarrow y = -\frac{3}{8} \Rightarrow \left(0, -\tfrac{3}{8}\right)$$

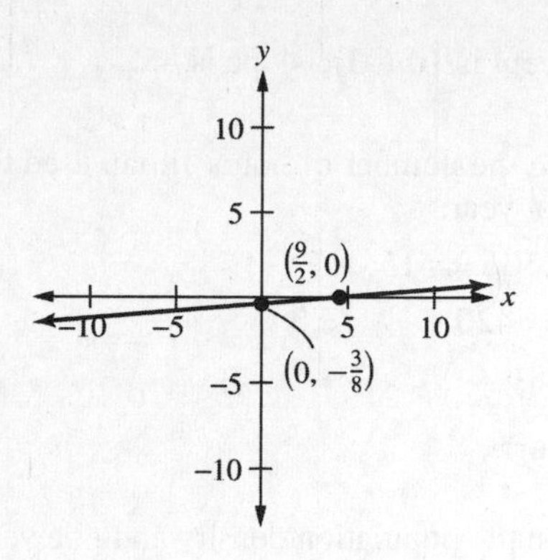

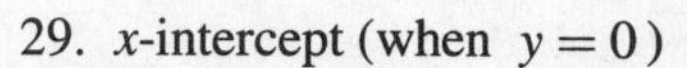

29. x-intercept (when $y = 0$)

$$0.2x = 4\cdot 0 + 1.6$$
$$0.2x = 0 + 1.6$$
$$0.2x = 1.6$$
$$x = 8 \Rightarrow (8, 0)$$

y-intercept (when $x = 0$)

$$0.2\cdot 0 = 4y + 1.6$$
$$0 = 4y + 1.6$$
$$-1.6 = 4y$$
$$-0.4 = y \Rightarrow (0, -0.4)$$

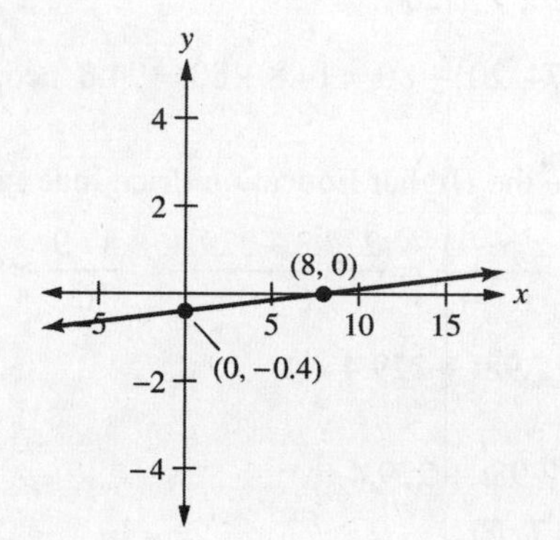

31. x-intercept (when $y = 0$)

$$0.4x - 0.3\cdot 0 = 1.2$$
$$0.4x - 0 = 1.2$$
$$x = 3 \Rightarrow (3, 0)$$

y-intercept (when $x = 0$)

$$0.4\cdot 0 - 0.3y = 1.2$$
$$0. - 0.3y = 1.2$$
$$y = -4 \Rightarrow (0, -4)$$

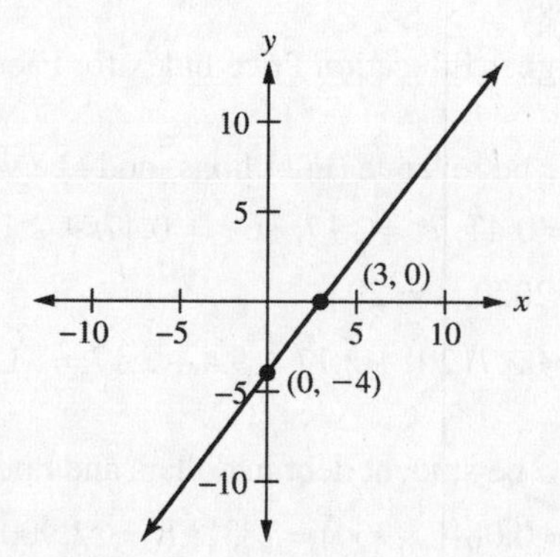

33. a

35. c

37. $m = \dfrac{y_2 - y_1}{x_2 - x_1} = \dfrac{8-5}{6-2} = \dfrac{3}{4}$

39. $m = \dfrac{y_2 - y_1}{x_2 - x_1} = \dfrac{2-6}{8-3} = \dfrac{-4}{5} = -\dfrac{4}{5}$

41. $m = \dfrac{y_2 - y_1}{x_2 - x_1} = \dfrac{1-(-4)}{5-3} = \dfrac{5}{2}$

43. a and d

45. f

47. The rise is zero.

49. y-intercept is $(0, -3)$; slope is 4

51. y-intercept is $(0,-3)$; slope is –5

53. Let h be the number of hours Jillian used the club last year.

$95+2.50h=515$

$2.50h=420$

$h=168$

168 hours

55. Let p be the number of pages.

$16+1.30(p-10)=44.60$

$16+1.30p-13=44.60$

$3+1.30p=44.60$

$1.30p=41.60$

$p=32$ pages

57. Let d be the population density and t be years since 2000.

$$m=\frac{d_2-d_1}{t_2-t_1}=\frac{87.4-80}{10-0}=\frac{7.4}{10}=0.74;\ b=80$$

so $d=0.74t+80$

59. For 2020, $t=20$;

$d=0.74(20)+80=14.8+80=94.8$ people per square mile.

61. Let y be the Higher Education Price Index for Faculty and t be years since 2003.

a) $m=\dfrac{d_2-d_1}{t_2-t_1}=\dfrac{277.3-229.4}{6-0}=\dfrac{47.9}{6}\approx 7.98;\ b=229.4$

b) $y=7.98t+229.4$

63. $300=7.98t+229.4$

$70.6=7.98t$

$t=\dfrac{70.6}{7.98}\approx 9$

The Higher Education Price Index for Faculty will be 300 by $2003+9=2012$.

65. a) Let r be revenue (in billions) and t be years since 2000.

$m=0.47;\ b=2.17$, so $r=0.47t+2.17$

b) For 2020, $t=20$;

$r=0.47(20)+2.17=9.4+2.17=\$11.57$ billion

67. a) Let d be student debt in dollars and t be years since 2011.

$m=0.08(22{,}900)=1832;\ b=22{,}900$, so $d=1832t+22{,}900$

b) For 2017, $t=6$;

$d=1832(6)+22{,}900=\$33{,}892$

69. Let d be the number of hours worked in the diner and p be the number of hours worked in the pet store.

$5.60d+7.35p=133$

71. Let t be the number of shares with Time-Warner and c be the number of shares with CDW.

$35t+55c=14{,}500$

73. Let e be the number of end tables to be constructed and c be the number of coffee tables to be constructed.

$9e+15c=342$

75. Let x be the number of tokens.

$$\begin{aligned} 9+0.70x &= 0.85x \\ 9 &= 0.15x \\ 60 &= x \end{aligned}$$

after 60 tokens

77 – 79. Answers will vary.

81. Let x be the number of computer systems Chuck sells.

$y=45x+225$ is the amount Chuck would receive from Buy More.

$y=20x+400$ is the amount Chuck would receive from Circuit Town.

To find when both deals are the same, we solve $45x+225=20x+400$.

$$\begin{aligned} 45x+225 &= 20x+400 \\ 25x+225 &= 400 \\ 25x &= 175 \\ x &= 7 \end{aligned}$$

If Chuck sells 7 computer systems, then the two deals are the same. Because he would receive more per computer system sold, if Chuck sells more than seven systems, then Buy More is better for him.

83. amount of interest on the CD: $5000(0.038)=190$

amount of tax paid to the government: $190(0.14)=26.60$

amount realized on the CD: $190-26.60=163.40$

Let r be the interest rate needed to be earned on the bonds.

$163.40=5000\cdot r$

$0.03268=r$

The bonds would need to earn 3.268%.

85. a) $y=105d$

b) $d=\dfrac{y}{105}$

87. $e=0.68d \Rightarrow \dfrac{e}{0.68}=d$

$p=2380d \Rightarrow \dfrac{p}{2380}=d$

$\dfrac{e}{0.68}=\dfrac{p}{2380} \Rightarrow p=\dfrac{2380}{0.68}e \Rightarrow$

$p=3{,}500e$

89. Examples will vary. The slope is the same regardless of which two points you choose.

91. The slope of $y = 3x + 5$ is 3, so $m = 3$.

$y=3x+b$

$1=3\cdot 2+b \Rightarrow 1=6+b \Rightarrow -5=b$

$y=3x-5$

93. The slope of the original line is $m=\dfrac{y_2-y_1}{x_2-x_1}=\dfrac{14-4}{4-2}=\dfrac{10}{2}=5.$

$y=mx+b$

$y=5x+b$

$3=5\cdot 2+b \Rightarrow 3=10+b \Rightarrow -7=b$

$y=5x-7$

95. $m = 0.08 = \frac{18}{x}$

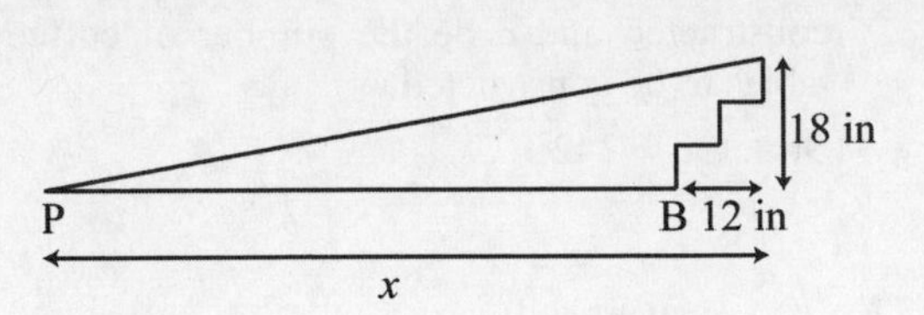

$x = \frac{18}{0.08} = 225$ inches

$225 - 12 = 213$ inches

$\frac{213}{12} = 17.75$ feet

Section 7.2 Modeling with Linear Equations

1. $y = mx + b$
   $y = 3x + b$
   $1 = 3 \cdot 2 + b \Rightarrow 1 = 6 + b \Rightarrow -5 = b$
   $y = 3x - 5$

3. $y = mx + b$
   $y = 4x + b$
   $6 = 4 \cdot (-2) + b \Rightarrow 6 = -8 + b \Rightarrow 14 = b$
   $y = 4x + 14$

5. $y = mx + b$
   $y = -2x + b$
   $3 = -2 \cdot 4 + b \Rightarrow 3 = -8 + b \Rightarrow 11 = b$
   $y = -2x + 11$

7. $y = mx + b$
   $y = -5x + b$
   $-1 = -5 \cdot (-6) + b \Rightarrow -1 = 30 + b \Rightarrow$
   $b = -31$
   $y = -5x - 31$

9. $m = \frac{y_2 - y_1}{x_2 - x_1} = \frac{9-3}{5-2} = \frac{6}{3} = 2$
   $y = mx + b$
   $y = 2x + b$
   $3 = 2 \cdot 2 + b \Rightarrow 3 = 4 + b \Rightarrow -1 = b$
   $y = 2x - 1$

11. $m = \frac{y_2 - y_1}{x_2 - x_1} = \frac{10-12}{9-17} = \frac{-2}{-8} = \frac{1}{4}$
    $y = mx + b$
    $y = \frac{1}{4}x + b$
    $10 = \frac{1}{4} \cdot 9 + b \Rightarrow 10 = \frac{9}{4} + b \Rightarrow \frac{31}{4} = b$
    $y = \frac{1}{4}x + \frac{31}{4}$

13. $m = \frac{y_2 - y_1}{x_2 - x_1} = \frac{2-(-4)}{-8-11} = \frac{6}{-19} = -\frac{6}{19}$
    $y = mx + b$
    $y = -\frac{6}{19}x + b$
    $2 = -\frac{6}{19} \cdot (-8) + b \Rightarrow 2 = \frac{48}{19} + b \Rightarrow$
    $b = -\frac{10}{19}$
    $y = -\frac{6}{19}x - \frac{10}{19}$

15. $m = \frac{y_2 - y_1}{x_2 - x_1} = \frac{-1-(-8)}{-4-(-6)} = \frac{7}{2}$
    $y = mx + b$
    $y = \frac{7}{2}x + b$
    $-1 = \frac{7}{2} \cdot (-4) + b$
    $-1 = -14 + b$
    $13 = b$
    $y = \frac{7}{2}x + 13$

17. Let x be the number of years after 2009 and y be the life expectancy in years.
    $y = 0.1x + 80.6$
    $y = 0.1(21) + 80.6 = 2.1 + 80.6 = 82.7$ years

19. a) Let t be the number of years after 2007 and y be the number of male students enrolled in college (in millions).
$y = 0.5t + 7.8$

b) For 2025, $t = 18$
$y = 0.5(18) + 7.8 = 9 + 7.8 = 16.8$
15,800,000 estimated male students enrolled

c) $15 = 0.5t + 7.8 \Rightarrow 7.2 = 0.5t \Rightarrow t = \dfrac{7.2}{0.5} \Rightarrow t \approx 15 \Rightarrow 2007 + 15 = 2022$

21. Let t be the number of years after 2009 and y be the price of tuition.
$y = 750t + 15{,}000$
For 2020, $t = 11$
$y = 750(11) + 15{,}000 = 8{,}250 + 15{,}000 = 23{,}250$
The estimated price of tuition is $23,250.

23. a) $m = \dfrac{v_2 - v_1}{t_2 - t_1} = \dfrac{429 - 481}{3 - 0}$
$= \dfrac{-52}{3} \approx -17.3$

b) $v = mt + b$
$v = -17.3t + b$
$481 = -17.3(0) + b$
$481 = 0 + b$
$481 = b$
$v = -17.3t + 481$

c) $v = -17.3(14) + 481 = -242.2 + 481$
$= 238.8 \approx 239$

25. Let t be the number of years after 2008 and y be the demand for nurses (in millions).
The given information corresponds to the two ordered pairs (0, 2.62) and (10, 3.20).

a) $m = \dfrac{y_2 - y_1}{t_2 - t_1} = \dfrac{3.20 - 2.62}{10 - 0} = 0.058$

$y = mt + b$
$y = 0.058t + b$
$2.62 = 0.058 \cdot 0 + b$
$2.62 = 0 + b$
$2.62 = b$
$y = 0.058t + 2.62$

b) The estimated demand for nurses in 2025 is 3.606 million
$y = 0.058 \cdot 17 + 2.62 = 3.606$

c) Demand will reach 4 million in 2032.
$4 = 0.058t + 2.62 \Rightarrow 1.38 = 0.058t \Rightarrow$
$t = \dfrac{1.38}{0.058} \Rightarrow t = 23.70 \Rightarrow$
$2008 + 23.79 = 2031.79$

27. Let x be the number of years after 1900 and y be the number of hours of sleep.
The given information corresponds to the two ordered pairs (0,8.5) and (102,6.9).

$$m=\frac{y_2-y_1}{x_2-x_1}=\frac{6.9-8.5}{104-0}=\frac{-1.6}{104}=-\frac{16}{1040}=-\frac{1}{65}\approx -0.0154$$

$$y=mx+b$$
$$y=-0.0154x+b$$
$$8.5=-0.0154\cdot 0+b$$
$$8.5=0+b$$
$$8.5=b$$
$$y=-0.0154x+8.5$$
$$0=-0.0154\cdot x+8.5$$
$$0.0154x=8.5$$
$$x\approx 552$$

In the year $1900+552=2452$, it is estimated that Americans won't be sleeping at all on a weeknight.

29. a) $m=\dfrac{l_2-l_1}{t_2-t_1}=\dfrac{151.3-146}{3-0}=\dfrac{5.3}{3}\approx 1.77$

b) $l=mt+b$

$l=1.77t+b$

$146=1.3\cdot 0+b$

$146=0+b$

$146=b$

$l=1.77t+146$

c) In 2009, $t=6$

$l=1.77t+146$

$l=1.77(6)+146=10.62+146=156.62$ million

d) The U.S. civilian labor force will reach 200 million in the year 2034.

$200=1.77t+146$

$54=1.77t$

$t=\dfrac{54}{1.77}\approx 31$

$2003+31=2034$

31. the slope of the line

33. Answers will vary.

35.

	2003 (year 0)	2004 (year 1)	2005 (year 2)
Actual data	146	147	149.2
Values predicted by your model	$l=1.77\cdot0+146$ $=0+146$ $=146$	$l=1.77\cdot1+146$ $=1.77+146$ $=147.77$	$l=1.77\cdot2+146$ $=3.54+146$ $=149.54$
Values predicted by line of best fit	$l=1.81\cdot0+145.66$ $=0+145.66$ $=145.66$	$l=1.81\cdot0+145.66$ $=1.81+145.66$ $=147.47$	$l=1.81\cdot2+145.66$ $=3.62+145.66$ $=149.28$

	2006 (year 3)	2006 (year 6)	
Actual data	151.3	154.1	
Values predicted by your model	$l=1.77\cdot3+146$ $=5.31+146$ $=151.31$	$l=1.77\cdot6+146$ $=10.62+146$ $=156.62$	
Values predicted by line of best fit	$l=1.81\cdot3+145.66$ $=5.43+145.66$ $=151.09$	$l=1.81\cdot6+145.66$ $=10.86+145.66$ $=156.52$	

Section 7.3 Modeling with Quadratic Equations

1. By using the quadratic formula, $x=\dfrac{-b\pm\sqrt{b^2-4ac}}{2a}$

$$x=\frac{-(-10)\pm\sqrt{(-10)^2-4(1)(16)}}{2(1)}=\frac{10\pm\sqrt{100-64}}{2}=\frac{10\pm\sqrt{36}}{2}=\frac{10\pm6}{2}$$

$$x=\frac{10+6}{2}=\frac{16}{2}=8 \text{ or } x=\frac{10-6}{2}=\frac{4}{2}=2$$

3. $$x=\frac{-(-5)\pm\sqrt{(-5)^2-4(2)(3)}}{2(2)}=\frac{5\pm\sqrt{25-24}}{4}=\frac{5\pm\sqrt{1}}{4}=\frac{5\pm1}{4}$$

$$x=\frac{5+1}{4}=\frac{6}{4}=\frac{3}{2} \text{ or } x=\frac{5-1}{4}=\frac{4}{4}=1$$

5. $$x=\frac{-(7)\pm\sqrt{(7)^2-4(3)(-6)}}{2(3)}=\frac{-7\pm\sqrt{49-(-72)}}{6}=\frac{-7\pm\sqrt{121}}{6}=\frac{-7\pm11}{6}$$

$$x=\frac{-7+11}{6}=\frac{4}{6}=\frac{2}{3} \text{ or } x=\frac{-7-11}{6}=\frac{-18}{6}=-3$$

7. $$x=\frac{-(-17)\pm\sqrt{(-17)^2-4(5)(-12)}}{2(5)}=\frac{17\pm\sqrt{289-(-240)}}{10}=\frac{17\pm\sqrt{529}}{10}=\frac{17\pm23}{10}$$

$$x=\frac{17+23}{10}=\frac{40}{10}=4 \text{ or } x=\frac{17-23}{10}=\frac{-6}{10}=-\frac{3}{5}$$

9. Opening: down since $a<0$

Vertex: $x=\frac{-b}{2a}=\frac{-6}{2(-1)}=\frac{-6}{-2}=3$. Substituting this value for *x*, we obtain

$y=-3^2+6\cdot3-8=-9+18-8=1$ as the *y*-coordinate of the vertex.

The vertex is therefore $(3,1)$.

x-intercepts: Set $y=0$. We get $0=-x^2+6x-8$ and by using the quadratic formula, we obtain

$$x=\frac{-(6)\pm\sqrt{(6)^2-4(-1)(-8)}}{2(-1)}=\frac{-6\pm\sqrt{36-32}}{-2}=\frac{-6\pm\sqrt{4}}{-2}=\frac{-6\pm2}{-2}$$

$$x=\frac{-6+2}{-2}=\frac{-4}{-2}=2 \text{ or } x=\frac{-6-2}{-2}=\frac{-8}{-2}=4$$

The *x*-intercepts of the graph are therefore $(2,0)$ and $(4,0)$.

y-intercept: Set $x=0$. Thus $y=-0^2+6\cdot0-8=-8$. The *y*-intercept is therefore $(0,-8)$.

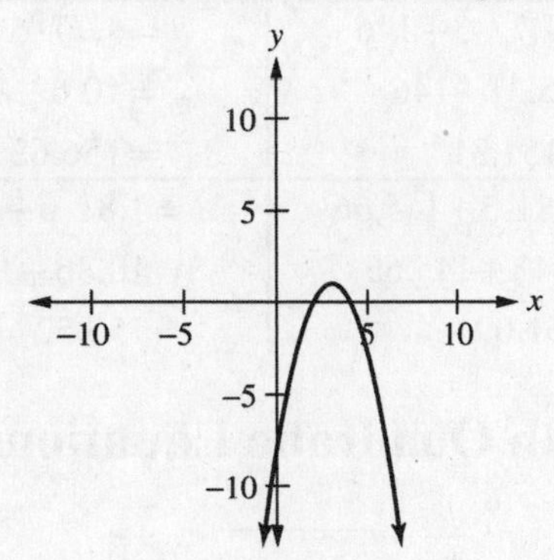

11. Opening: down since $a<0$

Vertex: $x=\frac{-b}{2a}=\frac{-8}{2(-4)}=\frac{-8}{-8}=1$. Substituting this value for *x*, we obtain

$y=-4\cdot1^2+8\cdot1+5=-4+8+5=9$ as the *y*-coordinate of the vertex.

The vertex is therefore $(1,9)$.

x-intercepts: Set $y=0$. We get $0=-4x^2+8x+5$ and by using the quadratic formula, we obtain

$$x=\frac{-(8)\pm\sqrt{(8)^2-4(-4)(5)}}{2(-4)}=\frac{-8\pm\sqrt{64+80}}{-8}=\frac{-8\pm\sqrt{144}}{-8}=\frac{-8\pm12}{-8}$$

$$x=\frac{-8+12}{-8}=\frac{4}{-8}=-\frac{1}{2} \text{ or } x=\frac{-8-12}{-8}=\frac{-20}{-8}=\frac{5}{2}$$

The *x*-intercepts of the graph are therefore $\left(-\frac{1}{2},0\right)$ and $\left(\frac{5}{2},0\right)$.

y-intercept: Set $x=0$. Thus $y=-4\cdot0^2+8\cdot0+5=5$. The *y*-intercept is therefore $(0,5)$.

11. (continued)

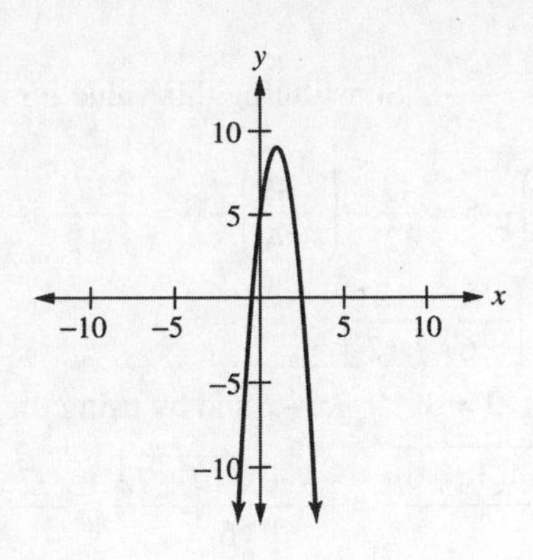

13. Opening: up since $a > 0$

Vertex: $x = \frac{-b}{2a} = \frac{-(-4)}{2(4)} = \frac{4}{8} = \frac{1}{2}$. Substituting this value for *x*, we obtain

$y = 4 \cdot \left(\frac{1}{2}\right)^2 - 4 \cdot \frac{1}{2} - 2 = 1 - 2 - 2 = -3$ as the *y*-coordinate of the vertex.

The vertex is therefore $\left(\frac{1}{2}, -3\right)$.

x-intercepts: Set $y = 0$. We get $0 = 4x^2 - 4x - 2$ and by using the quadratic formula, we obtain

$$x = \frac{-(-4) \pm \sqrt{(-4)^2 - 4(4)(-2)}}{2(4)} = \frac{4 \pm \sqrt{16 + 32}}{8} = \frac{4 \pm \sqrt{48}}{8} = \frac{4 \pm 4\sqrt{3}}{8} = \frac{1 \pm \sqrt{3}}{2}$$

$$x = \frac{1 + \sqrt{3}}{2} \approx 1.4 \text{ or } x = \frac{1 - \sqrt{3}}{2} = \approx -0.4$$

The *x*-intercepts of the graph are therefore $\left(\frac{1 + \sqrt{3}}{2}, 0\right)$ and $\left(\frac{1 - \sqrt{3}}{2}, 0\right)$.

y-intercept: Set $x = 0$. Thus $y = 4 \cdot 0^2 - 4 \cdot 0 - 2 = -2$. The *y*-intercept is therefore $(0, -2)$.

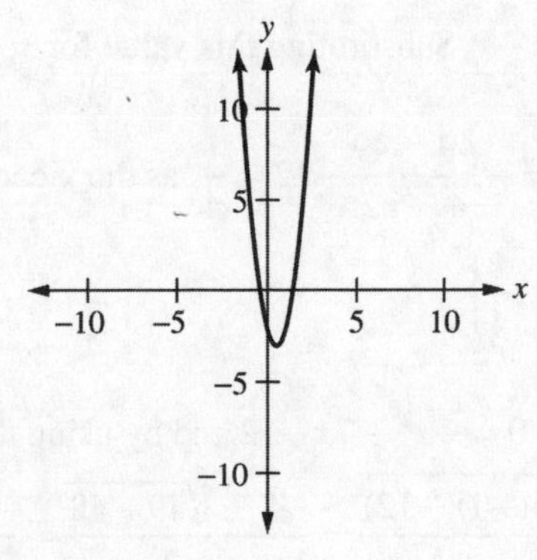

15. Opening: up since $a > 0$

Vertex: $x = \frac{-b}{2a} = \frac{-(7)}{2(3)} = \frac{-7}{6} = -\frac{7}{6}$. Substituting this value for x, we obtain

$$y = 3\cdot\left(-\frac{7}{6}\right)^2 + 7\cdot\left(-\frac{7}{6}\right) - 6 = \frac{49}{12} + \left(-\frac{49}{6}\right) - 6 = -\frac{121}{12}$$ as the y-coordinate of the vertex.

The vertex is therefore $\left(-\frac{7}{6}, -\frac{121}{12}\right)$.

x-intercepts: Set $y = 0$. We get $0 = 3x^2 + 7x - 6$ and by using the quadratic formula, we obtain

$$x = \frac{-(7) \pm \sqrt{(7)^2 - 4(3)(-6)}}{2(3)} = \frac{-7 \pm \sqrt{49 + 72}}{6} = \frac{-7 \pm \sqrt{121}}{6} = \frac{-7 \pm 11}{6} =$$

$$x = \frac{-7+11}{6} = \frac{4}{6} = \frac{2}{3} \text{ or } x = \frac{-7-11}{6} = -3$$

The x-intercepts of the graph are therefore $\left(\frac{2}{3}, 0\right)$ and $(-3, 0)$.

y-intercept: Set $x = 0$. Thus $y = 3\cdot 0^2 + 7\cdot 0 - 6 = -6$. The y-intercept is therefore $(0, -6)$.

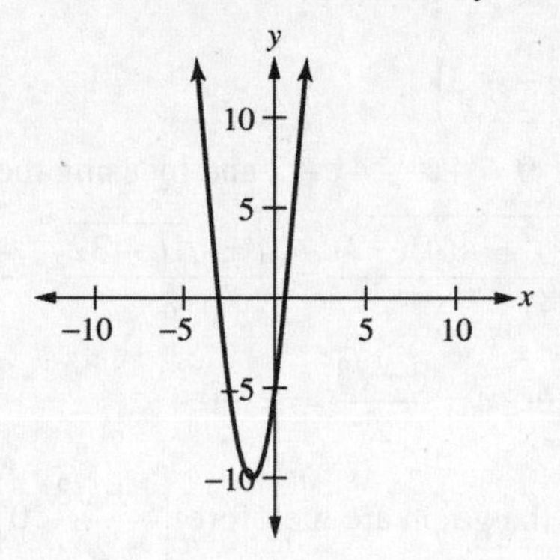

17. Opening: down since $a < 0$

Vertex: $x = \frac{-b}{2a} = \frac{-(7)}{2(-1)} = \frac{-7}{-2} = \frac{7}{2}$. Substituting this value for x, we obtain

$$y = -\left(\frac{7}{2}\right)^2 + 7\cdot\left(\frac{7}{2}\right) - 12 = -\frac{49}{4} + \frac{49}{2} - 12 = \frac{1}{4}$$ as the y-coordinate of the vertex.

The vertex is therefore $\left(\frac{7}{2}, \frac{1}{4}\right)$.

x-intercepts: Set $y = 0$. We get $0 = -x^2 + 7x - 12$ and by using the quadratic formula, we obtain

$$x = \frac{-(7) \pm \sqrt{(7)^2 - 4(-1)(-12)}}{2(-1)} = \frac{-7 \pm \sqrt{49 - 48}}{-2} = \frac{-7 \pm \sqrt{1}}{-2} = \frac{-7 \pm 1}{-2}$$

$$x = \frac{-7+1}{-2} = \frac{-6}{-2} = 3 \text{ or } x = \frac{-7-1}{-2} = \frac{-8}{-2} = 4$$

The x-intercepts of the graph are therefore $(3, 0)$ and $(4, 0)$.

y-intercept: Set $x = 0$. Thus $y = -1\cdot 0^2 + 7\cdot 0 - 12 = -12$. The y-intercept is therefore $(0, -12)$.

17. (continued)

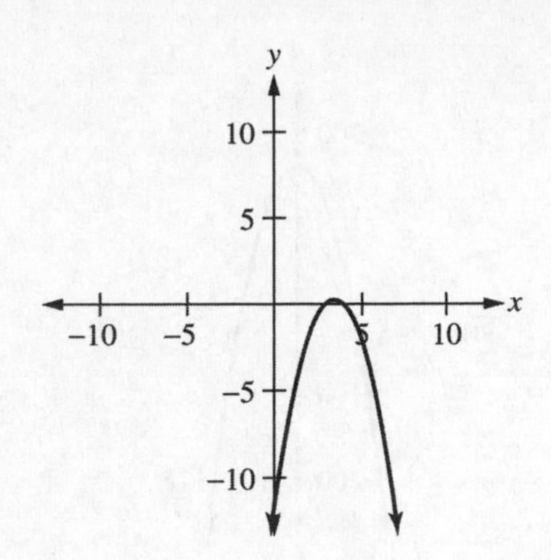

19. a) Week 2: $-0.125 \cdot 2^2 + 1.75 \cdot 2 + 2 = -0.5 + 3.5 + 2 = 5$;

Week 4: $-0.125 \cdot 4^2 + 1.75 \cdot 4 + 2 = -2 + 7 + 2 = 7$

Week 6: $-0.125 \cdot 6^2 + 1.75 \cdot 6 + 2 = -4.5 + 10.5 + 2 = 8$

b) We need to find the vertex of $A = -0.125x^2 + 1.75x + 2$. $x = \frac{-b}{2a} = \frac{-(1.75)}{2(-0.125)} = \frac{-1.75}{-0.25} = 7$

Substituting this value for x, we obtain

$A = -0.125 \cdot (7)^2 + 1.75 \cdot (7) + 2 = -6.125 + 12.25 + 2 = 8.125$.

21. a) We need to find the first component of the vertex of $S = -4n^2 + 16n - 12$

$n = \frac{-b}{2a} = \frac{-(16)}{2(-4)} = \frac{-16}{-8} = 2$ implies in week 2.

b) We need to find the second component of the vertex of $S = -4n^2 + 16n - 12$. Substituting $n = 2$, we obtain $S = -4 \cdot (2)^2 + 16 \cdot (2) - 12 = -16 + 32 - 12 = 4$. This implies \$4 million.

c) This corresponds to solving the equation $0 = -4n^2 + 16n - 12$ and by using the quadratic formula, we obtain

$$n = \frac{-(16) \pm \sqrt{(16)^2 - 4(-4)(-12)}}{2(-4)} = \frac{-16 \pm \sqrt{256 - 192}}{-8} = \frac{-16 \pm \sqrt{64}}{-8} = \frac{-16 \pm 8}{-8}$$

$$n = \frac{-16 + 8}{-8} = \frac{-8}{-8} = 1 \text{ or } n = \frac{-16 - 8}{-8} = \frac{-24}{-8} = 3$$

We can expect the sales to drop to zero in week 3.

23. a) Vertex: $t = \frac{-b}{2a} = \frac{-0}{2(-16)} = \frac{0}{-32} = 0$. Substituting this value for t, we obtain

$H = 160 - 16 \cdot 0^2 = 160 - 0 = 160$ as the second coordinate of the vertex, $(0,160)$

t-intercepts: We need to solve $0 = 160 - 16t^2$ and by using the quadratic formula, we obtain

$$t = \frac{-(0) \pm \sqrt{(0)^2 - 4(-16)(160)}}{2(-16)} = \frac{0 \pm \sqrt{10240}}{-32} = \frac{\pm 32\sqrt{10}}{-32} = \pm\sqrt{10} \approx \pm 3.16$$

The t-intercepts of the graph are therefore $\left(-\sqrt{10}, 0\right)$ and $\left(\sqrt{10}, 0\right)$.

23. (continued)

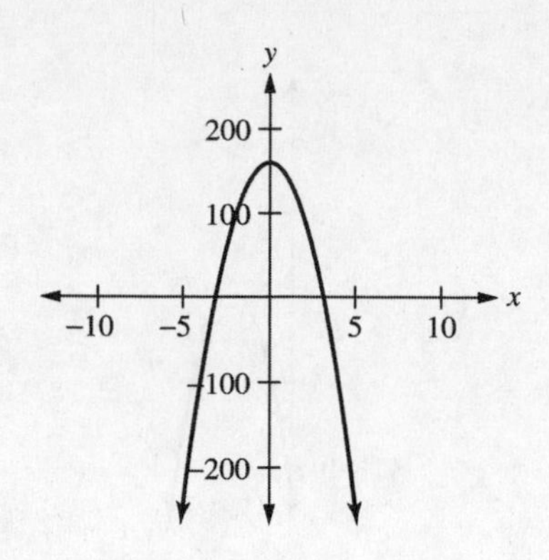

b) t (time) cannot be negative.

c) Approximately after 3.16 seconds

25. a) $-0.76(0)^2+30.79(0)+501.73=501.73$; $19.95(0)+518=518$; The parabola is closer.

b) $-0.76(5)^2+30.79(5)+501.73=636.68$; $19.95(5)+518=613.75$; The line is closer.

c) $-0.76(14)^2+30.79(14)+501.73=783.83$; $19.95(14)+518=793.3$; The parabola is closer.

27. Linear regression gives us the line of best fit for a set of data; quadratic regression gives the parabola of best fit.

29. The crate begins falling a point above the ground and picks up speed as it falls. A linear equation would not be appropriate, because the rate of change of the distance of the crate above the ground is not constant.

31. a) Vertex: $t=\frac{-b}{2a}=\frac{-8}{2(0.15)}=\frac{-8}{0.3}=-\frac{80}{3}\approx-26.7$. Substituting this value for t, we obtain

$d=0.15\left(-\frac{80}{3}\right)^2+8\cdot\left(-\frac{80}{3}\right)=\frac{320}{3}+\frac{-640}{3}=-\frac{320}{3}\approx-106.7$ as the second coordinate of the vertex, $\left(-\frac{80}{3},-\frac{320}{3}\right)$.

t-intercepts: We need to solve $0=0.15t^2+8t$ and by using the quadratic formula, we obtain

$$t=\frac{-(8)\pm\sqrt{(8)^2-4(0.15)(0)}}{2(0.15)}=\frac{-8\pm\sqrt{64-0}}{0.30}=\frac{-8\pm\sqrt{64}}{0.30}=\frac{-8\pm8}{0.30}$$

$$t=\frac{-8-8}{0.30}\approx-53.3 \text{ or } t=\frac{-8+8}{0.30}=0.$$

The t-intercepts of the graph are therefore approximately $(-53.3,0)$ and $(0,0)$

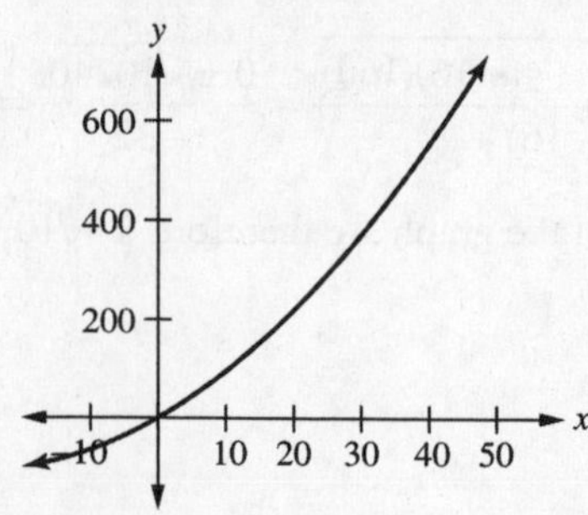

31. (continued)

b) t (time) cannot be negative.

c) Considering only non-negative values, Usain Bolt is running more slowly when t is close to 0.

d) We need to solve $100 = 0.15t^2 + 8t$ or $0 = 0.15t^2 + 8t - 100$ and by using the quadratic formula, we obtain

$$x = \frac{-(8) \pm \sqrt{(8)^2 - 4(0.15)(-100)}}{2(0.15)} = \frac{-8 \pm \sqrt{64+60}}{0.30} = \frac{-8 \pm \sqrt{124}}{0.30}$$

$$t = \frac{-8 - \sqrt{124}}{0.30} \approx -63.79 \text{ or } t = \frac{-8 + \sqrt{124}}{0.30} \approx 10.45$$

Since we disregard the negative solution, it would take approximately 10.45 seconds for Usain Bolt to finish the race.

33. No solution provided.

35. a)

	2003 (year 0)	2004 (year 1)	2005 (year 2)	2006 (year 3)	2009 (year 7)
Actual data	81.8	80.5	91.8	99.9	105.4
Values predicted by your model	81.8	87.73	93.86	99.89	124.01
Values predicted by line of best fit	78.66	85.22	91.78	98.34	124.58
Values Predicted by Parabola of Best Fit	$235 \cdot 0^2 - 0.49 \cdot 0 + 81.01 = 81.01$	$235 \cdot 1^2 - 0.49 \cdot 1 + 81.01 = 82.87$	$235 \cdot 2^2 - 0.49 \cdot 2 + 81.01 = 89.43$	$235 \cdot 3^2 - 0.49 \cdot 3 + 81.01 = 100.69$	$235 \cdot 7^2 - 0.49 \cdot 7 + 81.01 = 192.73$

b) The parabola gives better approximations until at least 2005. In 2009, the approximation is quite bad

Section 7.4 Exponential Equations and Growth

1. $A = 1{,}000(1+0.05)^1 = 1{,}000 \cdot 1.05 = \$1{,}050$

3. $A = 4{,}000(1+0.025)^1 = 4{,}000 \cdot 1.025 = \$4{,}100$

5. $A = 5{,}000(1+0.05)^5 = 5{,}000 \cdot 1.05^5 \approx \$6{,}381.40$

7. $A = 4{,}000(1+0.08)^2 = 4{,}000 \cdot 1.08^2 = \$4{,}665.60$

9. 3.24%

$$22000 = 15000(1+r)^{12}$$
$$\frac{22000}{15000} = (1+r)^{12}$$
$$(1+r)^{12} = \frac{22}{15}$$
$$1+r = \sqrt[12]{\frac{22}{15}}$$
$$r = \sqrt[12]{\frac{22}{15}} - 1 \approx 0.0324$$

11. 3.33%

$$13000 = 10000(1+r)^{8}$$
$$\frac{13000}{10000} = (1+r)^{8}$$
$$(1+r)^{8} = 1.3$$
$$1+r = \sqrt[8]{1.3}$$
$$r = \sqrt[8]{1.3} - 1 \approx 0.0333$$

13. $P(1+r)^n = 142(1+0.01566)^{2025-2011} = 142 \cdot 1.01566^{14} \approx 176.51$ million

15. $P(1+r)^n = 128(1+(-0.00278))^{2022-2011} = 128 \cdot 0.99722^{11} \approx 124.14$ million

17. $P(1+r)^n = 238(1+0.01069)^{2030-2011} = 238 \cdot 1.01069^{19} \approx 291.29$ million

19.
$$5^x = 20$$
$$\log 5^x = \log 20$$
$$x \log 5 = \log 20$$
$$x = \frac{\log 20}{\log 5} \approx \frac{1.30103}{0.69897} \approx 1.86$$

21.
$$10^x = 3.2$$
$$\log 10^x = \log 3.2$$
$$x \log 10 = \log 3.2$$
$$x = \frac{\log 3.2}{\log 10} \approx \frac{0.50515}{1} \approx 0.51$$

23.
$$(3.4)^x = 6.85$$
$$\log (3.4)^x = \log 6.85$$
$$x \log 3.4 = \log 6.85$$
$$x = \frac{\log 6.85}{\log 3.4} \approx \frac{0.83569}{0.53148} \approx 1.57$$

25. 1.32%

$$112 = 97(1+r)^{11}$$
$$\frac{112}{97} = (1+r)^{11}$$
$$\sqrt[11]{\frac{112}{97}} = 1+r$$
$$r = \sqrt[11]{\frac{112}{97}} - 1 \approx 0.0132$$

27. –0.25%

$$143 = 147(1+r)^{11}$$
$$\frac{143}{147} = (1+r)^{11}$$
$$\sqrt[11]{\frac{143}{147}} = 1+r$$
$$r = \sqrt[11]{\frac{143}{147}} - 1 \approx -0.0025$$

29.
$$P_{4+1} = [1+(0.03)\cdot(1-0.36)]\cdot 0.36$$
$$P_5 = [1+(0.03)\cdot 0.64]\cdot 0.36$$
$$P_5 = [1+0.0192]\cdot 0.36$$
$$P_5 = 1.0192 \cdot 0.36$$
$$P_5 \approx 0.3669$$

At the end of the fifth year, the population is at approximately 30.69% of its maximum capacity.

31.
$$P_{8+1} = [1+(0.045)\cdot(1-0.72)]\cdot 0.72$$
$$P_9 = [1+(0.045)\cdot 0.28]\cdot 0.72$$
$$P_9 = [1+0.0126]\cdot 0.72$$
$$P_9 = 1.0126 \cdot 0.72$$
$$P_9 \approx 0.7291$$

At the end of the ninth year, the population is at approximately 72.91% of its maximum capacity.

33. $P_{0+1} = [1+(0.08)\cdot(1-0.30)]\cdot 0.30$
$P_1 = [1+(0.08)\cdot 0.70]\cdot 0.30$
$P_1 = [1+0.056]\cdot 0.30$
$P_1 = 1.056\cdot 0.30$
$P_1 = 0.3168$

$P_{1+1} = [1+(0.08)\cdot(1-0.3168)]\cdot 0.3168$
$P_2 = [1+(0.08)\cdot 0.6832]\cdot 0.3168$
$P_2 = [1+0.054656]\cdot 0.3168$
$P_2 = 1.054656\cdot 0.3168$
$P_2 \approx 0.3341150208$

$P_{2+1} \approx [1+(0.08)\cdot(1-0.3341150208)]\cdot 0.3341150208$
$P_3 \approx [1+(0.08)\cdot 0.6658849792]\cdot 0.3341150208$
$P_3 \approx [1+0.053270798336]\cdot 0.3341150208$
$P_3 \approx 1.053270798336\cdot 0.3341150208$
$P_3 \approx 0.3519 \approx 35.2\%$

35. $P_{0+1} = [1+(0.10)\cdot(1-0.25)]\cdot 0.25$
$P_1 = [1+(0.10)\cdot 0.75]\cdot 0.25$
$P_1 = [1+0.075]\cdot 0.25$
$P_1 = 1.075\cdot 0.25$
$P_1 = 0.26875$

$P_{1+1} = [1+(0.10)\cdot(1-0.26875)]\cdot 0.26875$
$P_2 = [1+(0.10)\cdot 0.73125]\cdot 0.26875$
$P_2 = [1+0.073125]\cdot 0.26875$
$P_2 = 1.073125\cdot 0.26875$
$P_2 \approx 0.28840234375$

$P_{2+1} \approx [1+(0.10)\cdot(1-0.28840234375)]\cdot 0.28840234375$
$P_3 \approx [1+(0.10)\cdot 0.71159765625]\cdot 0.28840234375$
$P_3 \approx [1+0.071159765625]\cdot 0.28840234375$
$P_3 \approx 1.071159765625\cdot 0.28840234375$
$P_3 \approx 0.3089 \approx 30.9\%$

37. $P_{0+1} = [1+(0.045)\cdot(1-0.60)]\cdot 0.60$
$P_1 = [1+(0.045)\cdot 0.40]\cdot 0.60$
$P_1 = [1+0.018]\cdot 0.60$
$P_1 = 1.018\cdot 0.60$
$P_1 = 0.6108$

$P_{1+1} = [1+(0.045)\cdot(1-0.6108)]\cdot 0.6108$
$P_2 = [1+(0.045)\cdot 0.3892]\cdot 0.6108$
$P_2 = [1+0.017514]\cdot 0.6108$
$P_2 = 1.017514\cdot 0.6108$
$P_2 = 0.6214975512$

$P_{2+1} \approx [1+(0.045)\cdot(1-0.6214975512)]\cdot 0.6214975512$
$P_3 \approx [1+(0.045)\cdot 0.3785024488]\cdot 0.6214975512$
$P_3 \approx [1+0.017032610196]\cdot 0.6214975512$
$P_3 \approx 1.017032610196\cdot 0.6214975512$
$P_3 \approx 0.632 = 63.2\%$

39. $20{,}000 = 10{,}000(1+0.05)^n$

$$2 = 1.05^n$$
$$\log 2 = \log 1.05^n$$
$$\log 2 = n \cdot \log 1.05$$
$$\frac{\log 2}{\log 1.05} = n$$
$$n \approx \frac{0.30103}{0.02119} \approx 14.21 \text{ years}$$

41. $614 = 307(1+0.007)^n$

$$2 = 1.007^n$$
$$\log 2 = \log 1.007^n$$
$$\log 2 = n \cdot \log 1.007$$
$$\frac{\log 2}{\log 1.007} = n$$
$$n \approx \frac{0.30103}{0.00303} \approx 99.367 \text{ years}$$

By the year $2011 + 100 = 2111$, the population should double.

43. $\frac{70}{1.148} \approx 61$ years

45. $\frac{70}{2.614} \approx 27$ years

47. $7.4(1+(-0.00781))^{14} = 7.4(1+(-0.00781))^{2025-2011} = 7.4 \cdot 0.99219^{14} \approx 6.63$ million

49. $50 = 100(1+(-0.0035))^n$

$$\frac{1}{2} = 0.9965^n$$
$$\log\frac{1}{2} = \log 0.9965^n$$
$$\log\frac{1}{2} = n \cdot \log 0.9965$$
$$\frac{\log\frac{1}{2}}{\log 0.9965} = n$$
$$n \approx \frac{-0.30103}{-0.00152} \approx 197.70$$

after approximately 198 years

51. $500(1+(-0.15))^3 = 500 \cdot 0.85^3 \approx 307$ mg

53. After 5 hours: $500(1+(-0.15))^5 \approx 222$

After 6 hours: $500(1+(-0.15))^6 \approx 189$

After 7 hours: $500(1+(-0.15))^7 \approx 160$

After 8 hours: $500(1+(-0.15))^8 \approx 136$

Therefore after approximately 8 hours the patient should stop feeling numbness.

55. a) $c = 120(1.0299)^t$

Let c be the cost of the shoes.

Let t be the number of years since 2006.

$$c = a \cdot b^t$$
$$135 = 120 \cdot b^4$$
$$b^4 = \frac{135}{120}$$
$$b = \sqrt[4]{\frac{135}{120}} \approx 1.0299$$

b) $c = 120(1.0299)^{10} = \$161.11$

57. 80.55%

$$1712 = 4.65(1+r)^{10}$$
$$(1+r)^{10} = \frac{1712}{4.65}$$
$$1+r = \sqrt[10]{\frac{1712}{4.65}}$$
$$r = \sqrt[10]{\frac{1712}{4.65}} - 1 \approx 0.8055$$

59. 4.5%

$$1302 = 1000(1+r)^6$$
$$(1+r)^6 = \frac{1302}{1000}$$
$$1+r = \sqrt[6]{\frac{1302}{1000}}$$
$$r = \sqrt[6]{\frac{1302}{1000}} - 1 \approx 0.045$$

61. When the rate of growth of a quantity is proportional to the amount present.

63. $\frac{12,000(0.09)}{12} = \frac{1080}{12} = \$150$

65. You should double your interest rate.

Doubling investment: $A = 2000(1+0.04)^{20} = 2000(1.04)^{20} = \$4,382.25$

Doubling interest: $A = 1000(1+0.08)^{20} = 1000(1.08)^{20} = \$4,660.96$

67. $P_0 = \frac{200}{800} = 0.25$

$$P_{0+1} = [1+(0.18)\cdot(1-0.25)]\cdot 0.25$$
$$P_1 = [1+(0.18)\cdot 0.75]\cdot 0.25$$
$$P_1 = [1+0.135]\cdot 0.25$$
$$P_1 = 1.135\cdot 0.25$$
$$P_1 = 0.28375$$
$$\Rightarrow 0.28375\cdot 800 \approx 227 \text{ fish}$$

$$P_{1+1} = [1+(0.18)\cdot(1-0.28375)]\cdot 0.28375$$
$$P_2 = [1+(0.18)\cdot 0.71625]\cdot 0.28375$$
$$P_2 = [1+0.128925]\cdot 0.28375$$
$$P_2 = 1.128925\cdot 0.28375$$
$$P_2 \approx 0.32033246875$$
$$\Rightarrow 0.32033246875\cdot 800 \approx 256 \text{ fish}$$

$$P_{2+1} \approx [1+(0.18)\cdot(1-0.32033246875)]\cdot 0.32033246875$$
$$P_3 \approx [1+(0.18)\cdot 0.67966753125]\cdot 0.32033246875$$
$$P_3 \approx [1+0.122340155625]\cdot 0.32033246875$$
$$P_3 \approx 1.122340155625\cdot 0.32033246875$$
$$P_3 \approx 0.359521992829$$
$$\Rightarrow 0.359521992829\cdot 800 \approx 288 \text{ fish}$$

$$P_{3+1} \approx [1+(0.18)\cdot(1-0.359521992829)]\cdot 0.359521992829$$
$$P_4 \approx [1+(0.18)\cdot 0.640478007171]\cdot 0.359521992829$$
$$P_4 \approx [1+0.115286041291]\cdot 0.359521992829$$
$$P_4 \approx 1.11528604129\cdot 0.359521992829$$
$$P_4 \approx 0.400969860139$$
$$\Rightarrow 0.400969860139\cdot 800 \approx 321 \text{ fish}$$

The population will exceed 300 by the end of the fourth year. Fishing can begin toward the end of the fourth year.

69. Choose the one cent.
One cent: $1\cdot 2^{30}=1{,}073{,}741{,}824$ cents
One dollar: $100\cdot 2^{15}=3{,}276{,}800$ cents

Section 7.5 Proportions and Variation

1. Rewrite $24:x=18:3$ as $\frac{24}{x}=\frac{18}{3}$.

$$\begin{aligned}\frac{24}{x}&=\frac{18}{3}\\ 24\cdot 3&=x\cdot 18\\ 72&=18x\\ 4&=x\\ x&=4\end{aligned}$$

3.
$$\begin{aligned}\frac{50}{4}&=\frac{x}{5}\\ 50\cdot 5&=4\cdot x\\ 250&=4x\\ \frac{250}{4}&=x\\ x&=62.5\end{aligned}$$

5.
$$\begin{aligned}\frac{30}{40}&=\frac{27}{x}\\ 30\cdot x&=40\cdot 27\\ 30x&=1{,}080\\ x&=36\end{aligned}$$

7.
$$\begin{aligned}y&=kx\\ 37.5&=k\cdot 7.5\\ \tfrac{37.5}{7.5}&=k\\ k&=5\end{aligned}\qquad \begin{aligned}y&=5x\\ y&=5\cdot 13\\ y&=65\end{aligned}$$

9.
$$\begin{aligned}r&=\frac{k}{s}\\ 12&=\frac{k}{2/3}\\ 12\cdot\tfrac{2}{3}&=k\\ k&=8\end{aligned}\qquad \begin{aligned}r&=\frac{8}{s}\\ r&=\frac{8}{8}\\ r&=1\end{aligned}$$

11.
$$\begin{aligned}a&=kb^2\\ 16&=k\cdot 6^2\\ 16&=k\cdot 36\\ \frac{16}{36}&=k\\ k&=\tfrac{4}{9}\end{aligned}\qquad \begin{aligned}a&=\tfrac{4}{9}b^2\\ a&=\tfrac{4}{9}\cdot 15^2\\ a&=\tfrac{4}{9}\cdot 225\\ a&=100\end{aligned}$$

13.
$$\begin{aligned}D&=\frac{k}{C}\\ 3\cdot 2&=4k\\ 6&=4k\\ k&=\tfrac{6}{4}=\tfrac{3}{2}\end{aligned}\qquad \begin{aligned}D&=\frac{\frac{3}{2}}{C}\\ D&=\frac{\frac{3}{2}}{24}\\ D&=\tfrac{3}{2}\cdot\tfrac{1}{24}=\tfrac{1}{16}\end{aligned}$$

15.
$$\begin{aligned}r&=kxy\\ 12.5&=k\cdot 2\cdot 5\\ 12.5&=k\cdot 10\\ \frac{12.5}{10}&=k\\ k&=1.25\end{aligned}\qquad \begin{aligned}r&=1.25xy\\ r&=1.25\cdot 8\cdot 2.5\\ r&=25\end{aligned}$$

17.
$$\begin{aligned}y&=kwx^2\\ 504&=k\cdot 4\cdot 6^2\\ 504&=k\cdot 4\cdot 36\\ 504&=k\cdot 144\\ \frac{504}{144}&=k\\ k&=\tfrac{7}{2}\end{aligned}\qquad \begin{aligned}y&=\tfrac{7}{2}wx^2\\ 6{,}860&=\tfrac{7}{2}\cdot 10\cdot x^2\\ 6{,}860&=35\cdot x^2\\ x^2&=\frac{6{,}860}{35}=196\\ x&=\sqrt{196}=14\end{aligned}$$

19.
$$\begin{aligned}y&=k\frac{w}{x}\\ 4&=k\frac{6}{10}\\ 40&=6k\\ \frac{40}{6}&=k\\ k&=\tfrac{20}{3}\end{aligned}\qquad \begin{aligned}y&=\frac{20}{3}\cdot\frac{w}{x}\\ y&=\frac{20}{3}\cdot\frac{3}{15}\\ y&=\frac{4}{3}\end{aligned}$$

21.

$$y = k\frac{x^2 w}{z}$$
$$15 = k\frac{2.5^2 \cdot 8}{20}$$
$$300 = k(6.25 \cdot 8)$$
$$300 = 50k$$
$$\frac{300}{50} = k$$
$$k = 6$$

$$y = 6 \cdot \frac{x^2 w}{z}$$
$$y = 6 \cdot \frac{8^2 \cdot 7}{14}$$
$$y = 6 \cdot \frac{64 \cdot 7}{14}$$
$$y = 6 \cdot \frac{64}{2}$$
$$y = 6 \cdot 32 = 192$$

In the solutions to Exercises 23 – 43, your proportion could be set up differently.

23.

$$\frac{150}{6} = \frac{65}{x}$$
$$150 \cdot x = 6 \cdot 65$$
$$150x = 390$$
$$x = \frac{390}{150}$$
$$x = 2.6 \text{ mg}$$

25.

$$\frac{30}{48} = \frac{x}{56}$$
$$30 \cdot 56 = 48 \cdot x$$
$$1,680 = 48x$$
$$\frac{1,680}{48} = x$$
$$x = 35 \text{ miles}$$

27. The front lawn is $200 \cdot 650 = 130,000$ square feet.

$$\frac{1.5}{60,000} = \frac{x}{130,000}$$
$$1.5 \cdot 130,000 = 60,000 \cdot x$$
$$195,000 = 60,000x$$
$$\frac{195,000}{60,000} = x$$
$$x = 3.25 \text{ hours}$$

29.

$$\frac{\text{number of tagged bald eagles in population}}{\text{number of bald eagles in population}} = \frac{\text{number of tagged bald eagles in sample}}{\text{number of bald eagles in sample}}$$
$$\frac{400}{n} = \frac{8}{240}$$
$$400 \cdot 240 = n \cdot 8$$
$$96,000 = 8n$$
$$\frac{96,000}{8} = n$$
$$n = 12,000$$

31. $$\frac{\text{number of tagged grizzly bears in population}}{\text{number of grizzly bears in population}} = \frac{\text{number of tagged grizzly bears in sample}}{\text{number of grizzly bears in sample}}$$

$$\frac{55}{1{,}000} = \frac{n}{95}$$
$$55 \cdot 95 = 1{,}000 \cdot n$$
$$5{,}225 = 1{,}000n$$
$$\frac{5{,}225}{1{,}000} = n$$
$$n = 5.225$$

5 or 6 tagged grizzly bears could be expected.

33. Let t be the time of the trip.
Let s be the speed during the trip.

$$t = \frac{k}{s}$$
$$2 = \frac{k}{60}$$
$$2 \cdot 60 = k$$
$$k = 120$$

$$t = \frac{120}{s}$$
$$t = \frac{120}{65} = \frac{24}{13}$$
$$t \approx 1.8462 \text{ hours}$$

You would save $(2 - 1.8462) \times 60 \approx 9.23$ minutes by traveling at 65 miles per hour.

35. Let w be the width of the beam.
Let d be the depth of the beam.
Let l be the length of the beam.
Let s be the weight the beam can support.

$$s = k\frac{w \cdot d^2}{l}$$
$$672 = k\frac{6 \cdot 8^2}{4}$$
$$672 = k\frac{384}{4}$$
$$672 = k \cdot 96$$
$$k = \frac{672}{96} = 7$$

$$s = \frac{7 \cdot w \cdot d^2}{l}$$
$$s = \frac{7 \cdot 4 \cdot 6^2}{8}$$
$$s = \frac{7 \cdot 4 \cdot 36}{8}$$
$$s = \frac{1008}{8} = 126$$

The beam could support 126 pounds.

37. Let w be the width of the beam.
Let d be the depth of the beam.
Let l be the length of the beam.
Let s be the weight the beam can support.

$$s = k\frac{w \cdot d^2}{l}$$
$$1{,}280 = k\frac{4 \cdot 8^2}{12}$$
$$1{,}280 = k\frac{256}{12}$$
$$k = \frac{1{,}280}{1} \cdot \frac{12}{256} = 60$$
$$s = \frac{60 \cdot w \cdot d^2}{l}$$
$$540 = \frac{60 \cdot 3 \cdot 6^2}{l}$$
$$540 = \frac{6{,}480}{l}$$
$$l = \frac{6{,}480}{540} = 12$$

The beam would be 12 feet long.

39. Let E be the number of euros.

$$\frac{E}{500} = \frac{1}{1.2937}$$
$$E \cdot 1.2937 = 1 \cdot 500$$
$$1.2937E = 500$$
$$E = \frac{500}{1.2937} \approx 386.49$$

You could get 386.49 euros.

41. Let w be the amount of water Mario uses for irrigation in gallons.
Let r be the amount of rainfall in inches.

$$w = \frac{k}{r}$$
$$30{,}000 = \frac{k}{3}$$
$$30{,}000 \cdot 3 = k$$
$$k = 90{,}000$$

$$w = \frac{90{,}000}{r}$$
$$w = \frac{90{,}000}{5}$$
$$w = 18{,}000 \text{ gal}$$

43. Let d be the distance in feet.
Let t be the time in seconds.

$$d = kt^2$$
$$144 = k \cdot 3^2$$
$$144 = k \cdot 9$$
$$\frac{144}{9} = k$$
$$k = 16$$

$$d = 16t^2$$
$$d = 16 \cdot 5^2$$
$$d = 16 \cdot 25$$
$$d = 400 \text{ feet}$$

45. Let d be the braking distance of the car, in feet, and s be the speed of the car, in mph.

$$d = ks^2$$
$$43 = k(30)^2$$
$$43 = 900k$$
$$k = \frac{43}{900}$$

$$d = \frac{43s^2}{900}$$
$$d = \frac{43(60)^2}{900}$$
$$d = \frac{154800}{900}$$
$$d = 172 \text{ feet}$$

47. Let h be the amount of heating oil in gallons.
Let t be the temperature in degrees.

$$h = \frac{k}{t}$$
$$504 = \frac{k}{42}$$
$$504 \cdot 42 = k$$
$$k = 21{,}168$$

$$h = \frac{21{,}168}{t}$$
$$h = \frac{21{,}168}{36}$$
$$h = 588 \text{ gallons}$$

49. Let p be the pressure in pounds per square inch and v be the volume in cubic inches.

$$p = \frac{k}{v}$$
$$4 = \frac{k}{120}$$
$$4 \cdot 120 = k$$
$$k = 480$$

$$p = \frac{480}{v}$$
$$p = \frac{480}{75}$$
$$p = 6.4 \text{ psi}$$

51. y increases.

53. inversely

55. The strength doubles.

$$s = k\frac{(2w) \cdot d^2}{l} = 2\left(k\frac{w \cdot d^2}{l}\right)$$

57. The strength triples.

$$s = k\frac{w \cdot (3d)^2}{3l} = k\frac{9w \cdot d^2}{3l} = 3\left(k\frac{w \cdot d^2}{l}\right)$$

59. The total stopping distance of the car is 171 + 132 = 303 feet.

Let d be the braking distance of the car in feet and s be the speed of the car in mph.

$$d = ks^2$$
$$76 = k(40)^2$$
$$76 = 1600k$$
$$k = \frac{76}{1600}$$

$$d = \frac{76s^2}{1600}$$
$$d = \frac{76(60)^2}{1600}$$
$$d = \frac{273600}{1600}$$
$$d = 171 \text{ feet}$$

Let r be the reaction distance of the car in feet and s be the speed of the car in mph.

$$r = ks$$
$$88 = k(40)$$
$$k = \frac{88}{40} = \frac{11}{5}$$

$$r = \frac{11s}{5}$$
$$r = \frac{11(60)}{5}$$
$$r = \frac{660}{5}$$
$$r = 132 \text{ feet}$$

61. In $m:n$, we are comparing m objects with n objects. In $n:(m+n)$, we are comparing the n objects with the total of all the objects.

63. The ratio of PC's to all computers on campus is 7:9. Explanations and examples may vary.

Section 7.6 Modeling with Systems of Linear Equations and Inequalities

1. The solution is $(4,2)$.

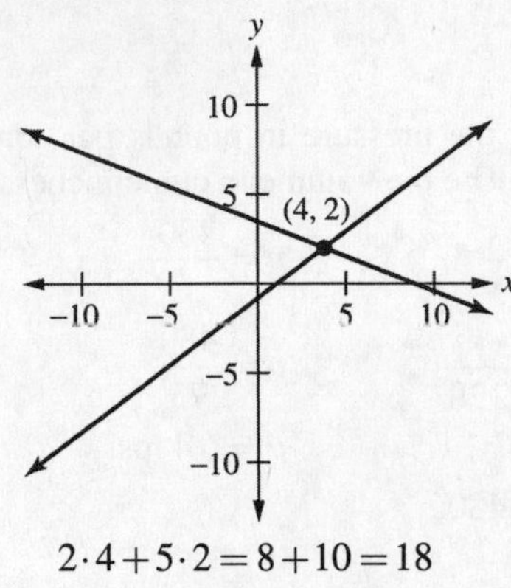

$$2\cdot 4+5\cdot 2=8+10=18$$
$$-3\cdot 4+4\cdot 2=-12+8=-4$$

3. The solution is $(4,-2)$.

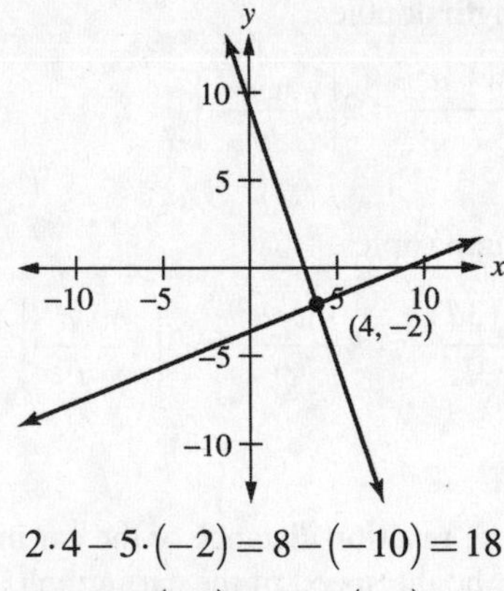

$$2\cdot 4-5\cdot(-2)=8-(-10)=18$$
$$3\cdot 4+1\cdot(-2)=12+(-2)=10$$

5. $(1,4)$

$$2\cdot[-3x+2y]=[5]\cdot 2$$
$$6x+4y=22$$
$$-6x+4y=10$$
$$6x+4y=22$$
$$0+8y=32$$
$$y=4$$

$$-3x+2\cdot 4=5$$
$$-3x+8=5$$
$$-3x=-3$$
$$x=1$$

7. $(-6,-2)$

$$3x-2y=-14$$
$$2\cdot[-4x+y]=[22]\cdot 2$$
$$3x-2y=-14$$
$$-8x+2y=44$$
$$-5x+0=30$$
$$x=-6$$

$$-6\cdot 3-2y=-14$$
$$-18-2y=-14$$
$$-2y=4$$
$$y=-2$$

9. no solution

$$8x-2y=-2$$
$$2\cdot[-4x+y]=[3]\cdot 2$$
$$8x-2y=-2$$
$$-8x+2y=6$$
$$0+0=4$$

11. infinite number of solutions

$$4\cdot[x-y]=[-2]\cdot 4$$
$$-4x+4y=8$$
$$4x-4y=-8$$
$$-4x+4y=8$$
$$0+0=0$$

13. no solution

$$2\cdot[6x-9y]=[8]\cdot 2$$
$$3\cdot[-4x+6y]=[10]\cdot 3$$
$$12x-18y=16$$
$$-12x+18y=30$$
$$0+0=46$$

15. $\left(\frac{1}{4}, \frac{1}{2}\right)$

$$12x - 8y = -1$$
$$6 \cdot [-2x + 5y] = [2] \cdot 6$$

$$12x - 8y = -1$$
$$-12x + 30y = 12$$

$$0 + 22y = 11$$
$$y = \tfrac{11}{22} = \tfrac{1}{2}$$

$$12x - 8 \cdot \tfrac{1}{2} = -1$$
$$12x - 4 = -1$$
$$12x = 3$$
$$x = \tfrac{3}{12} = \tfrac{1}{4}$$

17. $\left(\frac{1}{5}, \frac{2}{5}\right)$

$$4 \cdot [3x - 9y] = [-3] \cdot 4$$
$$3 \cdot [-4x + 2y] = [0] \cdot 3$$

$$12x - 36y = -12$$
$$-12x + 6y = 0$$

$$0 - 30y = -12$$
$$y = \tfrac{-12}{-30} = \tfrac{2}{5}$$

$$-4x + 2 \cdot \tfrac{2}{5} = 0$$
$$5 \cdot \left[-4x + \tfrac{4}{5}\right] = [0] \cdot 5$$
$$-20x + 4 = 0$$
$$-20x = -4$$
$$x = \tfrac{-4}{-20} = \tfrac{1}{5}$$

19. $(-1, 3)$

$$x - 2y = -7 \Rightarrow x = 2y - 7$$
$$3x + 5y = 12$$
$$3(2y - 7) + 5y = 12$$
$$6y - 21 + 5y = 12$$
$$11y - 21 = 12$$
$$11y = 33$$
$$y = 3$$
$$x = 2 \cdot 3 - 7 = 6 - 7 = -1$$

21. $(-1, 4)$

$$-2x + y = 6 \Rightarrow y = 2x + 6$$
$$-2x + 3y = 14$$
$$-2x + 3(2x + 6) = 14$$
$$-2x + 6x + 18 = 14$$
$$4x + 18 = 14$$
$$4x = -4$$
$$x = -1$$
$$y = 2 \cdot (-1) + 6 = -2 + 6 = 4$$

23. no solution

$$x + y = 6$$
$$2x = 8 - 2y \Rightarrow x = 4 - y$$
$$(4 - y) + y = 6$$
$$4 - y + y = 6$$
$$4 = 6$$ This is a false statement.

25. a and d

a) $3 \cdot 3 + 4 \cdot 5 \overset{?}{\geq} 2 \Rightarrow 9 + 20 \overset{?}{\geq} 2 \Rightarrow 29 \geq 2$; True

b) $3 \cdot 1 + 4 \cdot (-2) \overset{?}{\geq} 2 \Rightarrow 3 - 8 \overset{?}{\geq} 2 \Rightarrow -5 \geq 2$; False

c) $3 \cdot 0 + 4 \cdot 0 \overset{?}{\geq} 2 \Rightarrow 0 + 0 \overset{?}{\geq} 2 \Rightarrow 0 \geq 2$; False

d) $3 \cdot (-4) + 4 \cdot 6 \overset{?}{\geq} 2 \Rightarrow -12 + 24 \overset{?}{\geq} 2 \Rightarrow 12 \geq 2$; True

27. $3\cdot 0+4\cdot 0\overset{?}{\geq}12 \Rightarrow 0+0\overset{?}{\geq}12 \Rightarrow 0\geq 12$; False

Shade the side that does not contain $(0,0)$.

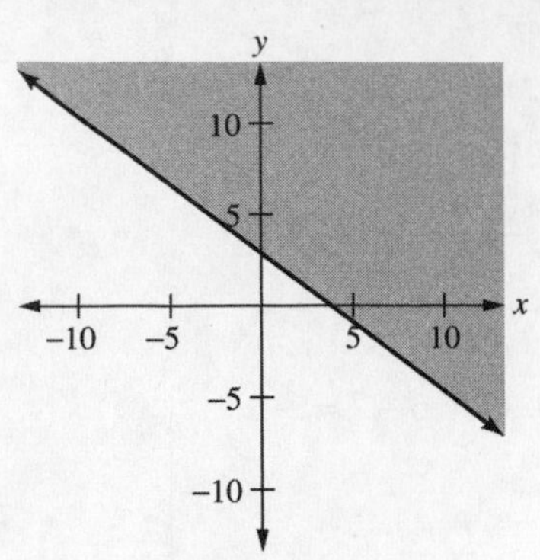

29. $2\cdot 0-4\cdot 0\overset{?}{<}12 \Rightarrow 0-0\overset{?}{<}12 \Rightarrow 0<12$; True

Shade the side that contains $(0,0)$.

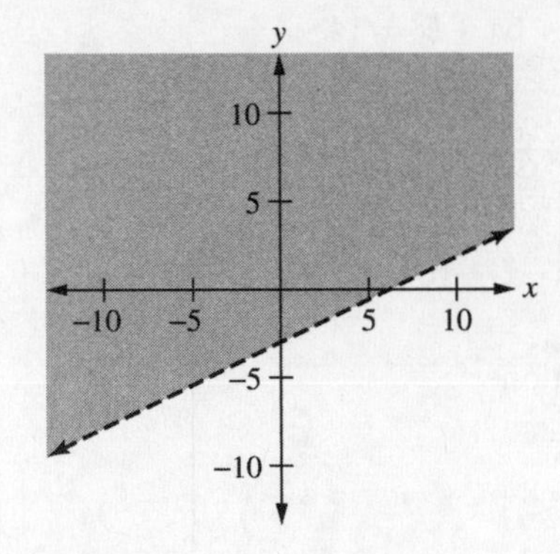

31. $0\overset{?}{\geq}3\cdot 0-9 \Rightarrow 0\overset{?}{\geq}0-9 \Rightarrow 0\geq -9$; True

Shade the side that contains $(0,0)$.

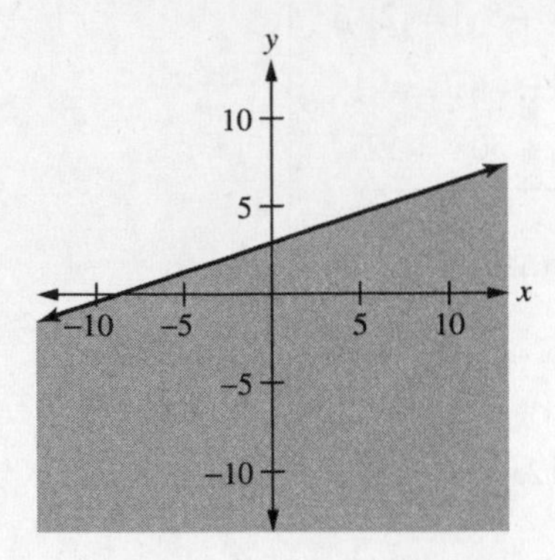

33. $4\cdot 0-8\overset{?}{<}2\cdot 0 \Rightarrow 0-8\overset{?}{<}0 \Rightarrow -8<0$; True

Shade the side that contains $(0,0)$.

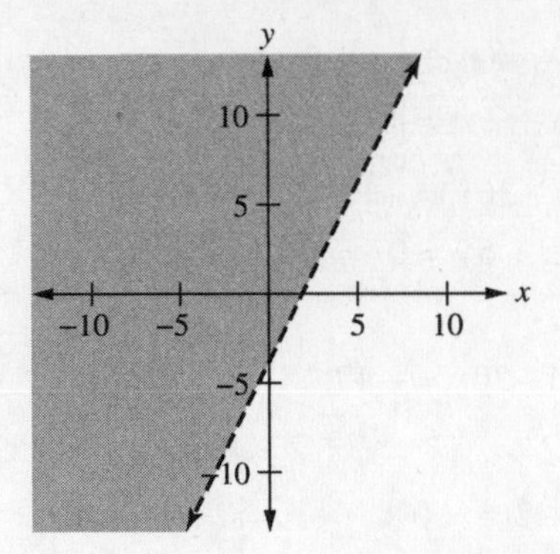

35. Testing $(0,0)$ in the first inequality, we get $2\cdot 0-3\cdot 0=0$ which is greater than -5, so $(0,0)$ is not a solution. Testing $(0,0)$ in the second inequality, we get $0-2\cdot 0=0$ which is greater than -8, so $(0,0)$ is not a solution. To solve for the corner point, we solve the system.

$$\begin{array}{c} 2x-3y=-5 \\ x-2y=-8 \end{array} \Rightarrow \begin{array}{c} 2x-3y=-5 \\ (-2)\cdot[x-2y]=[-8]\cdot(-2) \end{array} \Rightarrow \begin{array}{c} 2x-3y=-5 \\ -2x+4y=16 \end{array} \Rightarrow y=11$$

$$2x-3\cdot 11=-5 \Rightarrow 2x-33=-5 \Rightarrow 2x=28 \Rightarrow x=14$$

Thus the corner point is $(14,11)$.

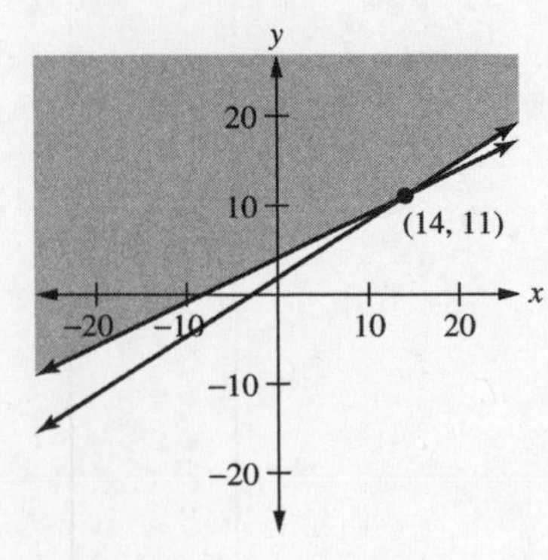

37. Testing $(0,0)$ in the first inequality, we get $2\cdot 0-0=0$ which is less than 3, so $(0,0)$ is not a solution. Testing $(0,0)$ in the second inequality, we get $0-0=0$ which is greater than -1, so $(0,0)$ is not a solution. To solve for the corner point, we solve the system.

$$\begin{array}{c} 2x-y=3 \\ x-y=-1 \end{array} \Rightarrow \begin{array}{c} 2x-y=3 \\ (-1)\cdot[x-y]=[-1]\cdot(-1) \end{array} \Rightarrow \begin{array}{c} 2x-y=3 \\ -x+y=1 \end{array} \Rightarrow x=4$$

$$4-y=-1 \Rightarrow -y=-5 \Rightarrow y=5$$

Thus the corner point is $(4,5)$.

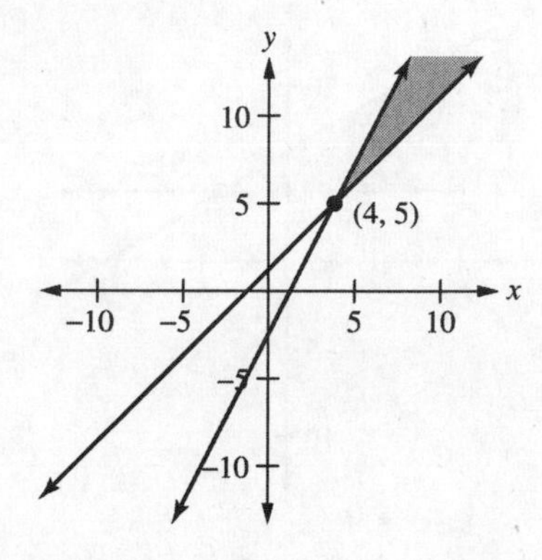

39. Testing $(0,0)$ in the first inequality, we get $-4\cdot 0+3\cdot 0=0$ which is less than 23, so $(0,0)$ is a solution. Testing $(0,0)$ in the second inequality, we get $3\cdot 0 \overset{?}{>} 19-5\cdot 0 \Rightarrow 0 \overset{?}{>} 19-0 \Rightarrow 0>19$ which is false, so $(0,0)$ is not a solution. To solve for the corner point, we solve the system.

$$\begin{array}{c} -4x+3y=23 \\ 3x=19-5y \end{array} \Rightarrow \begin{array}{c} 3\cdot[-4x+3y]=[23]\cdot 3 \\ 4\cdot[3x+5y]=[19]\cdot 4 \end{array} \Rightarrow \begin{array}{c} -12x+9y=69 \\ 12x+20y=76 \end{array} \Rightarrow 29y=145 \Rightarrow y=5$$

$$3x=19-5\cdot 5 \Rightarrow 3x=19-25 \Rightarrow 3x=-6 \Rightarrow x=-2$$

Thus the corner point is $(-2,5)$.

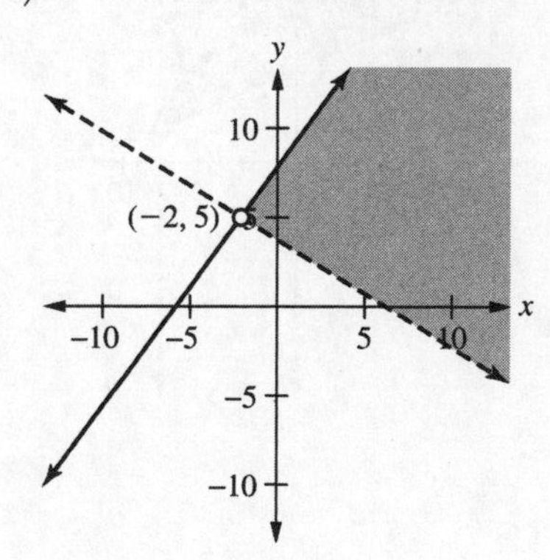

41. Testing $(0,0)$ in the first inequality, we get $3\cdot 0+5\cdot 0=0$ which is less than 32, so $(0,0)$ is a solution. Testing $(0,0)$ in the second inequality, we get $0\geq 4$ which is false, so $(0,0)$ is not a solution. To solve for the corner point, we solve the system.

$$3x+5y=32$$
$$y=4$$
$$3x+5\cdot 4=32 \Rightarrow 3x+20=32 \Rightarrow 3x=12 \Rightarrow x=4$$

Thus the corner point is $(4,4)$.

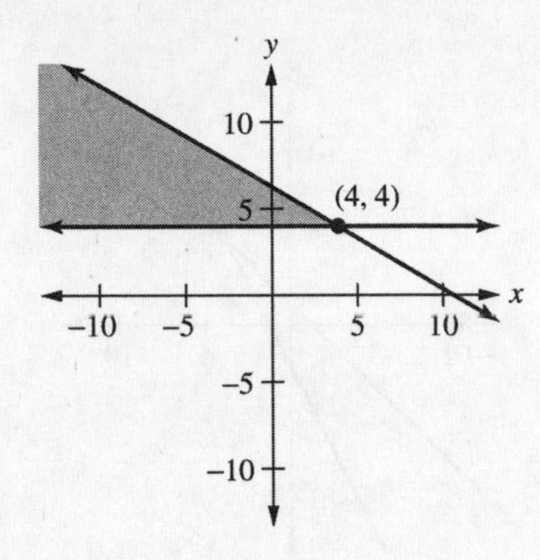

43. Testing $(0,0)$ in the first inequality, we get $2\cdot 0+5\cdot 0=0$ which is less than 26, so $(0,0)$ is not a solution. Testing $(0,0)$ in the second inequality, we get $0<8$ which is true, so $(0,0)$ is a solution. To solve for the corner point, we solve the system.

$$2x+5y=26$$
$$x=8$$
$$2\cdot 8+5y=26 \Rightarrow 16+5y=26 \Rightarrow 5y=10 \Rightarrow y=2$$

Thus the corner point is $(8,2)$.

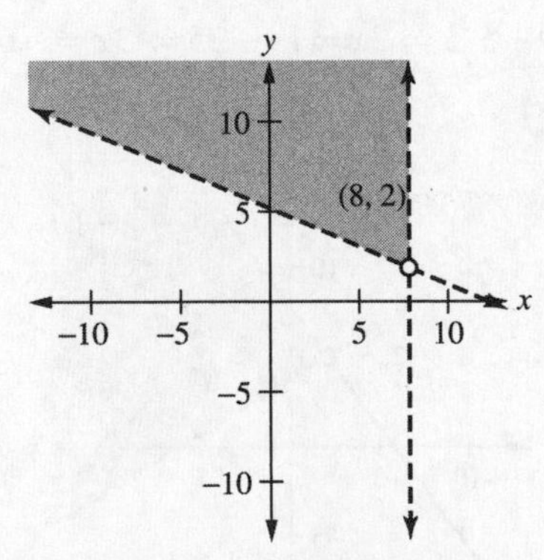

45. Let T be the number of points scored by Tennessee.
Let G be the number of points scored by Georgia.

$$T+G=116$$
$$T=G+30$$

$$(G+30)+G=116$$
$$2G+30=116$$
$$2G=86$$
$$G=43$$

$$T+43=116$$
$$T=73$$

Tennessee scored 73 points and Georgia scored 43 points.

47. 4 bagels and 5 ounces of cream cheese

Nutrition	Amount per Bagel (b)	Amount per Ounce of Cream Cheese (c)	Desired
Calcium	30 mg	25 mg	245 mg
Iron	2 mg	0.4 mg	10 mg

$$\begin{aligned}30b+25c&=245\\10\cdot[2b+0.4c]&=[10]\cdot 10\end{aligned} \Rightarrow \begin{aligned}2\cdot[30b+25c]&=[245]\cdot 2\\(-3)\cdot[20b+4c]&=[100]\cdot(-3)\end{aligned} \Rightarrow$$

$$\begin{aligned}60b+50c&=490\\-60b-12c&=-300\end{aligned} \Rightarrow 38c=190 \Rightarrow c=5$$

$$2b+0.4\cdot 5=10 \Rightarrow 2b+2=10 \Rightarrow 2b=8 \Rightarrow b=4$$

49. Let m be the number months and c be the total cost of each system.
$c=179+17.5m$ WorldCom
$c=135+21.5m$ Satellite, Inc

$$\begin{aligned}(-1)\cdot[-17.5m+c]&=[179]\cdot(-1)\\-21.5m+c&=135\end{aligned} \Rightarrow \begin{aligned}17.5m-c&=-179\\-21.5m+c&=135\end{aligned} \Rightarrow -4m=-44 \Rightarrow m=11$$

After 11 months, Maneka pays $c=179+17.5(11)=\$371.50$ with either plan. After that, Worldcom is the better deal.

51. Let A be the number of passengers at Atlanta's Hartsfield International (in millions).
Let B be the number of passengers at Beijing's Capital International (in millions).

$$\begin{aligned}A+B&=98\\A&=B+10\end{aligned} \Rightarrow \begin{aligned}A+B&=98\\A-B&=10\end{aligned} \Rightarrow 2A=108 \Rightarrow A=54 \text{ thus } 54+B=98 \Rightarrow B=44$$

Atlanta's Hartsfield International had 54 million and Beijing's Capital International had 44 million.

53. Let y be production in millions and x be years since 2000.

United States: $m=\dfrac{y_2-y_1}{x_2-x_1}=\dfrac{8-13}{10-0}=\dfrac{-5}{10}=-0.5$; $b=13$; so $y=-0.5x+13$

China: $m=\dfrac{y_2-y_1}{x_2-x_1}=\dfrac{18-2}{10-0}=\dfrac{16}{10}=1.6$; $b=2$; so $y=1.6x+2$

55. \$11 per hour

Supply

Price	Tutors Supplied
8	9
15	37

Demand

Price	Tutors Demanded
8	30
15	9

Supply:

$$\text{slope} = \frac{37-9}{15-8} = \frac{28}{7} = 4$$

$y = mx + b$
$y = 4x + b$
$9 = 4 \cdot 8 + b$
$9 = 32 + b \Rightarrow b = -23$
$y = 4x - 23 \Rightarrow -4x + y = -23$

Demand:

$$\text{slope} = \frac{9-30}{15-8} = \frac{-21}{7} = -3$$

$y = mx + b$
$y = -3x + b$
$30 = -3 \cdot 8 + b$
$30 = -24 + b \Rightarrow b = 54$
$y = -3x + 54 \Rightarrow 3x + y = 54$

We must now solve the following system.

$-4x + y = -23$ (supply)
$3x + y = 54$ (demand)

$$\begin{array}{c} -4x + y = -23 \\ (-1)\cdot[3x + y] = [54]\cdot(-1) \end{array} \Rightarrow \begin{array}{c} -4x + y = -23 \\ -3x - y = -54 \end{array} \Rightarrow -7x = -77 \Rightarrow x = 11$$

57. \$34 per book

Supply

Price	Books Supplied
30	60
42	120

Demand

Price	Books Demanded
30	95
42	50

Supply:

$$\text{slope} = \frac{120-60}{42-30} = \frac{60}{12} = 5$$

$y = mx + b$
$y = 5x + b$
$60 = 5 \cdot 30 + b$
$60 = 150 + b \Rightarrow b = -90$
$y = 5x - 90 \Rightarrow -5x + y = -90$

Demand:

$$\text{slope} = \frac{50-95}{42-30} = \frac{-45}{12} = -\frac{15}{4}$$

$y = mx + b$
$y = -\frac{15}{4}x + b$
$95 = -\frac{15}{4} \cdot 30 + b$
$95 = -\frac{225}{2} + b \Rightarrow b = \frac{415}{2}$
$y = -\frac{15}{4}x + \frac{415}{2} \Rightarrow 15x + 4y = 830$

We must now solve the following system.

$-5x + y = -90$ (supply)
$15x + 4y = 830$ (demand)

$$\begin{array}{c} (-4)\cdot[-5x + y] = [-90]\cdot(-4) \\ 15x + 4y = 830 \end{array} \Rightarrow \begin{array}{c} 20x - 4y = 360 \\ 15x + 4y = 830 \end{array} \Rightarrow 35x = 1190 \Rightarrow x = 34$$

59. Let x be the number of pairs of earbuds.

$C = 12x + 126000$

$R = 40x$

$R = C \Rightarrow 40x = 12x + 126000 \Rightarrow 28x = 126000 \Rightarrow x = 4500$

$C = R = 40(4500) = 180000$

The break even point occurs at 4500 earbud pairs with a cost or revenue of $180,000.

61. Since the number of entertainment centers cannot be negative, we will consider restrictions in the first quadrant given by the following table.

	Athens (a)	Barcelona (b)	Available	Inequalities
fancy molding in feet	4	15	360	$4a + 15b \le 360$
time in hours	4	3	120	$4a + 3b \le 120$

The origin, $(0,0)$, an a-intercept, $(30,0)$, and a b-intercept, $(0,24)$, are all corner points. To find the fourth corner point we solve the following system.

$$4a + 15b = 360$$
$$4a + 3b = 120$$

$$\begin{matrix} 4a + 15b = 360 \\ (-1)\cdot[4a + 3b] = [120]\cdot(-1) \end{matrix} \Rightarrow \begin{matrix} 4a + 15b = 360 \\ -4a - 3b = -120 \end{matrix} \Rightarrow 12b = 240 \Rightarrow b = 20$$

$4a + 3\cdot 20 = 120 \Rightarrow 4a + 60 = 120 \Rightarrow 4a = 60 \Rightarrow a = 15$ The fourth corner point is $(15, 20)$.

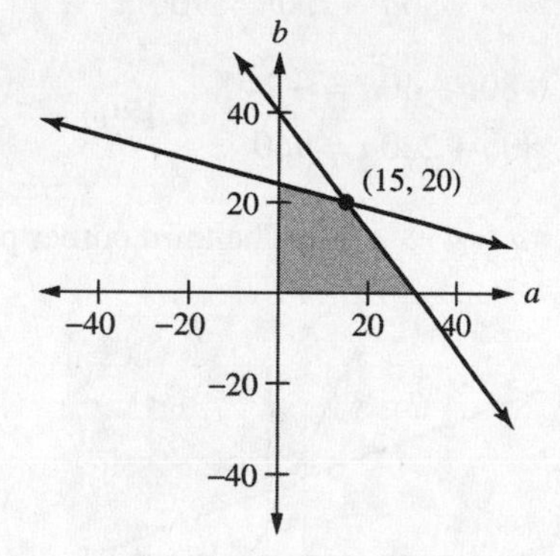

63. Let h be the number of hats and t be the number of T-shirts.

Since the number of items cannot be negative, we will consider restrictions in the first quadrant given by the following.

$$30h + 20t \le 600$$

The origin, $(0,0)$, the h-intercept, $(20,0)$, and the t-intercept, $(0,30)$, are all corner points.

63. (continued)

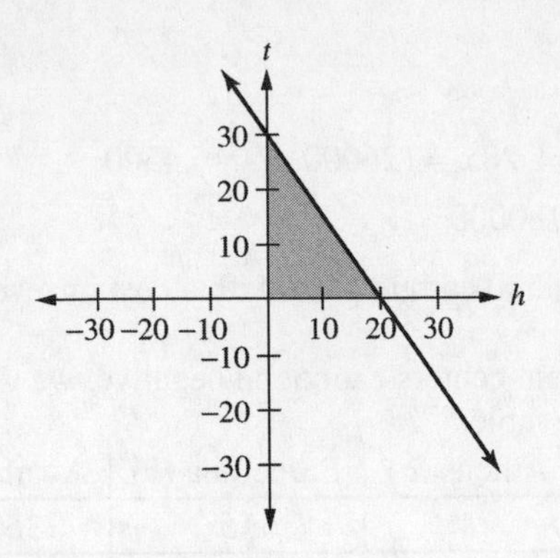

65. Since the amount of nutritional supplements cannot be negative, we will consider restrictions in the first quadrant given by the following table.

	Quantum (q)	NutraPlus (n)	Required	Inequalities
Niacin in mg	4	2	12	$4q+2n \geq 12$
Calcium in mg	80	220	960	$80q+220n \geq 960$

A q-intercept, $(12,0)$, and an n-intercept, $(0,6)$, are both corner points. To find the third corner point we solve the following system.

$$\begin{aligned} 4q+2n&=12 \\ 80q+220n&=960 \end{aligned}$$

$$\begin{aligned} (-20)[4q+2n]&=[12]\cdot(-20) \\ 80q+220n&=960 \end{aligned} \Rightarrow \begin{aligned} -80q-40n&=-240 \\ 80q+220n&=960 \end{aligned} \Rightarrow 180n=720 \Rightarrow n=4$$

$4q+2\cdot 4=12 \Rightarrow 4q+8=12 \Rightarrow 4q=4 \Rightarrow q=1$ The third corner point is $(1,4)$.

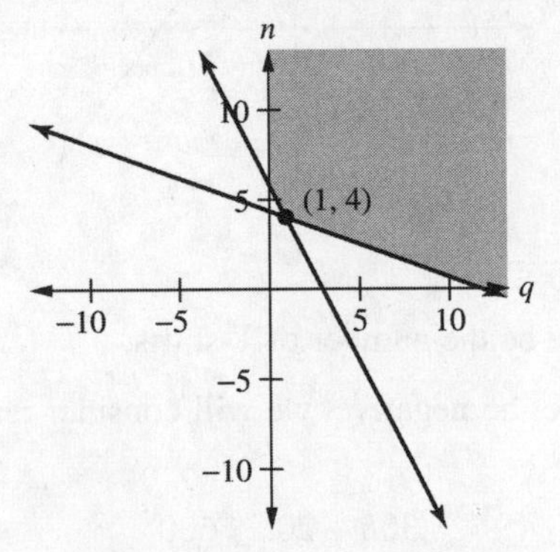

67. Let x = the number of Apple phones and y = the number of Android phones.
Since the number of phones cannot be negative, we will consider restrictions in the first quadrant given by the following.

$$\begin{aligned} x+y &\geq 30 \\ x+y &\leq 60 \\ x &\geq 2y \end{aligned}$$

The x-intercepts of $(30,0)$ and $(60,0)$ are both corner points.

67. (continued)

To find the third corner point we solve the following system.

$$x+y=30$$
$$x=2y$$

$(2y)+y=30 \Rightarrow 3y=30 \Rightarrow y=10$

$x=2\cdot 10=20$

The third corner point is $(20,10)$.

To find the fourth corner point we solve the following system.

$$x+y=60$$
$$x=2y$$

$(2y)+y=60 \Rightarrow 3y=60 \Rightarrow y=20$

$x=2\cdot 20=40$

The third corner point is $(40,20)$.

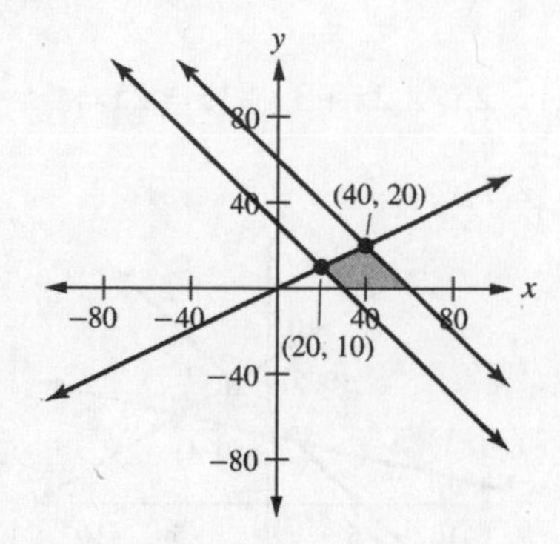

69. a) There can be one solution, an infinite number of solutions, or no solution.

b) If in solving we get a unique value for either x or y, then the lines intersect in a single point and the system has one solution. If in solving we get a false statement such as 2 + 2 = 0, then the lines are parallel and the system has no solution. If in solving we get a true statement that does not restrict either x or y, then both lines are the same, and we have an infinite number of solutions.

71. The solution will be a half-plane that can be determined by using the one-point test.

73. Answers will vary.

75. There are no points in which the regions given by the two inequalities intersect.

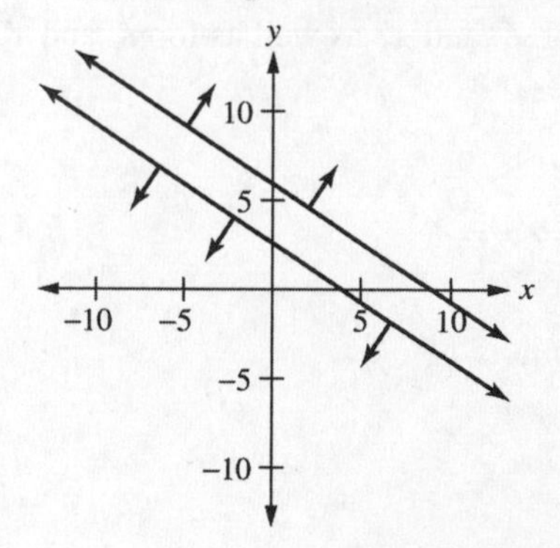

77. The constraints $x \geq 0$ and $y \geq 0$ imply that we are limited to quadrant I.

Let the equation that corresponds to the inequality $2x+3y \leq 25$ be ℓ_1.

Let the equation that corresponds to the inequality $5y \geq 20+x$ be ℓ_2.

Let the equation that corresponds to the inequality $y \leq x+5$ be ℓ_3.

77. (continued)

Two of the corner points are y-intercepts, namely $(0,4)$ and $(0,5)$. To find the other two corner points we need to find the point of intersection for ℓ_1 and ℓ_2 as well as for ℓ_1 and ℓ_3.

ℓ_1 and ℓ_2:

$$\begin{matrix} 2x+3y=25 \\ 5y=20+x \end{matrix} \Rightarrow \begin{matrix} 2x+3y=25 \\ 2\cdot[-x+5y]=[20]\cdot 2 \end{matrix} \Rightarrow \begin{matrix} 2x+3y=25 \\ -2x+10y=40 \end{matrix} \Rightarrow 13y=65 \Rightarrow y=5$$

$5\cdot 5=20+x \Rightarrow 25=20+x \Rightarrow 5=x$ A corner point is $(5,5)$.

ℓ_1 and ℓ_3:

$$\begin{matrix} 2x+3y=25 \\ y=x+5 \end{matrix} \Rightarrow 2x+3(x+5)=25 \Rightarrow 2x+3x+15=25 \Rightarrow 5x+15=25 \Rightarrow 5x=10 \Rightarrow x=2$$

$y=2+5=7$ A corner point is $(2,7)$.

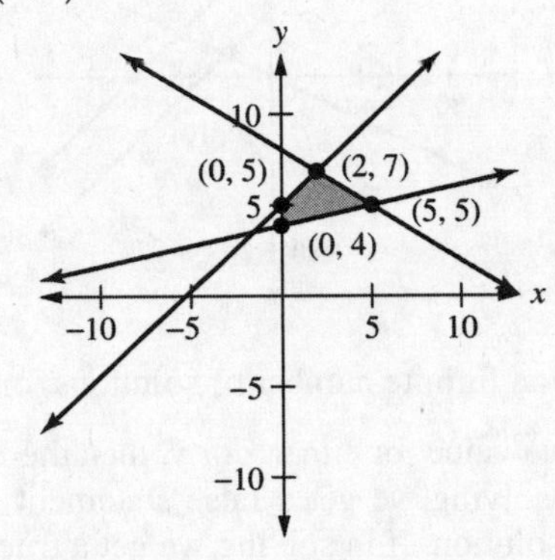

79. The constraints $x \ge 0$ and $y \ge 0$ imply that we are limited to quadrant I.

Let the equation that corresponds to the inequality $x-2y \le 8$ be ℓ_1.
Let the equation that corresponds to the inequality $2x+y \le 19$ be ℓ_2.
Let the equation that corresponds to the inequality $x-y \ge 5$ be ℓ_3.
One corner point is the x-intercept of ℓ_3, $(5,0)$. Another is the x-intercept of ℓ_1, $(8,0)$. To find the other two corner points we need to find the point of intersection for ℓ_1 and ℓ_2 as well as for ℓ_2 and ℓ_3.

ℓ_1 and ℓ_2:

$$\begin{matrix} x-2y=8 \\ 2x+y=19 \end{matrix} \Rightarrow \begin{matrix} x-2y=8 \\ 2\cdot[2x+y]=[19]\cdot 2 \end{matrix} \Rightarrow \begin{matrix} x-2y=8 \\ 4x+2y=38 \end{matrix} \Rightarrow 5x=46 \Rightarrow x=\tfrac{46}{5}$$

$\frac{46}{5}-2y=8 \Rightarrow -2y=8-\frac{46}{5}=\frac{40}{5}-\frac{46}{5}=-\frac{6}{5} \Rightarrow y=\frac{3}{5}$ A corner point is $\left(\frac{46}{5},\frac{3}{5}\right)$.

ℓ_2 and ℓ_3:

$$\begin{matrix} 2x+y=19 \\ x-y=5 \end{matrix} \Rightarrow 3x=24 \Rightarrow x=8$$

$8-y=5 \Rightarrow -y=-3 \Rightarrow y=3$ A corner point is $(8,3)$.

79. (continued)

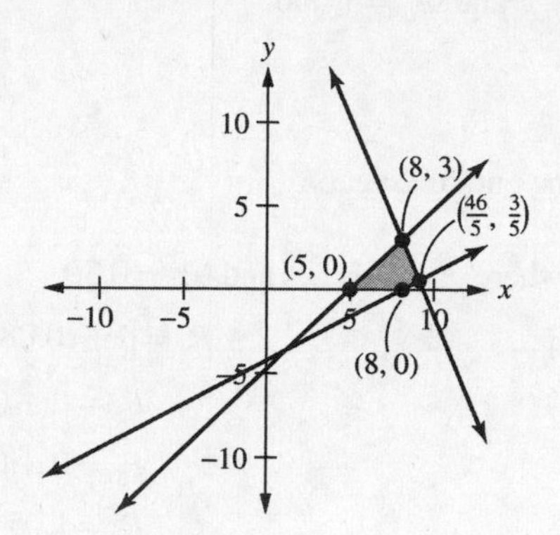

81. In Region 1, when we look at the location of the origin, $(0,0)$, Case A should be true, Case B should be false and Case C should be false. This yields the following set of inequalities.

$$3y \le 43 - 5x$$
$$11y \ge 27 - 2x$$
$$3y \ge 1 + 2x$$

83. In Region 3, when we look at the location of the origin, $(0,0)$, Case A should be true, Case B should be false and Case C should be true. This yields the following set of inequalities.

$$3y \le 43 - 5x$$
$$11y \ge 27 - 2x$$
$$3y \le 1 + 2x$$

Section 7.7 Looking Deeper: Dynamical Systems

1. $A_0 = 3$
   $A_1 = 2\cdot 3 - 1 = 6 - 1 = 5$
   $A_2 = 2\cdot 5 - 1 = 10 - 1 = 9$

3. $A_0 = -2$
   $A_1 = -3\cdot(-2) + 4 = 6 + 4 = 10$
   $A_2 = -3\cdot 10 + 4 = -30 + 4 = -26$
   $A_3 = -3\cdot(-26) + 4 = 78 + 4 = 82$

5. $A_0 = 4$
   $A_1 = 1.8\cdot 4 - 2 = 7.2 - 2 = 5.2$
   $A_2 = 1.8\cdot 5.2 - 2 = 9.36 - 2 = 7.36$
   $A_3 = 1.8\cdot 7.36 - 2 = 13.248 - 2 = 11.248$
   $A_4 = 1.8\cdot 11.248 - 2 = 20.2464 - 2 = 18.2464$

7. In $A_{n+1} = 2A_n + 3$, substitute a for both A_{n+1} and A_n.
   $a = 2a + 3$
   $-a = 3$
   $a = -3$
   Since $2 > 1$, the equilibrium value is unstable.

9. In $B_{n+1} = 0.25B_n + 4$, substitute a for both B_{n+1} and B_n.
   $a = 0.25a + 4$
   $0.75a = 4$
   $a = \frac{4}{0.75} = \frac{400}{75} = \frac{16}{3}$
   Since $-1 < 0.25 < 1$, the equilibrium value is stable.

11. $A_{n+1} = 1.05A_n$ where $n = 0, 1, 2, \ldots$ and $A_0 = 1{,}000$

$A_1 = 1.05 \cdot 1{,}000 = 1{,}050$

$A_2 = 1.05 \cdot 1{,}050 = 1{,}102.5$

You would have \$1,102.50 at the end of 2 years.

13. $P_{n+1} = [1 + (0.08) \cdot (1 - P_n)] \cdot P_n$ where $n = 0, 1, 2, \ldots$ and $P_0 = 0.30$

$P_1 = [1 + (0.08) \cdot (1 - 0.30)] \cdot 0.30$

$P_1 = [1 + (0.08) \cdot 0.70] \cdot 0.30$

$P_1 = [1 + 0.056] \cdot 0.30$

$P_1 = 1.056 \cdot 0.30$

$P_1 = 0.3168$

$P_2 = [1 + (0.08) \cdot (1 - 0.3168)] \cdot 0.3168$

$P_2 = [1 + (0.08) \cdot 0.6832] \cdot 0.3168$

$P_2 = [1 + 0.054656] \cdot 0.3168$

$P_2 = 1.054656 \cdot 0.3168$

$P_2 \approx 0.3341150208$

Approximately 33.4% of the lemur population's maximum capacity will be attained at the end of two years.

15. $D_{n+1} = 0.60 \cdot D_n + 250$ where $n = 0, 1, 2, \ldots$ and $D_0 = 0$

$D_1 = 0.60 \cdot 0 + 250 = 0 + 250 = 250$

$D_2 = 0.60 \cdot 250 + 250 = 150 + 250 = 400$

$D_3 = 0.60 \cdot 400 + 250 = 240 + 250 = 490$

490 milligrams of antibiotic will be in your bloodstream after three doses.

17. a) $D_{n+1} = 1.05 \cdot (D_n + 3000)$ where $n = 0, 1, 2, \ldots$ and $D_0 = 0$

b) $D_1 = 1.05 \cdot (0 + 3000) = 1.05 \cdot (3000) = 3150$

$D_2 = 1.05 \cdot (3150 + 3000) = 1.05 \cdot (6150) = 6457.50$

$D_3 = 1.05 \cdot (6457.50 + 3000) = 1.05 \cdot (9457.5) = 9930.375$

$D_4 = 1.05 \cdot (9930.38 + 3000) = 1.05 \cdot (12930.38) = 13576.89375$

There will be \$13,576.89 in the account.

19. There are equations relating A_1 to A_0, A_2 to A_1, A_3 to A_2, and so on.

21. If $-1 < m < 1$, then the equilibrium value will be stable.

23. $C_k = (0.99988)^k \cdot C_0 = 0.99988^k$ since $C_0 = 1$ where $k = 0, 1, 2, \ldots$

$$0.90 = 0.99988^k$$

$$\log 0.90 = \log 0.99988^k$$

$$\log 0.90 = k \cdot \log 0.99988$$

$$\frac{\log 0.90}{\log 0.99988} = k$$

$$k \approx \frac{-0.04575749}{-0.00005212} \approx 877.93$$

The fossilized bone is approximately 900 years old.

25. $50\left(\frac{1}{2}\right)^{\frac{50000}{24000}} \approx 11.8$ pounds

27. $P_0 = \dfrac{500}{1,000} = 0.5$ and 80 turkeys represents $\dfrac{80}{1,000} = 0.08$ or 8% of the maximum population.

$$\begin{aligned} P_1 &= [1+(0.10)\cdot(1-0.5)]\cdot 0.5 \\ P_1 &= [1+(0.10)\cdot 0.5]\cdot 0.5 \\ P_1 &= [1+0.05]\cdot 0.5 \\ P_1 &= 1.05\cdot 0.5 \\ P_1 &= 0.525 \\ &\Rightarrow 0.525-0.08 = 0.445 \end{aligned}$$

$$\begin{aligned} P_2 &= [1+(0.10)\cdot(1-0.445)]\cdot 0.445 \\ P_2 &= [1+(0.10)\cdot 0.555]\cdot 0.445 \\ P_2 &= [1+0.0555]\cdot 0.445 \\ P_2 &= 1.0555\cdot 0.445 \\ P_2 &= 0.4696975 \\ &\Rightarrow 0.4696975-0.08 = 0.3896975 \end{aligned}$$

$$\begin{aligned} P_3 &= [1+(0.10)\cdot(1-0.3896975)]\cdot 0.3896975 \\ P_3 &= [1+(0.10)\cdot 0.6103025]\cdot 0.3896975 \\ P_3 &= [1+0.06103025]\cdot 0.3896975 \\ P_3 &= 1.06103025\cdot 0.3896975 \\ P_3 &\approx 0.413480835849 \\ &\Rightarrow 0.413480835849-0.08 = 0.333480835849 \end{aligned}$$

There would be approximately 333 turkeys at the end of the three years.

Chapter Review Exercises

1. a)
$$\begin{aligned} \frac{2}{3}x+2 &= \frac{1}{6}x+4 \\ 6\cdot\left[\frac{2}{3}x+2\right] &= \left[\frac{1}{6}x+4\right]\cdot 6 \\ 4x+12 &= x+24 \\ 3x+12 &= 24 \\ 3x &= 12 \\ x &= 4 \end{aligned}$$

b)
$$\begin{aligned} 0.3x-2 &= 3.5x-0.4 \\ 10\cdot[0.3x-2] &= [3.5x-0.4]\cdot 10 \\ 3x-20 &= 35x-4 \\ -32x-20 &= -4 \\ -32x &= 16 \\ x &= \frac{16}{-32} = -\frac{1}{2} \end{aligned}$$

2.
$$\begin{aligned} A &= P(1+rt) \\ A &= P+\text{Pr}t \\ A-P &= Prt \\ \frac{A-P}{Pt} &= \frac{Prt}{Pt} = r \\ r &= \frac{A-P}{Pt} \end{aligned}$$

3. a) Let h be the number of hours Darryl works.
Let s be Darryl's salary for the week.

$$\begin{aligned} s &= 40\cdot 5+(h-40)(5)(2) \\ s &= 200+10(h-40) \end{aligned}$$

b) $s = 200+10(46-40) = 200+10\cdot 6 = 200+60 = \$260$

4. x-intercept (when $y=0$)

$3x+5\cdot 0=20$

$3x+0=20$

$3x=20$

$x=\frac{20}{3}\Rightarrow\left(6\frac{2}{3},0\right)$

y-intercept (when x=0)

$3\cdot 0+5y=20$

$0+5y=20$

$5y=20$

$y=4\Rightarrow(0,4)$

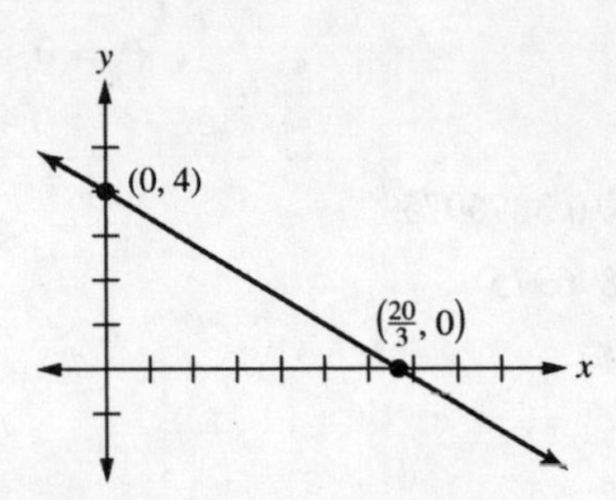

5. $m=\frac{y_2-y_1}{x_2-x_1}=\frac{8-5}{6-2}=\frac{3}{4}$

6. Line 1 is a; Line 2 is d; Line 3 is b; Line 4 is c.

7. Let m be the number of months.

$240+39m=350+28m$

$39m=110+28m$

$11m=110$

$m=10$

after ten months

8. $y=mx+b$

$y=-4x+b$

$5=-4\cdot 2+b$

$5=-8+b$

$13=b$

$y=-4x+13$

9. $m=\frac{y_2-y_1}{x_2-x_1}=\frac{9-4}{6-3}=\frac{5}{3}$

$y=mx+b$

$y=\frac{5}{3}x+b$

$4=\frac{5}{3}\cdot 3+b$

$4=5+b$

$-1=b$

$y=\frac{5}{3}x-1$

10. Let t be the number of years after 2002.
Let y be the amount spent by Americans on foreign travel (in billions).
The given information corresponds to the two ordered pairs (0, 81.8) and (3, 99.9).

a) $m=\frac{y_2-y_1}{t_2-t_1}=\frac{99.9-81.8}{3-0}=\frac{18.1}{3}=6.03$

$y=mt+b$

$y=6.03x+b$

$81.8=6.03\cdot 0+b$

$81.8=0+b$

$81.8=b$

$y=6.03t+81.8$

10. (continued)

b) $y = 6.03(13) + 81.8 = 78.39 + 81.8 = 160.19$

The amount spent by Americans on foreign travel is estimated to be \$160.19 billion in 2015.

11. It is the linear model that best approximates a set of data points.

12. When there is a constant rate of change between two variables.

13. We need to solve the equation $2x^2 + 7x = 4$ or $2x^2 + 7x - 4 = 0$ by using the quadratic formula, we obtain

$$x = \frac{-(7) \pm \sqrt{(7)^2 - 4(2)(-4)}}{2(2)} = \frac{-7 \pm \sqrt{49 + 32}}{4} = \frac{-7 \pm \sqrt{81}}{4} = \frac{-7 \pm 9}{4}$$

$$x = \frac{-7 - 9}{4} = \frac{-16}{4} = -4 \text{ or } x = \frac{-7 + 9}{4} = \frac{2}{4} = \frac{1}{2}$$

14. a) Opening: down since $a < 0$

b) Vertex: $x = \frac{-b}{2a} = \frac{-(-10)}{2(-2)} = \frac{10}{-4} = -\frac{5}{2}$. Substituting this value for *x*, we obtain

$y = -2 \cdot \left(-\frac{5}{2}\right)^2 - 10 \cdot \left(-\frac{5}{2}\right) - 8 = -\frac{25}{2} + 25 - 8 = \frac{9}{2}$ as the *y*-coordinate of the vertex.

The vertex is therefore $\left(-\frac{5}{2}, \frac{9}{2}\right)$.

c) *x*-intercepts: Set $y = 0$. We get $0 = -2x^2 - 10x - 8$ and by using the quadratic formula, we obtain

$$x = \frac{-(-10) \pm \sqrt{(-10)^2 - 4(-2)(-8)}}{2(-2)} = \frac{10 \pm \sqrt{100 - 64}}{-4} = \frac{10 \pm \sqrt{36}}{-4} = \frac{10 \pm 6}{-4}$$

$$x = \frac{10 + 6}{-4} = \frac{16}{-4} = -4 \text{ or } x = \frac{10 - 6}{-4} = \frac{4}{-4} = -1$$

The *x*-intercepts of the graph are therefore $(-4, 0)$ and $(-1, 0)$.

d) *y*-intercept: Set $x = 0$. Thus $y = -2 \cdot 0^2 - 10 \cdot 0 - 8 = -8$. The *y*-intercept is therefore $(0, -8)$.

e)

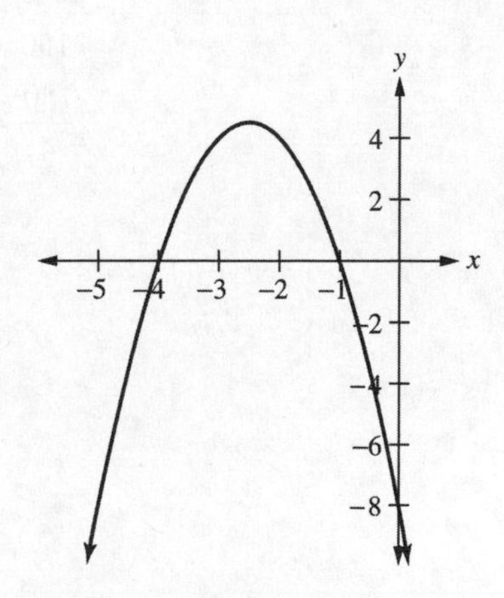

15. a technique by which we find a quadratic equation that bests models a set of data

16. We need to find the first component of the vertex of $A = -0.125x^2 + 2x + 1.125$.

$$x = \frac{-b}{2a} = \frac{-2}{2(-0.125)} = \frac{-2}{-0.25} = 8 \text{ implies in week 8.}$$

17. $A = 10{,}000(1+0.048)^5 = 10{,}000 \cdot 1.048^5 \approx \$12{,}641.72$

18.
$$20{,}000 = 10{,}000(1+0.048)^n$$
$$2 = 1.048^n$$
$$\log 2 = \log 1.048^n$$
$$\log 2 = n \cdot \log 1.048$$
$$\frac{\log 2}{\log 1.048} = n$$
$$n \approx \frac{0.30103}{0.02036} \approx 14.79$$

It will double by the 15th year.

19.
$$P_{3+1} = [1 + (0.03) \cdot (1 - 0.50)] \cdot 0.50$$
$$P_4 = [1 + (0.03) \cdot 0.50] \cdot 0.50$$
$$P_4 = [1 + 0.015] \cdot 0.50$$
$$P_4 = 1.015 \cdot 0.50$$
$$P_4 = 0.5075$$

20.
$$280 = 220(1+r)^4$$
$$(1+r)^4 = \frac{280}{220} = \frac{14}{11}$$
$$1 + r = \sqrt[4]{\frac{14}{11}}$$
$$r = \sqrt[4]{\frac{14}{11}} - 1 \approx 0.0621$$

$$A = 220(1+0.0621)^8$$
$$A = 220(1.0621)^8$$
$$A \approx \$356.24$$

21. In a logistic model we adjust the rate of growth as the population grows. The larger the population, the smaller the rate of growth.

22. a) Rewrite $25:8 = x:2$ as $\frac{25}{8} = \frac{x}{2}$.

$$\frac{25}{8} = \frac{x}{2}$$
$$25 \cdot 2 = 8 \cdot x$$
$$50 = 8x$$
$$\frac{50}{8} = x$$
$$x = \frac{25}{4} = 6.25$$

b)
$$\frac{30}{4} = \frac{x}{5}$$
$$30 \cdot 5 = 4 \cdot x$$
$$150 = 4x$$
$$\frac{150}{4} = x$$
$$x = \frac{75}{2} = 37.5$$

23. Your proportion could be set up differently.

$$\frac{3.5}{840}=\frac{x}{1{,}500}$$
$$3.5\cdot 1{,}500=840\cdot x$$
$$5{,}250=840x$$
$$\frac{5{,}250}{840}=x$$
$$x=6.25 \text{ gal}$$

24. $$\frac{\text{number of tagged penguins in population}}{\text{number of penguins in population}}=\frac{\text{number of tagged penguins in sample}}{\text{number of penguins in sample}}$$
$$\frac{180}{n}=\frac{12}{55}$$
$$180\cdot 55=n\cdot 12$$
$$9{,}900=12n$$
$$\frac{9{,}900}{12}=n$$
$$n=825$$

25.
$$y=\frac{k}{x}$$
$$14=\frac{k}{3}$$
$$14\cdot 3=k$$
$$k=42$$

$$y=\frac{42}{x}$$
$$y=\frac{42}{18}$$
$$y=\frac{7}{3}$$

26.
$$d=k\frac{a^2b}{c}$$
$$144=k\frac{3^2\cdot 7}{14}$$
$$2{,}016=k(9\cdot 7)$$
$$2{,}016=63k$$
$$\frac{2{,}016}{63}=k$$
$$k=32$$

$$d=32\cdot\frac{a^2b}{c}$$
$$d=32\cdot\frac{8^2\cdot 11}{16}$$
$$d=32\cdot\frac{64\cdot 11}{16}$$
$$d=32\cdot(4\cdot 11)$$
$$d=32\cdot 44=1{,}408$$

27. The strength of the beam stays the same.

$$s=k\frac{(2w)\cdot d^2}{2l}=k\frac{w\cdot d^2}{l}$$

28. a) $(3,-5)$

$$4x-3y=27$$
$$(-4)\cdot[x+2y]=[-7]\cdot(-4)$$

$$4x-3y=27$$
$$-4x-8y=28$$
$$0-11y=55$$
$$y=-5$$

$$x+2\cdot(-5)=-7$$
$$x-10=-7$$
$$x=3$$

b) The lines intersect at a single point.

29. a) infinite number of solutions

$$(4)\cdot\left[\frac{3}{4}x-\frac{1}{2}y\right]=\left[\frac{5}{4}\right]\cdot(4)$$
$$(3)\cdot\left[-x+\frac{2}{3}y\right]=\left[-\frac{5}{3}\right]\cdot(3)$$

$$3x-2y=5$$
$$-3x+2y=-5$$
$$0=0$$

b) The lines are the same.

30. a) no solutions

$$(3)\cdot[-4x+5y]=[3]\cdot(3)$$
$$12x-15y=-10$$

$$-12x+15y=9$$
$$\underline{12x-15y=-10}$$
$$0=-1$$

b) The lines are parallel.

31. Let u be the number of points scored by the U.S. team and p be the number of points by the opponent.

$$u+p=123$$
$$u=p+17$$

$$u+p=123$$
$$\underline{u-p=17}$$
$$2u+0=140$$
$$u=70$$

$$70+p=123$$
$$p=53$$

U.S.: 70; Opponent: 53

32. Let B be the number of Big Macs and C be the number of Chicken Sandwiches.

$5B+3C=3780$ Actual Meal

$3B+5C=3420$ Optional Meal

$$\begin{array}{l}(-3)\cdot[5B+3C]=[3780]\cdot(-3)\\(5)\cdot[3B+5C]=[3420]\cdot(5)\end{array}\Rightarrow\begin{array}{r}-15B-9C=-11340\\ \underline{15B+25C=17100}\\ 0+16B=5760\\ y=360\end{array}\Rightarrow\begin{array}{r}5B+3\cdot360=3780\\5B+1080=3780\\5B=2700\\B=540\end{array}$$

Big Mac's have 540 calories; McChicken sandwiches have 360 calories.

33. They should charge $18 per wreath so that supply and demand will both be 28 boxes.

Supply

Price	Wreaths Supplied
12	26
24	30

Demand

Price	Wreaths Demanded
12	32
24	24

Supply:

$$\text{slope}=\frac{30-26}{24-12}=\frac{4}{12}=\frac{1}{3}$$
$$y=mx+b$$
$$y=\tfrac{1}{3}x+b$$
$$26=\tfrac{1}{3}\cdot12+b$$
$$26=4+b\Rightarrow b=22$$
$$y=\tfrac{1}{3}x+22\Rightarrow -x+3y=66$$

Demand:

$$\text{slope}=\frac{24-32}{24-12}=\frac{-8}{12}=-\frac{2}{3}$$
$$y=mx+b$$
$$y=-\tfrac{2}{3}x+b$$
$$32=-\tfrac{2}{3}\cdot12+b$$
$$32=-8+b\Rightarrow b=40$$
$$y=-\tfrac{2}{3}x+40\Rightarrow 2x+3y=120$$

We must now solve the following system.

$$-x+3y=66\quad\text{(supply)}$$
$$2x+3y=120\quad\text{(demand)}$$

$$\begin{array}{r}(-1)\cdot[-x+3y]=[66]\cdot(-1)\\2x+3y=120\end{array}\Rightarrow\begin{array}{r}x-3y=-66\\2x+3y=120\end{array}\Rightarrow 3x=54\Rightarrow x=\frac{54}{3}=18\Rightarrow y=\tfrac{1}{3}\cdot18+22=28$$

34. $6\cdot 0+5\cdot 0\overset{?}{>}20 \Rightarrow 0+0\overset{?}{>}20 \Rightarrow 0>20$; False. Shade the side that does not contain $(0,0)$.

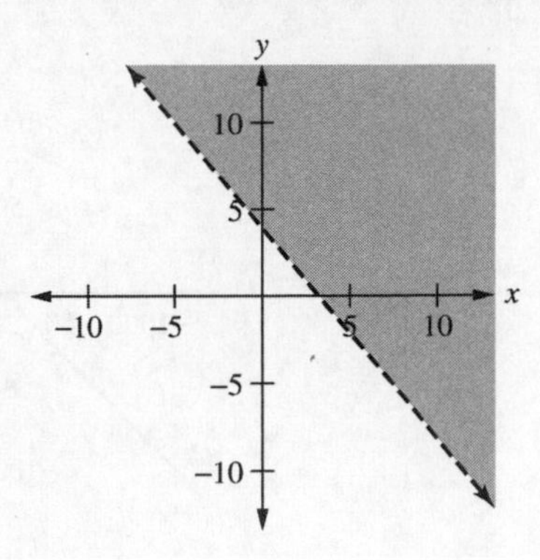

35. Testing $(0,0)$ in the first inequality, we get $2\cdot 0+5\cdot 0\overset{?}{\geq}24 \Rightarrow 0+0\overset{?}{\geq}24 \Rightarrow 0\geq 24$ which is false, so $(0,0)$ is not a solution. Testing $(0,0)$ in the second inequality, we get $2\cdot 0\overset{?}{\leq}2\cdot 0+18 \Rightarrow 0\overset{?}{\leq}0+18 \Rightarrow 0\leq 18$ which is true, so $(0,0)$ is a solution. To solve for the corner point, we solve the system.

$$2x+5y=24$$
$$2y=2x+18$$

$$\begin{array}{r} 2x+5y=24 \\ -2x+2y=18 \\ \hline 0+7y=42 \\ y=6 \end{array}$$

$$2x+5\cdot 6=24 \Rightarrow 2x+30=24 \Rightarrow$$
$$2x=-6 \Rightarrow x=-3$$

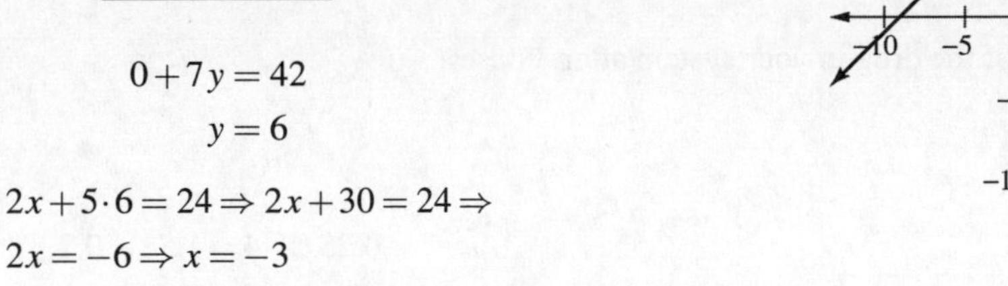

Thus the corner point is $(-3,6)$.

36. Let t be the number of T-shirts and h be the number of hats.

Since the number of hats and T-shirts cannot be negative, we will consider restrictions in the first quadrant given by the following.

$$4t+2h\leq 180$$
$$h\geq t+15$$

The h-intercepts, $(0,90)$ and $(0,15)$, are corner points. To find the third corner point we solve the following system.

$$4t+2h=180$$
$$h=t+15$$

36. (continued)

$$4t+2(t+15)=180$$
$$4t+2t+30=180$$
$$6t+30=180$$
$$6t=150$$
$$t=25$$

$h=25+15=40$

The third corner point is $(25,40)$.

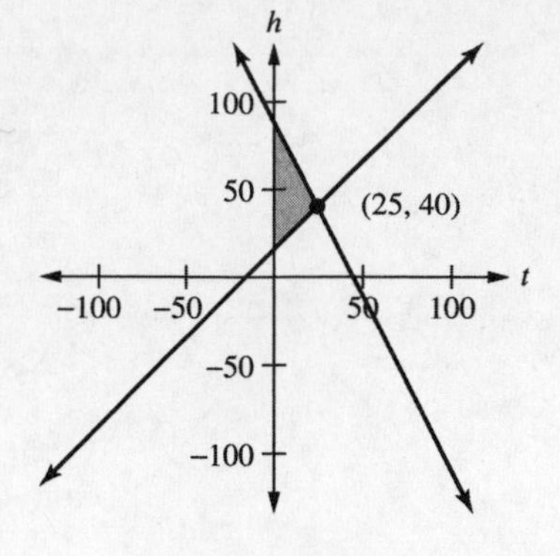

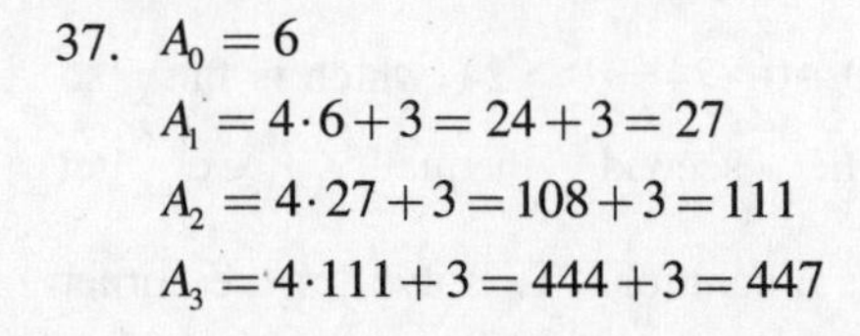

37. $A_0=6$
$A_1=4\cdot 6+3=24+3=27$
$A_2=4\cdot 27+3=108+3=111$
$A_3=4\cdot 111+3=444+3=447$

38. $A_{n+1}=4\cdot A_n+3$
$a=4a+3$
$-3a=3$
$a=\frac{-3}{3}=-1$

The value of a is unstable since $m=4>1$.

39. $D_{n+1}=0.50\cdot D_n+800$
$D_0=800$
$D_1=0.50\cdot 800+800=400+800=1200$
$D_2=0.50\cdot 1200+800=600+800=1400$

There will be 1400 mg of the drug in your system after 3 doses.

Chapter Test

1. a)

$$\frac{3}{4}x+5=\frac{2}{3}x+4$$
$$12\cdot\left[\frac{3}{4}x+5\right]=\left[\frac{2}{3}x+4\right]\cdot 12$$
$$9x+60=8x+48$$
$$x+60=48$$
$$x=-12$$

b)

$$0.25x+4=1.5x-0.2$$
$$100\cdot[0.25x+4]=[1.5x-0.2]\cdot 100$$
$$25x+400=150x-20$$
$$400=125x-20$$
$$420=125x$$
$$x=\frac{420}{125}=3.36$$

2.

$$X=a(1+b)$$
$$\frac{X}{a}=\frac{a(1+b)}{a}$$
$$\frac{X}{a}=1+b$$
$$b=\frac{X}{a}-1$$

or

$$X=a(1+b)$$
$$X=a+ab$$
$$X-a=ab$$
$$ab=X-a$$
$$b=\frac{X-a}{a}$$

3. $m=\frac{y_2-y_1}{x_2-x_1}=\frac{12-5}{8-3}=\frac{7}{5}$

4. x-intercept (when $y=0$)

$5x-4\cdot 0=10$

$5x-0=10$

$5x=10$

$x=\frac{10}{5}\Rightarrow (2,0)$

y-intercept (when x=0)

$5\cdot 0-4y=10$

$0-4y=10$

$-4y=10$

$y=\frac{10}{-4}\Rightarrow \left(-\frac{5}{2},0\right)$

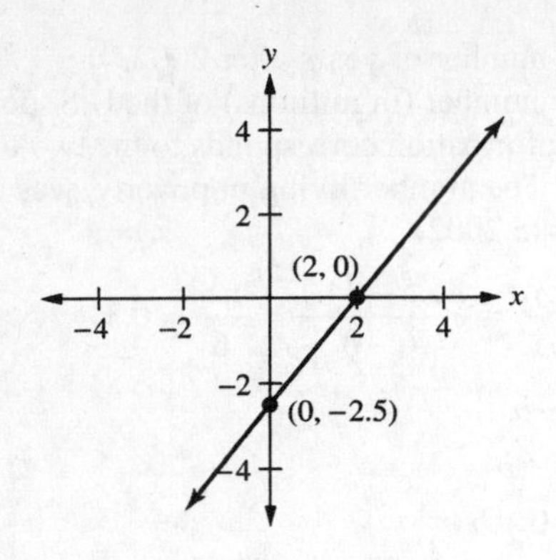

5. a) Let m be the number of minutes Brandon uses each month.
Let c be Brandon's cost for one month.
$c=30+0.045(m-1200)$

b) $c=30+0.045((1520)-1200))=30+0.045(320)=30+14.4=\$44.40$

6. $18.50=16(1+r)^6$

$(1+r)^6=\frac{18.50}{16}=1.15625$

$1+r=\sqrt[6]{1.15625}$

$r=\sqrt[6]{1.15625}-1\approx 0.024$

$A=16(1+0.024)^{13}$

$A=16(1.024)^{13}$

$A\approx \$21.78$

7. 1 and b, 2 and d, 3 and a, 4 and c

8. Let r be the number of rentals.

$40+1.5r=22+2r$

$18=0.5r$

$r=\frac{18}{0.5}$

$r=36$

after 36 rentals.

9. When the rate of change between the two variables is constant.

10. $m=\frac{y_2-y_1}{x_2-x_1}=\frac{8-5}{6-2}=\frac{3}{4}$

$y=mx+b$

$y=\frac{3}{4}x+b$

$5=\frac{3}{4}\cdot 2+b$

$5=\frac{3}{2}+b$

$\frac{7}{2}=b$

$y=\frac{3}{4}x+\frac{7}{2}$

11. Let p be the number of hours Shanaya works in the psychology lab.
Let d be the number of hours she works in the dining hall.

$8.25p+6.30d=137.70$

12. A technique to find a linear equation that best models a set of data.

13. Let t be the number of years after 2002.
Let n be the number (in millions) of the U.S. population that live in poverty.
The given information corresponds to the two ordered pairs (0, 34.7) and (6, 36.5).
(Note: The number living in poverty was up 1.8 million from 2002, so there are 36.5 – 1.8 = 34.7 million in 2002)

$$m = \frac{y_2 - y_1}{x_2 - x_1} = \frac{36.5 - 34.7}{6 - 0} = \frac{1.8}{6} = 0.3$$

$$n = mt + b$$
$$n = 0.3t + b$$
$$34.7 = 0.3 \cdot 0 + b$$
$$34.7 = 0 + b$$
$$34.7 = b$$
$$n = 0.3t + 34.7$$
$$y = 0.3 \cdot 18 + 34.7 = 5.4 + 34.7 = 40.1$$

The estimated number of people living in poverty in 2020 is 40.1 million.

14. $A_0 = 5$
$A_1 = 4 \cdot 5 + 2 = 20 + 2 = 22$
$A_2 = 4 \cdot 22 + 2 = 88 + 2 = 90$
$A_3 = 4 \cdot 90 + 2 = 360 + 2 = 362$

15. $A_{n+1} = 4 \cdot A_n + 2$
$a = 4a + 2$
$-3a = 2$
$a = -\frac{2}{3}$

The value of a is unstable since $m = 4 > 1$.

16. a) opening down since $a < 0$

b) Vertex: $x = \frac{-b}{2a} = \frac{-(11)}{2(-1)} = \frac{-11}{-2} = \frac{11}{2}$. Substituting this value for x, we obtain

$y = -\left(\frac{11}{2}\right)^2 + 11 \cdot \left(\frac{11}{2}\right) - 24 = -\frac{121}{4} + \frac{121}{2} - 8 = -\frac{121}{4} + \frac{242}{4} - \frac{96}{4} = \frac{25}{4}$ as the y-coordinate of the vertex.

The vertex is therefore $\left(\frac{11}{2}, \frac{25}{4}\right)$.

c) x-intercepts: Set $y = 0$. We get $0 = -x^2 + 11x - 24$ and by using the quadratic formula, we obtain

$$x = \frac{-(11) \pm \sqrt{(11)^2 - 4(-1)(-24)}}{2(-1)} = \frac{-11 \pm \sqrt{121 - 96}}{-2} = \frac{-11 \pm \sqrt{25}}{-2} = \frac{-11 \pm 5}{-2}$$

$$x = \frac{-11 + 5}{-2} = \frac{-6}{-2} = 3 \text{ or } x = \frac{-11 - 5}{-2} = \frac{-16}{-2} = 8$$

The x-intercepts of the graph are therefore $(3, 0)$ and $(8, 0)$.

d) y-intercept: Set $x = 0$. Thus $y = -(0)^2 + 11 \cdot 0 - 24 = -24$. The y-intercept is therefore $(0, -24)$.

16. (continued)

e)

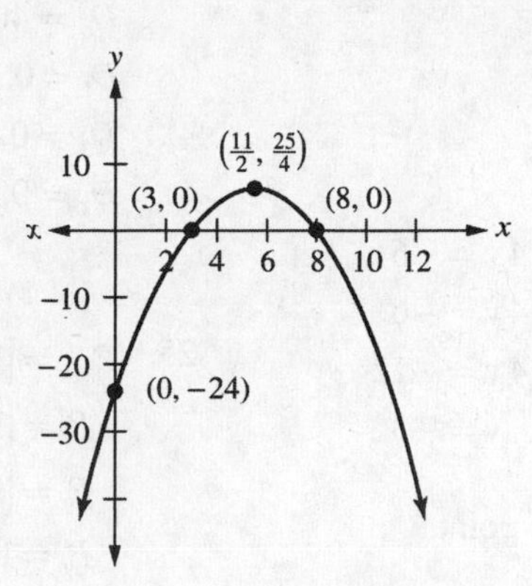

17. $A = 5{,}000(1+0.024)^8 = 5{,}000 \cdot 1.024^8 \approx \$6{,}044.63$

18.
$$10{,}000 = 5{,}000(1+0.024)^n$$
$$2 = 1.024^n$$
$$\log 2 = \log 1.024^n$$
$$\log 2 = n \cdot \log 1.024$$
$$\frac{\log 2}{\log 1.024} = n$$
$$n \approx \frac{0.30103}{0.103} \approx 29.2$$

It will double just after 29 years.

19. quadratic regression.

20. We need to find the first component of the vertex of $S = -1.2x^2 + 8x$.

$x = \frac{-b}{2a} = \frac{-(8)}{2(-1.2)} = \frac{-8}{-2.4} \approx 3.33$, which implies during week 4.

21. a) no solution

$$(4) \cdot [6x - 2y] = [-4] \cdot (4)$$
$$-24x + 8y = 10$$

$$24x - 8y = -16$$
$$-24x + 8y = 10$$
$$0 = -6$$

b) The lines are parallel.

22. a) infinite number of solutions

$$(-2) \cdot [4x - 3y] = [6] \cdot (-2)$$
$$8x - 6y = 12$$

$$-8x + 6y = -12$$
$$8x - 6y = 12$$
$$0 = 0$$

b) The lines are the same.

23. $(2,-3)$

$$(3)\cdot[3x+4y]=[-6]\cdot(3)$$
$$(4)\cdot[2x-3y]=[13]\cdot(4)$$

$$\begin{array}{r} 9x+12y=-18 \\ 8x-12y=\ \ 52 \\ \hline 17x+0=34 \\ x=2 \end{array} \qquad \begin{array}{r} 3(2)+4y=-6 \\ 6+4y=-6 \\ 4y=-12 \\ y=-3 \end{array}$$

b) The lines intersect in a single point.

24. $D_{n+1}=0.35\cdot D_n+400$

$D_0=400$

$D_1=0.35\cdot 400+400=140+400=540$

$D_2=0.35\cdot 540+400=189+400=589$

$D_3=0.35\cdot 589+400=206.15+400$

$=606.15$

25. $P_{8+1}=[1+(0.02)\cdot(1-0.40)]\cdot 0.40$

$P_9=[1+(0.02)\cdot 0.60]\cdot 0.40$

$P_9=[1+0.012]\cdot 0.40$

$P_9=1.012\cdot 0.40$

$P_9=0.4048$

26. $$\frac{\text{number of homeless in first sample}}{\text{number of homeless in population}}=\frac{\text{number of homeless from first sample in second sample}}{\text{total number of homeless in second sample}}$$

$$\frac{75}{x}=\frac{5}{120}$$
$$75\cdot 120=5\cdot x$$
$$9{,}000=5x$$
$$\frac{9{,}000}{5}=x$$
$$x=\frac{9{,}000}{5}=1{,}800$$

27. Let f be the number of hours Shandra worked at the fast food restaurant.
Let s be the number of hours she worked giving swimming lessons.

$$\begin{array}{c} s=f+5 \\ 5.65f+8.50s=212.30 \end{array} \Rightarrow \begin{array}{c} s=f+5 \\ 100\cdot[5.65f+8.50s]=[212.30]\cdot 100 \end{array} \Rightarrow \begin{array}{c} s=f+5 \\ 565f+850s=21{,}230 \end{array}$$

Using substitution, $565f+850(f+5)=21{,}230 \Rightarrow 565f+850f+4{,}250=21{,}230 \Rightarrow$

$1{,}415f+4{,}250=21{,}230 \Rightarrow 1{,}415f=16{,}980 \Rightarrow f=12$

Shandra worked 12 hours at the restaurant and $s=12+5=17$ hours giving swimming lessons.

28. He should charge $5 per box so that supply and demand will both be 23 boxes.

Supply		Demand	
Price	Donuts Supplied	Price	Donuts Demanded
3	17	3	27
8	32	8	17

Supply:

$$\text{slope} = \frac{32-17}{8-3} = \frac{15}{5} = 3$$

$$\begin{aligned} y &= mx+b \\ y &= 3x+b \\ 17 &= 3\cdot 3+b \\ 17 &= 9+b \Rightarrow b = 8 \\ y &= 3x+8 \Rightarrow -3x+y = 8 \end{aligned}$$

Demand:

$$\text{slope} = \frac{17-27}{8-3} = \frac{-10}{5} = -2$$

$$\begin{aligned} y &= mx+b \\ y &= -2x+b \\ 27 &= -2\cdot 3+b \\ 27 &= -6+b \Rightarrow b = 33 \\ y &= -2x+33 \Rightarrow 2x+y = 33 \end{aligned}$$

We must now solve the following system.

$$\begin{aligned} -3x+y &= 8 \quad \text{(supply)} \\ 2x+y &= 33 \quad \text{(demand)} \end{aligned}$$

$$\begin{array}{l} (-1)\cdot[-3x+y] = [8]\cdot(-1) \\ 2x+y = 33 \end{array} \Rightarrow \begin{array}{l} 3x-y = -8 \\ 2x+y = 33 \end{array} \Rightarrow 5x = 25 \Rightarrow x = \frac{25}{5} = 5 \Rightarrow y = -2\cdot 5+33 = 23$$

29. $4\cdot 0 - 3\cdot 0 \overset{?}{\geq} 8 \Rightarrow 0 - 0 \overset{?}{\geq} 8 \Rightarrow 0 \geq 8$; False. Shade the side that does not contain $(0,0)$.

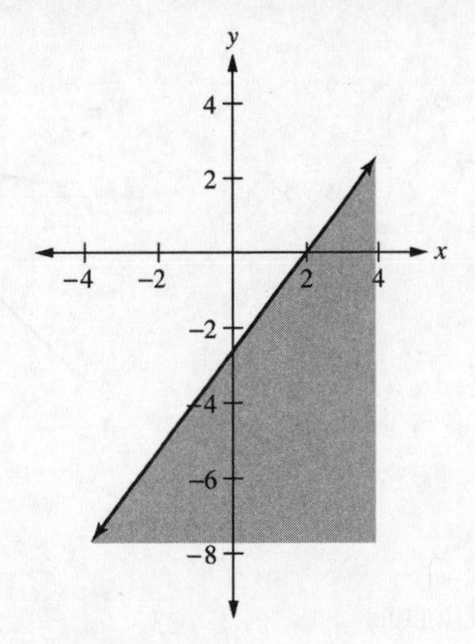

30. Your proportion could be set up differently.

$$\begin{aligned} \frac{x}{90} &= \frac{4}{35} \\ 35\cdot x &= 4\cdot 90 \\ 35x &= 360 \\ x &= \frac{360}{35} \approx 10.3 \end{aligned}$$

Cliff Lee could get 10 or 11 hits.

31. Testing $(0,0)$ in the first inequality yields $0\overset{?}{\leq}-(0)+14 \Rightarrow 0\overset{?}{\leq}14 \Rightarrow 0\leq 14$ which is true, so $(0,0)$ is a solution. Testing $(0,0)$ in the second inequality yields $2\cdot 0-5\cdot 0\overset{?}{\geq}-14 \Rightarrow 0\overset{?}{\geq}-14 \Rightarrow 0\geq 14$ which is true, so $(0,0)$ is a solution. Testing $(0,0)$ in the third inequality yields $5\cdot 0-2\cdot 0\overset{?}{\geq}7 \Rightarrow 0\overset{?}{\geq}7 \Rightarrow 0\geq 7$ which is false, so $(0,0)$ is not a solution. To solve for the corner point, we solve the following systems.

First and second

$$-2\cdot[x+y]=14\cdot -2$$
$$2x-5y=-14$$

$$\begin{array}{r} -2x-2y=-28 \\ 2x-5y=-14 \\ \hline -7y=-42 \\ y=6 \end{array} \qquad \begin{array}{l} x+6=14 \\ x=8 \end{array}$$

Corner point is (6,8)

First and third

$$2\cdot[x+y]=14\cdot 2$$
$$5x-2y=7$$

$$\begin{array}{r} 2x+2y=28 \\ 5x-2y=7 \\ \hline 7x=35 \\ x=5 \end{array} \qquad \begin{array}{l} 5+y=14 \\ y=9 \end{array}$$

Corner point is (5,9)

Second and third

$$-2\cdot[2x-5y]=-14\cdot -2$$
$$5\cdot[5x-2y]=7\cdot 5$$

$$\begin{array}{r} -4x+10y=28 \\ 25x-10y=35 \\ \hline 21x+0=63 \\ x=3 \\ 5\cdot 3-2y=7 \\ 15-2y=7 \\ -2y=-8 \\ y=4 \end{array}$$

Corner point is (3,4)

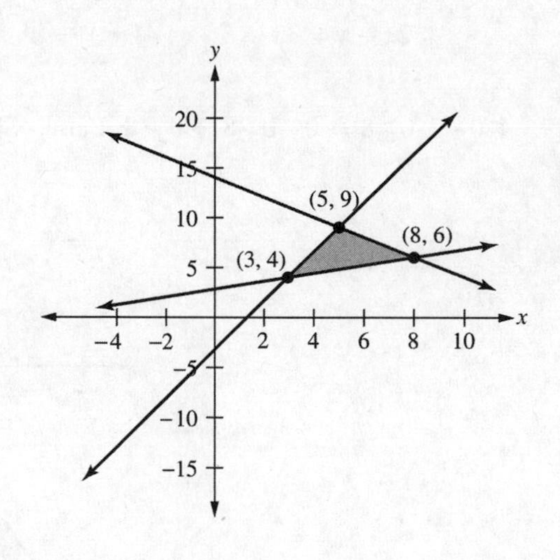

32. The strength of the beam would double.

$$s=k\frac{d^2}{l}$$

$$s_{new}=k\frac{(2d)^2}{(2l)}=k\frac{4d^2}{2l}=k\frac{2d^2}{l}=2\left(k\frac{d^2}{l}\right)$$

33.

$$y=\frac{k}{x}$$
$$8=\frac{k}{2/3}$$
$$8\cdot\left(\frac{2}{3}\right)=k$$
$$k=\frac{16}{3}$$

$$y=\frac{16/3}{x}=\frac{16}{3x}$$
$$y=\frac{16}{3(4)}$$
$$y=\frac{16}{12}=\frac{4}{3}$$

34.

$$s = \frac{kxy}{t}$$
$$16 = \frac{k(2)(4)}{(6)}$$
$$16 = \frac{8 \cdot k}{6}$$
$$k = \frac{16 \cdot 6}{8} = 12$$

$$s = \frac{12xy}{t}$$
$$s = \frac{12(3)(5)}{(9)} = 20$$

35. Let s be the number of small boxes.
Let l be the number of large boxes.

Since the number of boxes cannot be negative, we will consider restrictions in the first quadrant given by the following.

$2s + 6l \leq 90$ (Amount of wood)
$s + 2l \leq 40$ (Amount of time)

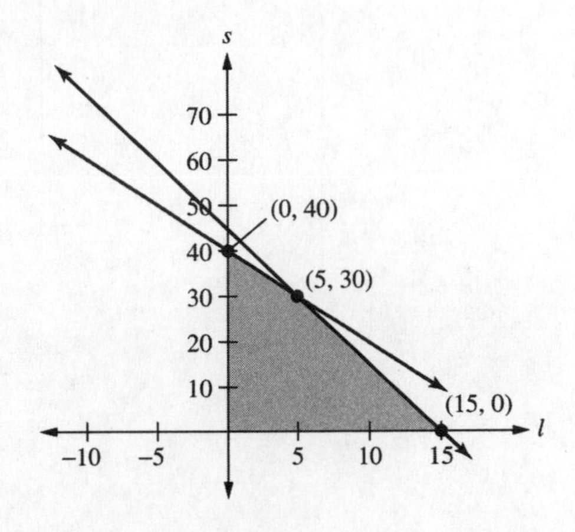

The points, $(40,0)$ and $(0,15)$, are corner points. To find the third corner point we solve the following system.

$$2s + 6l = 90$$
$$-2 \cdot [s + 2l] = 40 \cdot -2$$

$$2s + 6l = 90$$
$$\underline{-2s - 4l = -80}$$
$$2l = 10$$
$$l = 5$$

$$s + 2 \cdot 5 = 40$$
$$s + 10 = 40$$
$$s = 30$$

The third corner point is $(5,30)$.

Chapter 8

Consumer Mathematics: The Mathematics of Everyday Life

Section 8.1 Percents, Taxes, and Inflation

1. 0.78

3. 0.08

5. 0.2735

7. 0.0035

9. 43%

11. 36.5%

13. 145%

15. 0.2%

17. $\frac{3}{4} = 0.75 = 75\%$

19. $\frac{5}{16} = 0.3125 = 31.25\%$

21. $\frac{5}{2} = 2.5 = 250\%$

23. $percent \times 80 = 12$

    $percent = \frac{12}{80} = 0.15 = 15\%$

25. $0.22 \times base = 77$

    $base = \frac{77}{0.22} = 350$

27. $0.1225 \times 160 = amount$

    $19.6 = amount$

29. $percent \times 48 = 8.4$

    $percent = \frac{8.4}{48} = 0.175 = 17.5\%$

31. $0.2325 \times base = 29.76$

    $base = \frac{29.76}{0.2325} = 128$

33. $\frac{294.6}{3{,}124} \approx 0.094 = 9.4\%$

35. The average price of a new home in the United States in 2007 represents 100% of the 2000 price plus the 51.5% increase or $100\% + 51.5\% = 151.5\%$.

$$1.515 \times base = 314{,}000 \Rightarrow base = \frac{314{,}000}{1.515} \approx \$207{,}000$$

The average price for a new home in the United States in 2000 was approximately $207,000.

37. The number of radio stations in 2010 represents 100% of the number of stations in 2001 plus the 40.4% increase or $100\% + 40.4\% = 140.4\%$.

$$1.404 \times 314{,}000 = amount \Rightarrow amount \approx 806$$

The number of stations in 2010 in the United States in 2010 was approximately 806.

39. $percent\ increase = \dfrac{new\ amount - base\ amount}{base\ amount}$

$percent\ increase = \dfrac{11{,}818{,}500 - 37{,}000}{37{,}000} = \dfrac{11{,}781{,}500}{37{,}000}$

$percent\ increase \approx 318.42 = 31{,}842\%$

41. $\dfrac{900}{700+900+450+350+500} = \dfrac{900}{2900} \approx 0.31 = 31\%$

43. $\dfrac{500-350}{350} = \dfrac{150}{350} \approx 0.429 = 42.9\%$

45. The 81 authentic autographs represent 6% of all autographs in circulation.

$$0.06 \times base = 81 \Rightarrow base = \frac{81}{0.06} \approx 1{,}350$$

There are $1{,}350 - 81 = 1{,}269$ non-authentic Beatles autographs in circulation.

47. $\dfrac{307-281}{281} = \dfrac{26}{281} \approx 0.093 = 9.3\%$

49. $5.62 \times 13.3 = amount \Rightarrow amount \approx 76.4$ million music albums

51. a) $\dfrac{21{,}065 - 19{,}875}{19{,}875} = \dfrac{1{,}190}{19{,}875} = 0.06 = 6\%$

b) She divided by 21,065 instead of 19,875.

$\dfrac{21{,}065 - 19{,}875}{21{,}065} = \dfrac{1{,}190}{21{,}065} = 0.056 = 5.6\%$

53. $\dfrac{12{,}711 - 11{,}400}{11{,}400} = \dfrac{1{,}311}{11{,}400} = 0.115 = 11.5\%$

55. $\dfrac{150}{131+118+96+150+118} = \dfrac{150}{613} \approx 0.245 = 24.5\%$

57. $\dfrac{118-96}{96} = \dfrac{22}{96} \approx 0.229 = 22.9\%$

59. Amount of increase is $0.35 \times 28{,}000 = \$9{,}800$. Her new salary will be $28{,}000 + 9{,}800 = \$37{,}800$.

61. The sale price represents $100\% - 15\% = 85\%$ of the original price. $0.85 \times Base = \$578$. The original price was $Base = \dfrac{\$578}{0.85} = \$680$.

63. The amount that the fund is worth this quarter represents 100% of the amount the fund was worth last quarter minus the 12% decrease or $100\% - 12\% = 88\%$.

$$0.88 \times base = 11{,}264 \Rightarrow base = \frac{11{,}264}{0.88} = \$12{,}800$$

The fund was worth \$12,800 last quarter.

65. Using line 4, the Federal Income tax is \$35,771.50

$15{,}107.50+0.28(148{,}000-74{,}200)=15{,}107.50+0.28(73{,}800)=$

$15{,}107.50+20{,}664=35{,}771.5$

\$35,771.50 is owed.

67. Using line 3, the Federal Income tax is \$8,507.50

$4{,}220+0.25(47{,}800-30{,}650)=4{,}220+0.25(17{,}150)=$

$4{,}220+4{,}287.5=8{,}507.5$

\$8,507.50 is owed.

69. $\dfrac{196.6-164.4}{164.4}=\dfrac{32.2}{164.4}\approx 0.196=19.6\%$

71. $base(1+0.30)=218.2$

$base=\dfrac{218.2}{1.3}\approx 167.8$

73. – 75. No solution provided.

77. For each \$1,000 you will make \$20.10 more over three years with 3%, 1%, then 2%. Note that you will have the same final salary with either option.

With 3%, 1%, then 2% raises, in the three years, for each \$1,000 you will earn $1000\times 1.03+1000\times 1.03\times 1.01+1000\times 1.03\times 1.01\times 1.02=\$3{,}131.41$.

With 3%, 1%, then 2% raises, in the three years, for each \$1,000 you will earn $1000\times 1.02+1000\times 1.02\times 1.01+1000\times 1.02\times 1.01\times 1.03=3{,}111.31$.

The difference would be $3131.41-3111.31=\$20.10$ per \$1,000.

79. We must calculate the value for the 1st, 2nd, 3rd, then 4th year.

Value at the start of year		Depreciation	Value at the end of year
1	\$18,000	$0.12\times 18{,}000=\$2{,}160$	$18{,}000-2{,}160=\$15{,}840$
2	\$15,840	$0.12\times 15{,}840=\$1{,}900.80$	$15{,}840-1{,}900.80=\$13{,}939.20$
3	\$13,939.20	$0.12\times 13{,}939.20\approx \$1{,}672.70$	$13{,}939.20-1{,}672.70=\$12{,}266.50$
4	\$12,266.50	$0.12\times 12{,}266.50=\$1{,}471.98$	$12{,}266.50-1{,}471.98=\$10{,}794.52$

81. No; The price after the reduction will be less than the original price. Examples will vary.

Consider an original price of \$1,000. In general, after an increase of $x\%$, we have a new price of

$$1{,}000+0.01\cdot x\times 1{,}000=1{,}000+10x.$$

After a decrease of $x\%$, we have a new price of

$$(1{,}000+10x)-0.01\cdot x\times(1{,}000+10x)=$$
$$1{,}000+10x-10x-0.1\cdot x^2=$$
$$1{,}000-0.1x^2=$$
$$1{,}000-0.0001\cdot x^2\times 1{,}000$$

or

The final price is $(0.01\cdot x^2)\%$ less than the original price.

83. 8.3% increase

Compare the price per ounce: $\dfrac{\frac{3.89}{48}-\frac{4.79}{64}}{\frac{4.79}{64}}=\dfrac{0.08104-0.07484}{0.07484}\approx 0.083=8.3\%$

Section 8.2 Interest

1. $I=P\times r\times t$
$I=1,000\times 0.08\times 3$
$I=\$240$

3. $I=P\times r\times t$
$700=3,500\times r\times 4$
$700=14,000r$
$0.05=r$
$r=5\%$

5. $A=P(1+rt)$
$A=2,500(1+(0.08)3)$
$A=2,500(1+0.24)$
$A=2,500(1.24)$
$A=3,100$
$A=\$3,100$

7. $A=P(1+rt)$
$1,770=P(1+(0.06)3)$
$1,770=P(1+0.18)$
$1,770=P(1.18)$
$1,500=P$
$P=\$1,500$

9. $A=P(1+rt)$
$1,400=1,250(1+(r)2)$
$1,400=1,250(1+2r)$
$1,400=1,250+2500r$
$150=2500r$
$0.06=r$
$r=6\%$

11. $A=P\left(1+\dfrac{r}{m}\right)^n$
$A=5,000\left(1+\dfrac{0.05}{1}\right)^{1\cdot 5}$
$A=5,000(1+0.05)^5$
$A=5,000(1.05)^5$
$A\approx 6,381.41$
$A\approx \$6,381.41$

13. $A=P\left(1+\dfrac{r}{m}\right)^n$
$A=4,000\left(1+\dfrac{0.08}{4}\right)^{4\cdot 2}$
$A=4,000(1+0.02)^8$
$A=4,000(1.02)^8$
$A\approx 4,686.64$
$A\approx \$4,686.64$

15. $A=P\left(1+\dfrac{r}{m}\right)^n$
$A=20,000\left(1+\dfrac{0.08}{12}\right)^{12\cdot 2}$
$A=20,000\left(1+\dfrac{0.08}{12}\right)^{24}$
$A\approx 23,457.76$
$A\approx \$23,457.76$

17. $A = P\left(1+\frac{r}{m}\right)^{n}$

$A = 4{,}000\left(1+\frac{0.10}{365}\right)^{365\cdot 2}$

$A = 4{,}000\left(1+\frac{0.10}{365}\right)^{730}$

$A \approx 4{,}885.48$

$A \approx \$4{,}885.48$

19. $\left(1+\frac{0.075}{12}\right)^{12} - 1 \approx 0.0776 = 7.76\%$

21. $\left(1+\frac{0.06}{4}\right)^{4} - 1 \approx 0.0614 = 6.14\%$

23. You must compare $\left(1+\frac{r}{m}\right)^{n}$ for each case where m is the number of times the money compounds in a single year.

5% compounded yearly:

$\left(1+\frac{r}{m}\right)^{m} = \left(1+\frac{0.05}{1}\right)^{1}$

$= 1 + 0.05$

$= 1.05$

4.95% compounded quarterly:

$\left(1+\frac{r}{m}\right)^{m} = \left(1+\frac{0.0495}{4}\right)^{4} \approx 1.05043$

4.95% compounded quarterly is better.

25. $3^x = 10$

$\log 3^x = \log 10$

$x \log 3 = \log 10$

$x = \log 10/\log 3$

$x \approx 2.096$

27. $(1.05)^x = 2$

$\log(1.05)^x = \log 2$

$x \log(1.05) = \log 2$

$x = \log 2/\log(1.05)$

$x \approx 14.207$

29. a) $2 = 1.035^n$

$\log 2 = \log 1.035^n$

$\log 2 = n \cdot \log 1.035$

$\frac{\log 2}{\log 1.035} = n$

$n \approx \frac{0.30103}{0.01494} \approx 20.15$ years

b) $\frac{70}{3.5} = 20$ years

31. a) $2 = 1.03^n$

$\log 2 = \log 1.03^n$

$\log 2 = n \cdot \log 1.03$

$\frac{\log 2}{\log 1.03} = n$

$n \approx \frac{0.30103}{0.01284} \approx 23.45$ years

b) $\frac{70}{3} = 23.3$ years

33. $A = P(1+r)^t$

$2500 = 2000(1+r)^5$

$1.25 = (1+r)^5$

$(1.25)^{1/5} = \left((1+r)^5\right)^{1/5}$

$(1.25)^{1/5} = 1 + r$

$(1.25)^{1/5} - 1 = r$

$r \approx 0.0456 = 4.56\%$

35.
$$\begin{aligned} A &= P(1+r)^t \\ 1{,}500 &= 1{,}000(1+0.04)^t \\ 1.5 &= (1.04)^t \\ \log(1.5) &= \log(1.04)^t \\ \log(1.5) &= t\cdot\log(1.04) \\ \log(1.5)/\log(1.04) &= t \\ t &\approx 10.34 \end{aligned}$$

37. a) $36\cdot 136 = \$4{,}896$

b) $4{,}896 - 3{,}600 = \$1{,}296$

39. $I = P\times r\times t$

$I = 10{,}000\times 0.08\times\frac{1}{12}$

$I = \$66.67$

41.
$$\begin{aligned} I &= P\times r\times t \\ 300 &= 1{,}200\times r\times 2 \\ 300 &= 2{,}400r \\ 0.125 &= r \\ r &= 12.5\% \end{aligned}$$

43.
$$\begin{aligned} I &= P\times r\times t \\ 425-400 &= 400\times r\times\frac{1}{12} \\ 25 &= 400\times r\times\frac{1}{12} \\ 25 &= \frac{100}{3}r \\ 0.75 &= r \\ r &= 75\% \end{aligned}$$

45.
$$\begin{aligned} A &= P\left(1+\frac{r}{m}\right)^n \\ 30{,}000 &= P\left(1+\frac{0.06}{4}\right)^{4\cdot 15} \\ 30{,}000 &= P(1+0.015)^{60} \\ 30{,}000 &= P(1.015)^{60} \\ \frac{30{,}000}{1.015^{60}} &= P \\ P &\approx \$12{,}278.88 \end{aligned}$$

47. a) $\frac{218.1-38.8}{38.8}\approx 4.621 = 462.1\%$

b) The price of the shoes will increase by $100+462.1 = 562.1\%$, so the shoes will cost $(33)(5.621) = \$185.50$ in 2010.

49. a) $\frac{218.1-82.4}{82.4}\approx 1.647 = 164.7\%$

b) The price of the car has increased by $100+164.7 = 264.7\%$, so the car will cost $\frac{17{,}000}{2.647} = \$6422.36$ in 1980.

51. $4.65(1+2.26)^5 = 4.65(3.26)^5 \approx \$1{,}712.15$

53. a) She earned $2,450-2,375=\$75$ in interest.

$$I=P\times r\times t$$
$$75=2,375\times r\times\frac{8}{12}$$
$$75=\frac{4750}{3}\times r$$
$$\frac{9}{190}=r$$

$r\approx 0.04736842$ or approximately 4.74%

b)
$$A=P\left(1+\frac{r}{m}\right)^n$$
$$2,450=2,375\left(1+\frac{r}{12}\right)^{12\cdot\frac{8}{12}}$$
$$\frac{98}{95}=\left(1+\frac{r}{12}\right)^8$$
$$\left(\frac{98}{95}\right)^{1/8}=\left(\left(1+\frac{r}{12}\right)^8\right)^{1/8}$$
$$\left(\frac{98}{95}\right)^{1/8}=1+\frac{r}{12}$$
$$\left(\frac{98}{95}\right)^{1/8}-1=\frac{r}{12}$$

$r\approx 4.67\%$

55.
$$A=P\left(1+\frac{r}{m}\right)^n$$
$$20,000=9,420\left(1+\frac{0.075}{12}\right)^{12t}$$
$$20,000=9,420(1+0.00625)^{12t}$$
$$\frac{1,000}{471}=(1.00625)^{12t}$$
$$\log\left(\frac{1,000}{471}\right)=\log(1.00625)^{12t}$$
$$\log\left(\frac{1,000}{471}\right)=12t\cdot\log(1.00625)$$
$$t\approx\frac{\log\left(\frac{1,000}{471}\right)/\log(1.00625)}{12}$$

$t\approx 10.07$ years

57.
$$A=P\left(1+\frac{r}{m}\right)^{nt}$$
$$1,200,000=P\left(1+\frac{0.06}{365}\right)^{365(35)}$$
$$P=\frac{1,200,000}{\left(1+\frac{0.06}{365}\right)^{12775}}=\$146,973.08$$

59.
$$A=P\left(1+\frac{r}{m}\right)^{nt}$$
$$A=24\left(1+\frac{0.05}{1}\right)^{1(389)}$$
$$A=24(1.05)^{389}\approx\$4,196,126,573$$

61. 68 cents;
$$A=P(1+r)^t$$
$$1=P(1.04)^{10}$$
$$P=\frac{1}{(1.04)^{10}}\approx\$0.68$$

63. Simple interest only pays interest on the principle. Compound interest pays interest on the principle and previously earned interest.

65. when $m=1$

67. between 1% and 3% of \$350, or between \$3.50 and \$10.50

69. $\left(1+\frac{0.10}{100,000}\right)^{100,000}-1\approx 0.105170863=10.5170863\%$

71. $e^r-1\approx 2.718281828^{0.10}-1\approx 0.105170918=10.5170918\%$

73. $\dfrac{15.2\times10^{12}}{312\times10^{6}} = \$48,717.95$

75. There are $366\times24\times60 = 527,040$ minutes in a leap year, so the debt will be increasing at

$\dfrac{\$7.6\times10^{11}}{527,040min} = \$1,442,015.79$ per minute.

Section 8.3 Consumer Loans

1. $I = Prt = 900(0.12)(2) = 216$ Monthly Payment $= \dfrac{P+I}{n} = \dfrac{900+216}{2\cdot12} = \dfrac{1,116}{24} = \$46.50$

3. $I = Prt = 1,360(0.08)(4) = 435.20$

Monthly Payment $= \dfrac{P+I}{n} = \dfrac{1,360+435.20}{4\cdot12} = \dfrac{1,795.20}{48} = \$37.40$

5. $I = Prt = 1,280(0.095)(2) = \$243.20$

Monthly Payment $= \dfrac{P+I}{n} = \dfrac{1,280+243.20}{2\cdot12} = \dfrac{1,523.20}{24} \approx \$63.47$

7. $I = Prt = 6,480(0.1165)(4) = \$3,019.68$

Monthly Payment $= \dfrac{P+I}{n} = \dfrac{6,480+3,019.68}{4\cdot12} = \dfrac{9,499.68}{48} = \$197.91$

9. $I = Prt = 900(0.12)(2) = 216$ Monthly Payment $= \dfrac{P+I}{n} = \dfrac{900+216}{2\cdot12} = \dfrac{1,116}{24} = \$46.50$

$\dfrac{900}{24} = \$37.50$ of the loan amount is paid each month 46.50-37.50 $= \$9$ in interest each month

$\dfrac{9}{37.50} = 0.24 = 24\%$ per month or $12\cdot24\% = 288\%$ annual interest rate

11. $I = Prt = 1360(0.08)(4) = 435.20$ Monthly Payment $= \dfrac{P+I}{n} = \dfrac{1,360+435.20}{4\cdot12} = \dfrac{1,795.20}{48} = \$37.40$

$\dfrac{1,360}{48} \approx \$28.33$ of the loan amount is paid each month 37.40-28.33 $= \$9.07$ in interest each month

$\dfrac{9.07}{28.33} \approx 0.32 = 32\%$ per month or $12\cdot32\% = 384\%$ annual interest rate

13. a) $I = Prt = 11,000(0.092)(4) = 4,048$

Monthly Payment $= \dfrac{P+I}{n} = \dfrac{11,000+4048}{4\cdot12} = \dfrac{15,048}{48} = \$313.50$

b) $I = Prt = 9,000(0.092)(4) = 3,312$

Monthly Payment $= \dfrac{P+I}{n} = \dfrac{9,000+3,312}{4\cdot12} = \dfrac{12,312}{48} = \$256.50$

His monthly payments will be reduced by 313.50 – 256.50 = \$57.00

13. (continued)

c) Monthly Payment $= 200 = \dfrac{P+I}{48} \Rightarrow P+I = 9{,}600 = P+P(0.092)(4) = P+0.368P = 1.368P \Rightarrow$
$9{,}600 = 1.368P \Rightarrow P \approx \$7{,}017.54$ This represents the amount financed, so
$11{,}000 - 7{,}017.54 = \$3{,}982.46$ should be the down payment.

15. a) $I = Prt = 15{,}000(0.096)(3) = 4{,}320$

Monthly Payment $= \dfrac{P+I}{n} = \dfrac{15{,}000+4{,}320}{3 \cdot 12} = \dfrac{19{,}320}{36} \approx \$536.67$

b) $I = Prt = 12{,}000(0.096)(3) = 3{,}456$

Monthly Payment $= \dfrac{P+I}{n} = \dfrac{12{,}000+3{,}456}{3 \cdot 12} = \dfrac{15{,}456}{36} \approx \$429.34$

Her monthly payments will be reduced by 536.67 – 429.34=$107.33

c) Monthly Payment $= 300 = \dfrac{P+I}{36} \Rightarrow P+I = 10{,}800 = P+P(0.096)(3) = P+0.288P = 1.288P \Rightarrow$
$10{,}800 = 1.288P \Rightarrow P \approx \$8{,}385.09$ This represents the amount financed, so
$15{,}000 - 8{,}385.09 = \$6{,}614.91$ should be the down payment.

17. $I = Prt = 132{,}000(0.20)(3) = 79{,}200$

Monthly Payment $= \dfrac{P+I}{n} = \dfrac{132{,}000+79{,}200}{3 \cdot 12} = \dfrac{211{,}100}{36} \approx \$5{,}866.67$

19. $I = Prt = 354{,}000(0.12)(3) = 127{,}440$

Monthly Payment $= \dfrac{P+I}{n} = \dfrac{354{,}000+127{,}440}{3 \cdot 12} = \dfrac{481{,}440}{36} \approx \$13{,}373.34$ (round up)

21.

Method	Finance Charge = (Last Month's Balance)*rt*	*P*	*r*	*t*	Finance Charge = *I*=*Prt*
Unpaid Balance	$475 \times (0.18/12) \approx 7.13$	Last Month's Balance + Finance – Payment + New Charges – Returned Charges = 475 + 7.13 – 225 + 180 – 145 = 292.13	18%	$\frac{1}{12}$	$292.13 \times (0.18/12)$ $\approx$ **$4.38**

23.

Method	Finance Charge = (Last Month's Balance)*rt*	*P*	*r*	*t*	Finance Charge = *I*=*Prt*
Unpaid Balance	$640 \times (0.165/12) = 8.80$	Last Month's Balance + Finance – Payment + New Charges = 640 + 8.80 – 320 + 140 + 35 + 75 = 578.80	16.5%	$\frac{1}{12}$	$578.80 \times (0.165/12)$ $\approx$ **$7.96**

25.

Method	Finance Charge = (Last Month's Balance)rt	P	r	t	Finance Charge = $I=Prt$
Unpaid Balance	$460\times(0.188/12)\approx 7.21$	Last Month's Balance + Finance – Payment + New Charges = 460 + 7.21 – 300 + 140 + 135 + 175 = 617.21	18.8%	$\frac{1}{12}$	$617.21\times(0.188/12)$ $\approx$ **\$9.67**

27. a) $I = Prt = 1,200(0.08)(30) = \$2,880$

b) Total cost of computer $= \$1,200 + \$2,880 = \$4,080$

29.

Day	Balance	Number of days × Balance
1,2,3,4	\$280	4 × 280 = 1,120
5,6,7,8,9,10,11,12,13,14	\$205	10 × 205 = 2,050
15,16,17,18,19,20	\$340	6 × 340 = 2,040
21,22,23	\$356	3 × 356 = 1,068
24,25,26,27,28,29,30,31	\$382	8 × 382 = 3,056

Average daily balance $\dfrac{1,120+2,050+2,040+1,068+3,056}{31} = \dfrac{9,334}{31} \approx 301.10$

Finance charge $I = Prt = 301.10(.21)(31/365) \approx \$5.37$

31.

Day	Balance	Number of days × Balance
1,2	\$240	2 × 240 = 480
3,4,5,6,7,8,9,10,11,12	\$375	10 × 375 = 3,750
13,14,15,16,17,18,19,20,21,22	\$225	10 × 225 = 2,250
23,24,25,26,27	\$255	5 × 255 = 1,275
28,29,30	\$283	3 × 283 = 849

Average daily balance $\dfrac{480+3,750+2,250+1,275+849}{30} = \dfrac{8,604}{30} = 286.80$

Finance charge $I = Prt = 286.80(.21)(30/365) \approx \$4.95$

33. Finance charge on unpaid balance $I=Prt = 280\cdot 0.21\cdot(1/12) = \$4.90$

Method	P	r	t	Finance Charge = $I = Prt$
Unpaid Balance	Last month's balance + finance charge + purchases – payment = 280 + 4.90 + 135 + 16 + 26 – 75 = 386.90	21%	$\frac{1}{12}$	$386.90\times(0.21/12)$ $\approx$ **\$6.77**

35. Finance charge on unpaid balance $I = Prt = 240 \cdot 0.21 \cdot (1/12) = \$4.20$

Method	P	r	t	Finance Charge = $I = Prt$
Unpaid Balance	Last month's balance + finance charge + purchases – payment = 240 + 4.20 + 135 + 30 + 28 – 150 = 287.20	21%	$\frac{1}{12}$	$287.20 \times (0.21/12)$ **≈ $5.03**

37. Add-on Interest: $I = Prt = 1{,}000(0.105)\left(\frac{10}{12}\right) = 87.50$

Unpaid Balance Method

Beginning of Month	Finance Charge = (Last Month's Balance)rt	Last Month's Balance + Finance Charge – Payment
2	1,000 × (0.18/12) = 15.00	1,000 + 15.00 – (100 + 15.00) = 900
3	900 × (0.18/12) = 13.50	900 + 13.50 – (100 + 13.50) = 800
4	800 × (0.18/12) = 12.00	800 + 12.00 – (100 + 12.00) = 700
5	700 × (0.18/12) = 10.50	700 + 10.50 – (100 + 10.50) = 600
6	600 × (0.18/12) = 9.00	600 + 9.00 – (100 + 9.00) = 500
7	500 × (0.18/12) = 7.50	500 + 7.50 – (100 + 7.50) = 400
8	400 × (0.18/12) = 6.00	400 + 6.00 – (100 + 6.00) = 300
9	300 × (0.18/12) = 4.50	300 + 4.50 – (100 + 4.50) = 200
10	200 × (0.18/12) = 3.00	200 + 3.00 – (100 + 3.00) = 100
11	100 × (0.18/12) = 1.50	100 + 1.50 – (100 + 1.50) = 0

The total interest paid for the unpaid balance method is $82.50, so the unpaid balance method would accumulate less finance charges.

39. Unpaid Balance Method

Beginning of Month	Finance Charge = (Last Month's Balance)r	Last Month's Balance + Finance Charge
2	1,150 × (0.0175) ≈ 20.13	1,150 + 20.13 = 1,170.13
3	1,170.13 × (0.0175) ≈ 20.48	1,170.13 + 20.48 = 1,190.61
4	1,190.61 × (0.0175) ≈ 20.84	1,190.61 + 20.84 = 1,211.45

The total interest accumulated with the unpaid balance method is $61.45 – or – By using the formula $A = P(1+r)^n$ where P=1,150, r=0.0175, and n=3 we get

$$A = 1{,}150(1+0.0175)^3 = 1{,}150 \cdot 1.0175^3 \approx 1{,}211.44$$
$$1{,}211.44 - 1{,}150 = \$61.44$$

41. The simple interest for the length of the loan is added to the principal and the total is divided by the number of months of the loan to determine the payments.

43. The interest due the first month is $I = Prt = 35{,}000(0.085)(1/12) = \$247.92$. If you make a payment of $800, the amount going to reduce the principle is $\$800 - \$247.92 = \$552.08$. So you will still owe $\$35{,}000 - \$552.08 = \$34{,}447.92$.

45. – 49. No solution provided.

Section 8.4 Annuities

1. $\dfrac{x^8-1}{x-1}$

Your calculations may differ slightly from ours due to rounding at intermediate steps.

3. $\dfrac{r}{m}=\dfrac{0.06}{12}=0.005$; $n=7$; $A=100\left(\dfrac{(1+0.005)^7-1}{0.005}\right)\approx \$710.59$

5. $\dfrac{r}{m}=\dfrac{0.03}{12}$; $n=12\cdot 8=96$; $A=200\left(\dfrac{\left(1+\frac{0.03}{12}\right)^{96}-1}{0.03/12}\right)\approx \$21,669.48$

7. $\dfrac{r}{m}=\dfrac{0.09}{12}$ and $n=12\cdot 4=48$; $A=400\left(\dfrac{\left(1+\frac{0.09}{12}\right)^{48}-1}{0.09/12}\right)\approx \$23,008.28$

9. $\dfrac{r}{m}=\dfrac{0.095}{12}$ and $n=12\cdot 8=96$; $A=600\left(\dfrac{\left(1+\frac{0.095}{12}\right)^{96}-1}{0.095/12}\right)\approx \$85,785.11$

11. $\dfrac{r}{m}=\dfrac{0.08}{4}$ and $n=4\cdot 5=20$; $A=500\left(\dfrac{\left(1+\frac{0.08}{4}\right)^{20}-1}{0.08/4}\right)\approx \$12,148.68$

13. $\dfrac{r}{m}=\dfrac{0.036}{4}$; $n=4\cdot 6=24$; $A=280\left(\dfrac{\left(1+\frac{0.036}{4}\right)^{24}-1}{0.036/4}\right)\approx \$7,463.67$

15. $\dfrac{r}{m}=\dfrac{0.06}{12}=0.005$ and $n=12\cdot 1=12$

$$2,000=R\left(\frac{(1+0.005)^{12}-1}{0.005}\right)$$

$$2,000\approx R(12.3355623729)$$

$$R\approx\frac{2,000}{12.3355623729}\approx \$162.14$$

17. $\dfrac{r}{m}=\dfrac{0.075}{12}=0.00625$ and $n=12\cdot 2=24$

$$5,000=R\left(\frac{(1+0.00625)^{24}-1}{0.00625}\right)$$

$$5,000\approx R(25.8067228988)$$

$$R\approx\frac{5,000}{25.8067228988}\approx \$193.75$$

19.
$$3^x = 20$$
$$\log 3^x = \log 20$$
$$x \cdot \log 3 = \log 20$$
$$x = \frac{\log 20}{\log 3} \approx 2.7268$$

21.
$$\frac{8^x + 2}{5} = 12$$
$$8^x + 2 = 60$$
$$8^x = 58$$
$$\log 8^x = \log 58$$
$$x \cdot \log 8 = \log 58$$
$$x = \frac{\log 58}{\log 8} \approx 1.9527$$

23.
$$10,000 = 200\left(\frac{\left(1+\frac{0.09}{12}\right)^n - 1}{0.09/12}\right)$$
$$50 = \left(\frac{(1.0075)^n - 1}{0.0075}\right)$$
$$0.375 = 1.0075^n - 1$$
$$1.375 = 1.0075^n$$
$$\log 1.375 = \log 1.0075^n$$
$$\log 1.375 = n \cdot \log 1.0075$$
$$\frac{\log 1.375}{\log 1.0075} = n$$
$$n \approx 42.62$$

25.
$$5,000 = 150\left(\frac{\left(1+\frac{0.06}{12}\right)^n - 1}{0.06/12}\right)$$
$$\frac{100}{3} = \left(\frac{(1.005)^n - 1}{0.005}\right)$$
$$0.1666667 \approx 1.005^n - 1$$
$$1.1666667 \approx 1.005^n$$
$$\log 1.1666667 \approx \log 1.005^n$$
$$\log 1.1666667 \approx n \cdot \log 1.005$$
$$\frac{\log 1.1666667}{\log 1.005} \approx n$$
$$n \approx 30.91$$

27.
$$6,000 = 250\left(\frac{\left(1+\frac{0.075}{12}\right)^n - 1}{0.075/12}\right)$$
$$24 = \left(\frac{(1.00625)^n - 1}{0.00625}\right)$$
$$0.15 = 1.00625^n - 1$$
$$1.15 = 1.00625^n$$
$$\log 1.15 = \log 1.00625^n$$
$$\log 1.15 = n \cdot \log 1.00625$$
$$\frac{\log 1.15}{\log 1.00625} = n$$
$$n \approx 22.43$$

29. $\frac{r}{m} = \frac{0.065}{12}$ and $n = 30$; $A = 75\left(\frac{\left(1+\frac{0.065}{12}\right)^{30} - 1}{0.065/12}\right) \approx \$2,435.99$

31. $\frac{r}{m}=\frac{0.102}{12}$ and $n=12\cdot 3=36$; $A=150\left(\frac{\left(1+\frac{0.102}{12}\right)^{36}-1}{0.102/12}\right)\approx \$6,286.36$

33. $\frac{r}{m}=\frac{0.065}{12}$ and $n=12\cdot 15=180$; $A=400\left(\frac{\left(1+\frac{0.065}{12}\right)^{180}-1}{0.065/12}\right)\approx \$121,417.91$

35. $1,000,000=200\frac{\left(1+\frac{0.035}{12}\right)^{n}-1}{\frac{0.035}{12}}$

$2917=200\left[(1.002917)^{n}-1\right]$

$15.585=(1+0.002917)^{n}$

$\log(15.585)=n\log(1.002917)$

$n=\frac{\log(15.585)}{\log(1.002917)}\approx 942.86$ months

It will take $\frac{942.86}{12}\approx 79$ years.

37. $1,000,000=800\frac{\left(1+\frac{0.02}{12}\right)^{n}-1}{\frac{0.02}{12}}$

$1670=800\left[(1.00167)^{n}-1\right]$

$3.0875=(1.00167)^{n}$

$\log(3.0875)=n\log(1.00167)$

$n=\frac{\log(3.0875)}{\log(1.00167)}=675.63$ months

It will take $\frac{675.63}{12}\approx 56$ years.

39. $n=12\cdot 8=96$

$14,000=R\left(\frac{(1+0.007)^{96}-1}{0.007}\right)$

$14,000\approx R(136.224411498)$

$R\approx\frac{14,000}{136.2244114988}\approx \$102.78$

41. $\frac{r}{m}=\frac{0.082}{12}$ and $n=6$

$600=R\left(\frac{\left(1+\frac{0.082}{12}\right)^{6}-1}{0.082/12}\right)$

$600\approx R(6.10343868814)$

$R\approx\frac{600}{6.10343868814}\approx \$98.31$

43. a)

tax-deferred plan

$$\frac{r}{m} = \frac{0.06}{12} = 0.005,\ n = 12 \cdot 30 = 360;$$

$$R = 300$$

$$A = 300\left(\frac{(1.005)^{360} - 1}{0.005}\right) \approx \$301,354.51$$

non-deferred plan

$$\frac{r}{m} = \frac{0.06}{12} = 0.005,\ n = 12 \cdot 30 = 360;$$

$$R = 300 \cdot 0.75 = 225$$

$$A = 225\left(\frac{(1.005)^{360} - 1}{0.005}\right) \approx \$226,015.88$$

b)

$$\begin{aligned} I &= 301,354.51 - 300 \cdot 360 \\ &= 301,354.51 - 108,000 \\ &= \$193,354.51 \end{aligned}$$

$$\begin{aligned} I &= 226,015.88 - 225 \cdot 360 \\ &= 226,015.88 - 81,000 \\ &= \$145,015.88 \end{aligned}$$

c)

When you withdraw your money, you will have $301,354.51(0.82) = \$247,110.70$.

When you withdraw your money, you will have $81,000 + 145,015.88(0.82) =$ \$199,913.02.

The tax-deferred plan earns \$47,197.68 more.

45. a)

tax-deferred plan

$$\frac{r}{m} = \frac{0.04}{12},\ n = 12 \cdot 20 = 240;$$

$$R = 400$$

$$A = 400\left(\frac{\left(1 + \frac{0.04}{12}\right)^{240} - 1}{0.04/12}\right) \approx \$146,709.85$$

non-deferred plan

$$\frac{r}{m} = \frac{0.04}{12},\ n = 12 \cdot 20 = 240;$$

$$R = 400 \cdot 0.70 = 280$$

$$A = 280\left(\frac{\left(1 + \frac{0.04}{12}\right)^{240} - 1}{0.04/12}\right) \approx \$102,696.90$$

b)

$$\begin{aligned} I &= 146,709.85 - 400 \cdot 240 \\ &= 146,709.85 - 96,000 \\ &= \$50,709.85 \end{aligned}$$

$$\begin{aligned} I &= 102,696.90 - 280 \cdot 240 \\ &= 102,696.90 - 67,200 \\ &= \$35,496.90 \end{aligned}$$

c)

When you withdraw your money, you will have $146,709.85(0.70) = \$102,696.90$.

When you withdraw your money, you will have $67,200 + 35,496.90(0.70) =$ \$92,047.83.

The tax-deferred plan earns \$10,649.07 more.

47. a)

tax-deferred plan

$\frac{r}{m}=\frac{0.034}{12}$, $n=12\cdot 35=420$;

$R=500$

$$A=500\left(\frac{\left(1+\frac{0.034}{12}\right)^{420}-1}{0.034/12}\right)\approx \$402,627.32$$

non-deferred plan

$\frac{r}{m}=\frac{0.034}{12}$, $n=12\cdot 35=420$;

$R=500\cdot 0.75=375$

$$A=375\left(\frac{\left(1+\frac{0.034}{12}\right)^{420}-1}{0.034/12}\right)\approx \$301,970.49$$

b)

$$\begin{aligned}I&=402,627.32-500\cdot 420\\&=402,627.32-210,000\\&=\$192,627.32\end{aligned}$$

$$\begin{aligned}I&=301,970.49-375\cdot 420\\&=301,970.49-157,500\\&=\$144,470.49\end{aligned}$$

c)

When you withdraw your money, you will have $402,627.32(0.70)=\$281,839.12$.

When you withdraw your money, you will have $157,500+144,470.49(0.70)=$ \$258,629.34.

The tax-deferred plan earns \$23,209.78 more.

49.

$$400,000=5,000\left(\frac{\left(1+\frac{0.108}{12}\right)^{n}-1}{0.108/12}\right)$$

$$80=\left(\frac{(1+0.009)^{n}-1}{0.009}\right)$$

$$0.72=1.009^{n}-1$$

$$1.72=1.009^{n}$$

$$\log 1.72=\log 1.009^{n}$$

$$\log 1.72=n\cdot\log 1.009$$

$$\frac{\log 1.72}{\log 1.009}=n$$

$$n\approx 61 \text{ months}$$

51.

$$30,000=550\left(\frac{\left(1+\frac{0.078}{12}\right)^{n}-1}{0.078/12}\right)$$

$$\frac{600}{11}=\left(\frac{(1+0.0065)^{n}-1}{0.0065}\right)$$

$$0.3545455\approx 1.0065^{n}-1$$

$$1.3545455\approx 1.0065^{n}$$

$$\log 1.3545455\approx\log 1.0065^{n}$$

$$\log 1.3545455\approx n\cdot\log 1.0065$$

$$\frac{\log 1.3545455}{\log 1.0065}\approx n$$

$$n\approx 47 \text{ months}$$

53. Julio

$\frac{r}{m} = \frac{0.06}{1} = 0.06$ and $n = 15$

$$A = 1000\left(\frac{(1+0.06)^{15} - 1}{0.06}\right) \approx \$23,275.97$$

Max

$\frac{r}{m} = \frac{0.06}{1} = 0.06$ and $n = 25$

$$A = 2000\left(\frac{(1+0.06)^{25} - 1}{0.06}\right) \approx \$109,729.02$$

$$A = P\left(1 + \frac{r}{m}\right)^n$$

$$A = 23,275.97\left(1 + \frac{0.06}{1}\right)^{30}$$

$$A = 23,275.97(1.06)^{30}$$

$$A \approx 133,685.32$$

55. $\frac{r}{m} = \frac{0.06}{1} = 0.06$ and $n = 25$

$$133,685.32 = R\left(\frac{(1+0.06)^{25} - 1}{0.06}\right)$$

$$133,685.32 \approx R(54.864512)$$

$$R \approx \frac{133,685.32}{54.864512} \approx \$2436.65$$

57. To find the future value we solve for A; to find the payments we solve for R.

59. Answers will vary.

61. Think of this as making investments rather than payments. This is the same problem as investing \$10,000 for 5 years or making ordinary annuity payments for 5 years. At 8%, you pay less by making payments. At 3%, paying cash is better.

At 3%:

Cash payment

$$A = 10000(1.03)^5 \approx \$11,592.74$$

Monthly payment

$$A = 2500\frac{(1.03)^5 - 1}{0.03} \approx \$13,272.84$$

At 8%:

Cash payment

$$A = 10000(1.08)^5 \approx \$14,693.28$$

Monthly payment

$$A = 2500\frac{(1.08)^5 - 1}{0.08} \approx \$14,666.50$$

63. a) Think of an annuity due as an ordinary annuity that earns one more month of interest. So if A is the future value of an ordinary annuity, $A\left(1+\frac{r}{m}\right)$ is the value of the corresponding annuity due.

b) The annuity from Example 2 was worth \$1,966.81. The annuity due would be worth $1966.81\left(1+\frac{0.06}{12}\right)=\$1,976.64$

Section 8.5 Amortization

1. $\frac{r}{m}=\frac{0.10}{12}$ and $n=12\cdot 4=48$

a) $A=5000\left(1+\frac{0.10}{12}\right)^{48}\approx 7446.77049$

b) $B=\left(\frac{\left(1+\frac{0.10}{12}\right)^{48}-1}{\frac{0.10}{12}}\right)\approx 58.72249$

c) $7446.77049=58.72249R$
$R\approx \$126.81$

3. $\frac{r}{m}=\frac{0.08}{12}$ and $n=12\cdot 10=120$

a) $A=8000\left(1+\frac{0.08}{12}\right)^{120}\approx 17757.12188$

b) $B=\left(\frac{\left(1+\frac{0.08}{12}\right)^{120}-1}{\frac{0.08}{12}}\right)\approx 182.94604$

c) $17757.12188=182.94604R$
$R\approx \$97.06$

5. $\frac{r}{m}=\frac{0.06}{12}=0.005$ and $n=12\cdot 30=360$

a) $A=120000(1+0.005)^{360}=722709.02547$

b) $B=\left(\frac{(1+0.005)^{360}-1}{0.005}\right)=1004.515042$

c) $722709.02547=1004.515042R$
$R\approx \$719.46$

7. From Table 8.6, the payment would be \$24.41 per \$1,000, so the payment would be $4\times 24.41=\$97.64$ per month.

9. From Table 8.6, the payment would be \$29.53 per \$1,000, so the payment would be $8.5\times 29.53=251.00$ per month.

11. From Table 8.6, the payment would be \$14.35 per \$1,000, so the payment would be $100\times 14.35=\$1435.00$ per month.

13.

Payment	Interest Paid	Paid on Principal	Balance
\$126.82	\$35.06	\$91.76	\$4,115.81

15.

Payment	Interest Paid	Paid on Principal	Balance
\$246.01	\$30.54	\$215.47	\$4,147.02

17. a) From Table 8.6, the payment would be \$6.00 per \$1,000, so the payment would be $100 \times 6.00 = \$600.00$ per month.

b)

Payment	Interest Paid	Paid on Principal	Balance
			\$100,000.00
\$600.00	\$500.00	\$100.00	\$99,900.00
\$600.00	\$499.50	\$100.50	\$99,799.50
\$600.00	\$499.00	\$101.00	\$99,698.50

c)

Payment	Interest Paid	Paid on Principal	Balance
			\$100,000.00
\$700.00	\$500.00	\$200.00	\$99,800.00
\$700.00	\$499.00	\$201.00	\$99,599.00
\$700.00	\$498.00	\$202.00	\$99,397.00

19. From Table 8.6, the payment would be \$26.33 per \$1,000, so the payment for a $13,500 - 2000 = \$11,500.00$ loan would be $11.5 \times 26.33 \approx \$302.80$ per month.

$48 \cdot 302.80 = \$14,534.40$ and $14,534.40 - 11,500.00 = \$3,034.40$

This represents the approximate amount of interest paid. The actual amount is slightly lower due to the fact that we round our payments up to the next penny and our final payment is slightly less than the other 47 payments.

21. From Table 8.6, the payment would be \$10.12 per \$1,000, so the payment for a $13,500 - 2500 = \$11,000.00$ loan would be $11.0 \times 25.36 \approx \$278.96$ per month.

$48 \cdot 278.96 = \$13,390.08$ and $13,390.08 - 11,000.00 = \$2,390.08$

This represents the approximate amount of interest paid. The actual amount is slightly lower due to the fact that we round our payments up to the next penny and our final payment is slightly less than the other 47 payments.

23. a) 1st year: From Table 8.6, the payment would be \$4.47 per \$1,000, so the payment would be $200 \times 4.77 = \$954$ per month.

b) 3rd year: $r = 0.04 + 0.02 + 0.02 = 0.08$; From Table 8.6, the payment would be \$7.34 per \$1,000, so the payment would be $200 \times 7.34 = \$1468$ per month.

25. a) 1st year: From Table 8.6, the payment would be \$5.37 per \$1,000, so the payment would be $220 \times 5.37 = \$1,181.40$ per month.

b) 3rd year: $r = 0.05 + 0.02 + 0.01 = 0.08$; From Table 8.6, the payment would be \$7.34 per \$1,000, so the payment would be $220 \times 7.34 = \$1,614.80$ per month.

27. $A = 50{,}000\left(\frac{(1+0.10)^{20}-1}{0.10}\right) \approx \$2{,}863{,}749.97$

$$A = P\left(1+\frac{r}{m}\right)^n$$
$$2{,}863{,}749.97 = P(1+0.10)^{20}$$
$$2{,}863{,}749.97 = P(1.10)^{20}$$
$$\frac{2{,}863{,}749.97}{1.10^{20}} = P$$
$$P \approx \$425{,}678.19$$

The present value of the annual payments is slightly better than the lump sum of $425,000.

29. $\frac{r}{m} = \frac{0.108}{12} = 0.009$ and $n = 12 \cdot 4 = 48$

$$P(1+0.009)^{48} = 350\left(\frac{(1+0.009)^{48}-1}{0.009}\right)$$
$$P(1.53736142389) \approx 20{,}897.3887067$$
$$P \approx \frac{20{,}897.3887067}{1.53736142389} \approx \$13{,}593.02$$

31. $\frac{r}{m} = \frac{0.09}{12} = 0.0075$ and $n = 12 \cdot 20 = 240$

$$A = 350\left(\frac{(1+0.0075)^{240}-1}{0.0075}\right) \approx \$233{,}760.40$$
$$A = P\left(1+\frac{r}{n}\right)^{nt}$$
$$233{,}760.40 = P(1+0.0075)^{240}$$
$$233{,}760.40 = P(1.0075)^{240}$$
$$\frac{233{,}760.40}{1.0075^{240}} = P$$
$$P \approx \$38{,}900.73$$

The present value of the annual payments is slightly worse than the lump sum of $40,000.

33. You have 24 months left on the loan. $246.20 \times 24 = \$5{,}908.80$, so you will save $5{,}908.80 - 5{,}416 = \$492.80$ in interest.

35. From Table 8.6, the original payment would be $7.34 per $1,000, so the original payment would be $120 \times 7.34 = \$880.80$ per month. The new payment would be $7.16 per $1,000, so the new payment would be $105.218 \times 7.16 \approx \$753.36$ per month.

    The interest saved would be $240 \times 880.80 - 240 \times 753.36 = \$30{,}585.60$.

37. From Table 8.6, the original payment would be $25.36 per $1,000, so the original payment would be $18 \times 25.36 = \$456.48$ per month. The new payment would be $29.97 per $1,000, so the new payment would be $14.404 \times 29.97 \approx \$431.69$ per month.

    The interest saved would be $36 \times 456.48 - 36 \times 431.69 = \$892.44$.

39. a) From Table 8.6, the payment would be \$9.65 per \$1,000, so the original payment would be $(160-32)\times 8.78=\$1,123.84$ per month and the new payment would be $(160-40)\times 8.78=\$1,053.60,$ per month, so the payment would be reduced by $1,123.84-1,053.60=\$70.24$ per month.

    b) The amount saved would be $360\times 70.24=\$25,286.40$.

41. a) From Table 8.6, the payment would be \$4.77 per \$1,000, so the original payment would be $(240-50)\times 4.77=\$906.30$ per month and the new payment would be $(240-75)\times 4.77=\$787.05$ per month, so the payment would be reduced by $906.30-787.05=\$119.25$ per month.

    b) The amount saved would be $360\times 119.25=\$42,930$.

43. You must pay interest on whatever you borrow now. If you borrow \$12,000, your monthly payments would exceed \$200.

45. If there are fees to refinance that exceed the amount that you will save.

47. The initial lower rate makes the mortgage affordable at first; however, as the rate increases, the payments may become excessive.

49. Option B has the highest total costs.
    Option A: From Table 8.6, the payment would be \$5.37 per \$1,000, so the payment would be $140\times 5.37=\$751.80$ per month.

$$\begin{aligned}\text{total cost} &= \text{points + closing costs + total mortgage payments}\\ &= 4500+0.02\times 140000+360\times 751.80\\ &= \$4,500+\$2,800+\$270,648=\$277,948\end{aligned}$$

    Option B: From Table 8.6, the payment would be \$6.00 per \$1,000, so the payment would be $140\times 6.00=\$840$ per month.

$$\begin{aligned}\text{total cost} &= \text{points + closing costs + total mortgage payments}\\ &= 2500+0.01\times 140000+360\times 840\\ &= \$2,500+\$1400+\$302,400=\$306,300\end{aligned}$$

51. Examples will vary. You payments will be smaller if you refinance earlier.

Section 8.6 Looking Deeper: Annual Percentage Rate

1. $$\begin{aligned}&I=P\times r\times t\\ &I=6,000\times 0.08\times 3\\ &I=\$1,440.00\end{aligned}$$

$$\begin{aligned}&6,000\times a\times 1+4,000\times a\times 1+2,000\times a\times 1=1,440\\ &12,000a=1,440\\ &a=\frac{1,440}{12,000}=0.12=12\%\end{aligned}$$

3. $$\frac{\text{finance charge}}{\text{amount borrowed}}\times 100=\frac{270}{1,800}\times 100=\$15.00$$

5. $\dfrac{\text{finance charge}}{\text{amount borrowed}} \times 100 = \dfrac{260}{2{,}000} \times 100 = \$13.00$

7. $\dfrac{\text{finance charge}}{\text{amount borrowed}} \times 100 = \dfrac{420}{3{,}000} \times 100 = \$14.00$

From Table 8.7, we see that for a 24-payment loan, this corresponds to an APR of roughly 13%.

9. $\dfrac{\text{finance charge}}{\text{amount borrowed}} \times 100 = \dfrac{165}{4{,}000} \times 100 \approx \$4.13$

From Table 8.7, we see that for a 6-payment loan, this corresponds to an APR of roughly 14%.

11. 12.45%

$$\text{APR} \approx \frac{2 \times 36 \times 0.064}{36+1} = \frac{4.608}{37} = 0.1245$$

13. 13.67%

$$\text{APR} \approx \frac{2 \times 42 \times 0.07}{42+1} = \frac{5.88}{43} = 0.1367$$

15. Finance charge = $24 \cdot 485 - 10{,}000 = 11{,}640 - 10{,}000 = 1{,}640$

$\dfrac{\text{finance charge}}{\text{amount borrowed}} \times 100 = \dfrac{1{,}640}{10{,}000} \times 100 = \$16.40$

From Table 8.7, we see that for a 24-payment loan, this corresponds to an APR of roughly 15%.

17. Finance charge = $48 \cdot 116.50 - 4{,}500 = 5{,}592 - 4{,}500 = 1{,}092$

$\dfrac{\text{finance charge}}{\text{amount borrowed}} \times 100 = \dfrac{1{,}092}{4{,}500} \times 100 \approx \$24.27$

From Table 8.7, we see that for a 48-payment loan, this corresponds to an APR of roughly 11%.

19. $I = Prt = 2{,}000 \times 0.08 \times 2 = 320$

$\dfrac{\text{finance charge}}{\text{amount borrowed}} \times 100 = \dfrac{320}{2{,}000} \times 100 = \$16.00$

From Table 8.7, we see that for a 24-payment loan, this corresponds to an APR of roughly 15%.

21. $I = Prt = P \times 0.082 \times 3 = 0.246P$

$\dfrac{\text{finance charge}}{\text{amount borrowed}} \times 100 = \dfrac{0.246P}{P} \times 100 = \$24.60$

From Table 8.7, we see that for a 36-payment loan, this corresponds to an APR of roughly 15%.

23. A is better

A: An amortized loan at 10% for 3 years:

From Table 8.6, the original payment would be \$32.27 per \$1,000, so the original payment would be $5 \times 32.27 = \$161.35$ per month.

Finance charge = $36 \cdot 161.35 - 5000 = 5808.60 - 5000 = 808.60$

$\dfrac{\text{finance charge}}{\text{amount borrowed}} \times 100 = \dfrac{808.60}{5000} \times 100 = \$16.17$

From Table 8.7, we see that this corresponds to an APR of roughly 10%.

23. (continued)

B: An add-on interest loan at 6% for three years:

$I = Prt = 5000 \times 0.06 \times 3 = 900$

$$\frac{\text{finance charge}}{\text{amount borrowed}} \times 100 = \frac{900}{5000} \times 100 = \$18$$

From Table 8.7, we see that for a 36-payment loan, this corresponds to an APR of roughly 11%.

25. B is better

A: An amortized loan at 8% for 4 years:

From Table 8.6, the original payment would be $24.41 per $1,000, so the original payment would be $5 \times 24.41 = \$122.05$ per month.

Finance charge = $48 \cdot 122.05 - 5000 = 5858.40 - 5000 = 858.40$

$$\frac{\text{finance charge}}{\text{amount borrowed}} \times 100 = \frac{858.40}{5000} \times 100 = \$17.17$$

From Table 8.7, we see that this corresponds to an APR of less than 10%.

B: 48 payments of $120:

Finance charge = $48 \cdot 120 - 5000 = 5760 - 5000 = 760$

$$\frac{\text{finance charge}}{\text{amount borrowed}} \times 100 = \frac{760}{5000} \times 100 = \$15.20$$

From Table 8.7, we see that for a 48-payment loan, this also corresponds to an APR less than 10%, but the finance charge per $100 is smaller than option A.

27. 24 payments of $18.75:

Finance charge = $24 \cdot 18.75 - 375 = 450 - 375 = 75$

$$\frac{\text{finance charge}}{\text{amount borrowed}} \times 100 = \frac{75}{375} \times 100 = \$20.00$$

From Table 8.7, we see that for a 24-payment loan, this corresponds to an APR of over 16%.

29. The amortized loan is better because with the add-on loan you are paying interest on the entire amount of the loan for the whole length of the loan.

31. The monthly percentage rate is $\frac{110-100}{100} = 10\%$. The lender would probably claim the APR is $12 \times 10\% = 120\%$. After one year, you would owe $100(1.10)^{12} = \$313.84$, so the APR would be $\frac{313.84}{100} \approx 314\%$.

33. No solution provided.

Chapter Review Exercises

1. 12.45%

2. 0.01365

3. $\frac{11}{16} = 0.6875 = 68.75\%$

4. $percent \times 3,400 = 2,890$

$$percent = \frac{2,890}{3,400} = 0.85 = 85\%$$

5. The \$238.4 million represents 13.2% of the total spent on chocolate candy in 2007.

$$0.132 \times base = 238.4 \Rightarrow base = \frac{238.4}{0.132} \approx 1,806$$

Total amount spent on chocolate candy in 2007 was \$1,806 million.

6. Using line 3, the Federal Income tax is \$10,657.50

$4,220 + 0.25(56,400 - 30,650) = 4,220 + 0.25(25,750) =$

$4,220 + 6,437.5 = 10,657.50$

Maribel owes \$10,657.50

7. $A = P(1+r)^t$

$A = 130(1+0.043)^5$

$A = 130(1.043)^5 \approx \$160.46$

8. $I = P \times r \times t$

$I = 1,500 \times 0.09 \times 2$

$I = 270$

$1,500 + 270 = \$1,770$

9. $I = 400 \cdot 24 - 8,000 = 1,600$

$I = P \times r \times t$

$1,600 = 8,000 \times r \times 2$

$1,600 = 16,000r$

$0.10 = r$

$r = 10\%$

10. $A = P\left(1 + \frac{r}{m}\right)^n$

$A = 11,400\left(1 + \frac{0.18}{12}\right)^{12\left(\frac{1}{2}\right)}$

$A = 11,400(1+0.015)^6$

$A \approx 12,465.25$

$A \approx \$12,465.25$

11. $A = P\left(1 + \frac{r}{m}\right)^n$

$10,000 = P\left(1 + \frac{0.06}{12}\right)^{12 \cdot 10}$

$10,000 = P(1+0.005)^{120}$

$10,000 = P(1.005)^{120}$

$\frac{10,000}{1.005^{120}} = P$

$P \approx \$5,496.33$

12. $A = P\left(1 + \frac{r}{m}\right)^n$

$2,000 = 1,000\left(1 + \frac{0.064}{12}\right)^n$

$2 \approx (1.00533333333)^n$

$\log(2) \approx \log(1.00533333333)^n$

$\log(2) \approx n \cdot \log(1.00533333333)$

$n \approx \frac{\log(2)}{\log(1.00533333333)} \approx 130.311$

$n \approx 131$ months

13. $A = P(1+r)^t$

$1,400 = 1,200(1+r)^5$

$\frac{7}{6} = (1+r)^5$

$\left(\frac{7}{6}\right)^{1/5} = \left((1+r)^5\right)^{1/5}$

$\left(\frac{7}{6}\right)^{1/5} = 1 + r$

$\left(\frac{7}{6}\right)^{1/5} - 1 = r$

$r \approx 0.0313 = 3.13\%$

14. $\frac{70}{5.5} \approx 12.7$ years

15. $I = Prt = 1,320(0.0825)(3) = \$326.70$

$$\text{Monthly Payment} = \frac{P+I}{n} = \frac{1,320+326.70}{3 \cdot 12} = \frac{1,646.70}{36} \approx \$45.75$$

16.

Method	Finance Charge = (Last Month's Balance)*rt*	*P*	*r*	*t*	Finance Charge = *I=Prt*
Unpaid Balance	$1,350 \times (0.21/12) \approx 23.63$	Last Month's Balance + Finance – Payment + New Charges – Returned Charges = 1,350 + 23.63 – 375 + 120 – 140 = 978.63	21%	$\frac{1}{12}$	$978.63 \times (0.21/12)$ $\approx$ **$17.13**

17.

Day	Balance	Number of days × Balance
1,2,3,4,5	$275	5 × 275 = 1,375
6,7,8,9,10,11	$200	6 × 200 = 1,200
12,13,14,15,16,17,18	$315	7 × 315 = 2,205
19,20,21,22,23	$335	5 × 335 = 1,675
24,25,26,27,28,29,30,31	$351	8 × 351 = 2,808

$$\text{Average daily balance } \frac{1,375+1,200+2,205+1,675+2,808}{31} = \frac{9,263}{31} \approx 298.81$$

Finance charge $I = Prt = 298.81(0.18)(31/365) \approx \$4.57$

18. $r = \frac{0.0935}{12}$ and $n = 10 \cdot 12 = 120$

$$A = 175\left(\frac{\left(1+\frac{0.0935}{12}\right)^{120} - 1}{0.0935/12}\right) \approx \$34,543.31$$

19. $r = \frac{0.06}{12} = 0.005$

$$2,000 = R\left(\frac{(1+0.005)^{36} - 1}{0.005}\right)$$

$$2,000 \approx R(39.3361049646)$$

$$R \approx \frac{2,000}{39.3361049646} \approx \$50.85$$

20. $\frac{3^x - 4}{2} = 10$

$$3^x - 4 = 20$$
$$3^x = 24$$
$$\log 3^x = \log 24$$
$$x \cdot \log 3 = \log 24$$
$$x = \frac{\log 24}{\log 3} \approx 2.8928$$

21. $\frac{r}{m}=\frac{0.09}{12}=0.0075$

$$10{,}000=300\left(\frac{(1+0.0075)^n-1}{0.0075}\right)$$
$$\frac{100}{3}=\left(\frac{(1.0075)^n-1}{0.0075}\right)$$
$$0.25=1.0075^n-1$$
$$1.25=1.0075^n$$
$$\log 1.25=\log 1.0075^n$$
$$\log 1.25=n\cdot\log 1.0075$$
$$\frac{\log 1.25}{\log 1.0075}=n$$
$$n\approx 30 \text{ months}$$

22. a)

tax-deferred plan

$\frac{r}{m}=\frac{0.042}{12}=0.0035$, $n=12\cdot 30=360$;

$R=350$

$$A=350\left(\frac{(1.0035)^{360}-1}{0.0035}\right)\approx \$251{,}767.45$$

b)

$$I=251{,}767.45-350\cdot 360$$
$$=251{,}767.45-126{,}000$$
$$=\$125{,}767.45$$

c)

When you withdraw your money, you will have $251{,}767.45(0.82)=\$206{,}449.31$.

non-deferred plan

$\frac{r}{m}=\frac{0.042}{12}=0.0035$, $n=12\cdot 30=360$;

$R=350\cdot 0.75=262.50$

$$A=262.5\left(\frac{(1.0035)^{360}-1}{0.0035}\right)\approx \$188{,}825.59$$

$$I=188{,}825.59-262.50\cdot 360$$
$$=188{,}825.59-94{,}500$$
$$=\$94{,}325.59$$

When you withdraw your money, you will have $94{,}500+94{,}325.59(0.82)=$ \$171,846.98

The tax-deferred plan earns \$34,602.33 more than the non-deferred plan.

23. $\frac{r}{n}=\frac{0.10}{12}$ and $n=12\cdot 4=48$

$$5{,}000\left(1+\frac{0.10}{12}\right)^{48}=R\left(\frac{\left(1+\frac{0.10}{12}\right)^{48}-1}{0.10/12}\right)$$

$$7{,}446.77049303\approx R(58.7224918326)$$

$$R\approx\frac{7{,}446.77049303}{58.722491826}\approx \$126.82$$

24. We need to determine the monthly payments where $\frac{r}{m}=\frac{0.08}{12}$ and $n=12\cdot 20=240$.

$$100,000\left(1+\frac{0.08}{12}\right)^{240}=R\left(\frac{\left(1+\frac{0.08}{12}\right)^{240}-1}{0.08/12}\right)$$

$$492,680.277085\approx R(589.020415627)$$

$$R\approx\frac{492,680.277085}{589.020415627}\approx \$836.45$$

Payment Number	Amount of Payment	Interest Payment	Applied to Principal	Balance
				$100,000.00
1	$836.45	$666.67	$169.78	$ 99,830.22
2	$836.45	$665.53	$170.92	$ 99,659.30

25. $A=50,000\left(\frac{(1+0.08)^{20}-1}{0.08}\right)\approx \$2,288,098.21$

$$A=P\left(1+\frac{r}{m}\right)^{n}$$

$$2,288,098.21=P(1+0.08)^{20}$$

$$2,288,098.21\approx P(4.66095714385)$$

$$\frac{2,288,098.21}{4.66095714385}\approx P$$

$$P\approx \$490,907.37$$

The present value is $490,907.37; the lump sum of $500,000 is a better deal.

26. For this exercise, we will use the same method as described in the footnote for Exercises 23 – 26 in Section 8.5.

a) 1st year: $R=\frac{180000\left(1+\frac{0.045}{12}\right)^{360}}{\left(\frac{\left(1+\frac{0.045}{12}\right)^{360}-1}{0.045/12}\right)}=\frac{692585.649}{759.3861467}\approx \$912.04$

b) In year 5: $r=0.125$

$$R=\frac{180000\left(1+\frac{0.125}{12}\right)^{360}}{\left(\frac{\left(1+\frac{0.125}{12}\right)^{360}-1}{0.125/12}\right)}=\frac{7506767.182}{3907.609164}\approx \$1,921.07$$

27. From Table 8.6, the original payment would be \$14.35 per \$1,000, so the original payment would be $235\times 14.35 = \$3{,}372.25$ per month. The new payment would be \$24.41 per \$1,000, so the new payment would be $127.96\times 24.41 \approx \$3{,}123.50$ per month.

    The difference in monthly payments is $\$3{,}372.25 - \$3{,}123.50 = \$248.75$.

28. Finance charge = $12\cdot 163 - 1{,}800 = 1{,}960.80 - 1{,}800 = 160.80$

    $$\frac{\text{finance charge}}{\text{amount borrowed}}\times 100 = \frac{160.80}{1{,}800}\times 100 = \$8.93$$

    From Table 8.7, we see that for a 24-payment loan, this corresponds to an APR of roughly 16%.

29. 15.2%

    $$\text{APR} \approx \frac{2\times 20\times 0.08}{20+1} = \frac{3.2}{21} = 0.152$$

Chapter Test

1. 36.24%

2. 0.2345

3. $\frac{7}{16} = 0.4375 = 43.75\%$

4.

Method	Finance Charge = (Last Month's Balance)*rt*	*P*	*r*	*t*	Finance Charge = *I*=*Prt*
Unpaid Balance	$950\times(0.24/12)\approx 19.00$	Last Month's Balance + Finance – Payment + New Charges = 950 + 19 – 270 + 217 + 23 = 292.13	24%	$\frac{1}{12}$	$939\times(0.24/12)$ ≈ **\$18.78**

5. $I = P\times r\times t$

   $I = 3{,}400\times 0.025\times 3$

   $I = 255$

   $3{,}400 + 255 = \$3{,}655$

6. $percent \times 2{,}840 = 994$

   $$percent = \frac{994}{2{,}840} = 0.35 = 35\%$$

7. The amount of decrease in price is $\$169.99 - \$149.99 = \$20$.

   The percent reduction in price is $\frac{\$20}{\$169.99} \approx 0.1177 = 11.77\%$.

8. Finance charge = $24\cdot 75.00 - 1{,}600 = 1{,}800 - 1{,}600 = 200$

   $$\frac{\text{finance charge}}{\text{amount borrowed}}\times 100 = \frac{200}{1{,}600}\times 100 = \$12.50$$

   From Table 8.7, we see that for a 24-payment loan, this corresponds to an APR of roughly 12%.

9. $I = 162.50 \cdot 24 - 3{,}000 = 900$

$$I = P \times r \times t$$
$$900 = 3{,}000 \times r \times 2$$
$$900 = 6{,}000r$$
$$0.15 = r$$
$$r = 15\%$$

10. $A = P\left(1+\frac{r}{m}\right)^n$

$$A = 4{,}000\left(1+\frac{0.036}{12}\right)^{12\cdot 4}$$
$$A = 4{,}000(1+0.003)^{48}$$
$$A \approx 4{,}618.54$$
$$A \approx \$4{,}618.54$$

11. $A = P\left(1+\frac{r}{n}\right)^n$

$$2{,}000 = 1{,}000\left(1+\frac{0.048}{12}\right)^n$$
$$2 \approx (1.004)^n$$
$$\log(2) \approx \log(1.004)^n$$
$$\log(2) \approx n \cdot \log(1.004)$$
$$n \approx \frac{\log(2)}{\log(1.004)} \approx 173.633$$
$$n \approx 174 \text{ months}$$

12. For this exercise, we will use the same method as described in the footnote for Exercises 23 – 26 in Section 8.5.

a) 1st year: $R = \dfrac{220000\left(1+\frac{0.032}{12}\right)^{360}}{\dfrac{\left(\left(1+\frac{0.032}{12}\right)^{360}-1\right)}{0.032/12}} = \dfrac{573839.5441}{603.1355866} \approx \$951.43$

b) In year 5: $r = 0.032 + 0.015 + 0.015 + 0.015 + 0.015 = 0.92$

$$R = \dfrac{220000\left(1+\frac{0.092}{12}\right)^{360}}{\dfrac{\left(\left(1+\frac{0.092}{12}\right)^{360}-1\right)}{0.092/12}} = \dfrac{3439568.522}{1908.835092} \approx \$1{,}801.93$$

13. $\frac{70}{8.5} \approx 8.235 = 8.3$ years

(Round up to guarantee doubling.)

14. $A = P\left(1+\frac{r}{m}\right)^n$

$$15{,}000 = P\left(1+\frac{0.042}{12}\right)^{12\cdot 18}$$
$$15{,}000 = P(1+0.0035)^{216}$$
$$15{,}000 = P(1.0035)^{216}$$
$$\frac{15{,}000}{1.0035^{216}} = P$$
$$P \approx \$7{,}052.42$$

15. $I = Prt = 1{,}560(0.105)(2) = \$327.60$

$$\text{Monthly Payment} = \frac{P+I}{n} = \frac{1{,}560 + 327.60}{2\cdot 12} = \frac{1{,}887.60}{24} = \$78.65$$

16.
$$A = P(1+r)^t$$
$$2{,}400 = 2{,}100(1+r)^3$$
$$\frac{8}{7} = (1+r)^3$$
$$\left(\frac{8}{7}\right)^{1/3} = \left((1+r)^3\right)^{1/3}$$
$$\left(\frac{8}{7}\right)^{1/3} = 1+r$$
$$r = \left(\frac{8}{7}\right)^{1/3} - 1 \approx 0.0455 = 4.55\%$$

17. $\dfrac{257-250}{250} = \dfrac{7}{250} = 0.028 = 2.8\%$

18.

Day	Balance	Number of days × Balance
1,2,3	$425	3 × 425 = 1,275
4,5,6,7,8,9	$340	6 × 340 = 2,040
10,11,12,13,14	$365	5 × 365 = 1,825
15,16,17,18,19,20,21,22,23,24	$380	10 × 380 = 3,800
25,26,27,28,29,30	$460	6 × 460 = 2,760

$$\text{Average daily balance } \frac{1{,}275 + 2{,}040 + 1{,}825 + 3{,}800 + 2{,}760}{30} = \frac{11{,}700}{30} = 390.00$$

Finance charge $I = Prt = 390.00(0.21)(30/365) \approx \$6.74$

19. $r = \dfrac{0.0515}{12}$ and $n = 8\cdot 12 = 96$

$$A = 200\left(\frac{\left(1+\frac{0.0515}{12}\right)^{96} - 1}{0.0515/12}\right) \approx \$23{,}697.27$$

20.
$$\frac{5^x - 4}{3} = 10$$
$$5^x - 4 = 30$$
$$5^x = 34$$
$$\log 5^x = \log 34$$
$$x\cdot \log 5 = \log 34$$
$$x = \frac{\log 34}{\log 5} \approx 2.1911$$

21. 15.2%

$$\text{APR} \approx \frac{2\times 20\times 0.08}{20+1} = \frac{3.2}{21} = 0.152$$

22. $r = \dfrac{0.04}{12} = 0.0033333333$

$$1{,}800 = R\left(\frac{(1+0.0033333333)^{36} - 1}{0.0033333333}\right)$$
$$1{,}800 \approx R(38.181562)$$
$$R \approx \frac{1{,}800}{38.181562} \approx \$47.15$$

23. $\frac{r}{m}=\frac{0.09}{12}=0.0075$ and $n=12\cdot 8=96$

$$20{,}000(1+0.0075)^{96}=R\left(\frac{(1+0.0075)^{96}-1}{0.0075}\right)$$

$$40{,}978.4245646\approx R(139.85613674)$$

$$R\approx\frac{40{,}978.4245646}{139.85613674}\approx \$293.01$$

24. $\frac{r}{m}=\frac{0.0375}{12}=0.003125$

$$9{,}000=450\left(\frac{(1+0.003125)^{n}-1}{0.003125}\right)$$

$$20=\left(\frac{(1.003125)^{n}-1}{0.003125}\right)$$

$$0.0625=1.003125^{n}-1$$

$$1.0625=1.003125^{n}$$

$$\log 1.0625=\log 1.003125^{n}$$

$$\log 1.0625=n\cdot\log 1.003125$$

$$\frac{\log 1.0625}{\log 1.003125}=n$$

$$n\approx 20 \text{ months}$$

25. From Table 8.6, the original payment would be \$12.13 per \$1,000, so the original payment would be $130\times 12.13=\$1{,}576.90$ per month. The new payment would be \$23.49 per \$1,000, so the new payment would be $64.608\times 23.49\approx \$1{,}517.64$ per month.

The difference in monthly payments is $\$1{,}576.90-\$1{,}517.64=\$59.26$.

26. Using line 3, the Federal Income tax is \$8,707.50

$4{,}220+0.25(48{,}600-30{,}650)=4{,}220+0.25(17{,}950)=$
$4{,}220+4{,}487.5=8{,}707.50$

27. $A=100{,}000\left(\frac{(1+0.04)^{10}-1}{0.04}\right)\approx \$1{,}200{,}610.71$

$$A=P\left(1+\frac{r}{m}\right)^{n}$$

$$1{,}200{,}610.71=P(1+0.04)^{10}$$

$$1{,}200{,}610.71\approx P(1.480244285)$$

$$\frac{1{,}200{,}610.71}{1.480244285}\approx P$$

$$P\approx \$811{,}089.58$$

The annuity is the better deal because its present value is \$811,089.58.

28. We need to determine the monthly payments where $\frac{r}{m} = \frac{0.075}{12}$ and $n = 12 \cdot 20 = 240$.

$$140{,}000\left(1+\frac{0.075}{12}\right)^{240} = R\left(\frac{\left(1+\frac{0.075}{12}\right)^{240}-1}{0.075/12}\right)$$

$$624{,}514.384397 \approx R(553.730725)$$

$$R \approx \frac{624{,}514.384397}{553.730725} \approx \$1127.84$$

Payment Number	Amount of Payment	Interest Payment	Applied to Principal	Balance
				$140,000.00
1	$1,127.84	$875.00	$252.84	$139,747.16
2	$1,127.84	$873.42	$254.42	$139,492.74

29. $A = P(1+r)^t$

$A = 450(1+0.034)^5$

$A = 450(1.034)^5 \approx \$531.89$

30. $347.34

$\frac{r}{m} = \frac{0.096}{12} = 0.008$ and $n = 12 \cdot 5 = 60$

$$16{,}500(1+0.008)^{60} = R\left(\frac{(1+0.008)^{60}-1}{0.008}\right)$$

$$26{,}614.350421808 \approx R(76.6238668)$$

$$R \approx \frac{26{,}614.350421808}{76.6238668} \approx \$347.34$$

Chapter 9
Geometry: Ancient and Modern Mathematics Embrace

Section 9.1 Lines, Angles, and Circles

1. 7 and 9; 8 and 10; 2 and 5; 1 and 6; 3 and 4

3. 3 and 7

5. 10; 8

7. 9 and 10; 7 and 10; 7 and 8; 8 and 9; 3 and 10; 4 and 10; 4 and 8; 3 and 8

9. false; Complementary angles are two angles whose sum of the measures is 90 degrees. The two angles do not each have to measure 45 degrees.

11. true

13. false; For example, a $30°$ angle is the complement of a $60°$ angle and the supplement of a $150°$ angle.

15. e and d

17. b and f; c and g

19. complement: $90° - 30° = 60°$; supplement: $180° - 30° = 150°$

21. complement: none; supplement: $180° - 120° = 60°$

23. complement: $90° - 51.2° = 38.8°$; supplement: $180° - 51.2° = 128.8°$

25. $m\angle a = 180° - 36° = 144°$; $m\angle b = 36°$; $m\angle c = 144°$

27. $m\angle a = 45°$; $m\angle b = 180° - 45° = 135°$; $m\angle c = 45°$

29. $m\angle a = 90° - 38° = 52°$; $m\angle b = 90°$; $m\angle c = 180° - 52° = 128°$

31.
$$\frac{\text{measure of central angle}}{360°} = \frac{\text{arc length}}{\text{circumference}}$$
$$\frac{90°}{360°} = \frac{a}{24 \text{ feet}}$$
$$90 \cdot 24 = 360a$$
$$2{,}160 = 360a$$
$$a = \frac{2{,}160}{360} = 6$$
Arc AB has length 6 feet.

33.
$$\frac{\text{measure of central angle}}{360°} = \frac{\text{arc length}}{\text{circumference}}$$
$$\frac{m}{360°} = \frac{4 \text{ m}}{12 \text{ m}}$$
$$12m = 4 \cdot 360$$
$$12m = 1{,}440$$
$$m = 120$$
The measure of the central angle, $\angle ACB$, is $120°$.

35. $$\frac{\text{measure of central angle}}{360°} = \frac{\text{arc length}}{\text{circumference}}$$
$$\frac{30°}{360°} = \frac{100 \text{ mm}}{c}$$
$$30c = 360 \cdot 100$$
$$30c = 36{,}000$$
$$c = 1{,}200$$
The circumference of the circle is 1,200 millimeters.

37. $$\frac{\text{measure of central angle}}{360°} = \frac{\text{arc length}}{\text{circumference}}$$
$$\frac{180° - 60°}{360°} = \frac{a}{18 \text{ ft}}$$
$$\frac{120°}{360°} = \frac{a}{18 \text{ ft}}$$
$$360a = 120 \cdot 18$$
$$360a = 2{,}160$$
$$a = 6$$
The arc length is 6 feet.

39. $$\frac{\text{measure of central angle}}{360°} = \frac{\text{arc length}}{\text{circumference}}$$
$$\frac{m}{360°} = \frac{15 \text{ in} - 3 \text{ in}}{30 \text{ in}}$$
$$\frac{m}{360°} = \frac{12 \text{ in}}{30 \text{ in}}$$
$$30m = 360 \cdot 12$$
$$30m = 4{,}320$$
$$m = 144°$$
$m\angle BCD$ is 144°.

41. $$2x + x = 180°$$
$$3x = 180°$$
$$x = 60°$$

43. $$3x + 2x = 90°$$
$$5x = 90°$$
$$x = 18°$$

45. $$m\angle D = 2 \times m\angle B$$
$$m\angle D = 2 \times 24°$$
$$m\angle D = 48°$$
$$\frac{m\angle D}{360°} = \frac{\text{arc length}}{\text{circumference}}$$
$$\frac{48°}{360°} = \frac{x}{30 \text{ in.}}$$
$$360 \cdot x = 48 \cdot 30$$
$$360x = 1440$$
$$x = 4$$
The length of arc AC is 4 in.

47. $$\frac{m\angle D}{360°} = \frac{\text{arc length}}{\text{circumference}}$$
$$\frac{m\angle D}{360°} = \frac{20 \text{ in.}}{60 \text{ in.}}$$
$$60 \cdot m\angle D = 20 \cdot 360$$
$$60 \cdot m\angle D = 7200$$
$$m\angle D = 120°$$
$$m\angle D = 2 \times m\angle B$$
$$120° = 2 \times m\angle B$$
$$m\angle B = 60°$$

49. 12

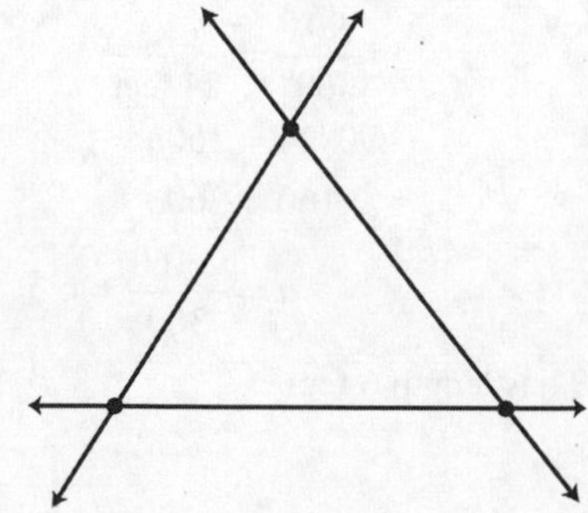

51. $$\frac{\text{measure of central angle}}{360^\circ}=\frac{\text{arc length}}{\text{circumference}}$$
$$\frac{m}{360^\circ}=\frac{1{,}000\text{ miles}}{25{,}000\text{ miles}}$$
$$25{,}000m=360\cdot 1{,}000$$
$$25{,}000m=360{,}000$$
$$m=14.4$$
The measure of the angle should have been 14.4°.

53. The figure below shows the equal angles, from both the incoming and outgoing angles of the ball's path and one set of alternate interior angles.. From the triangle in the upper left, $x+50^\circ=90^\circ$, so $x=40^\circ$.

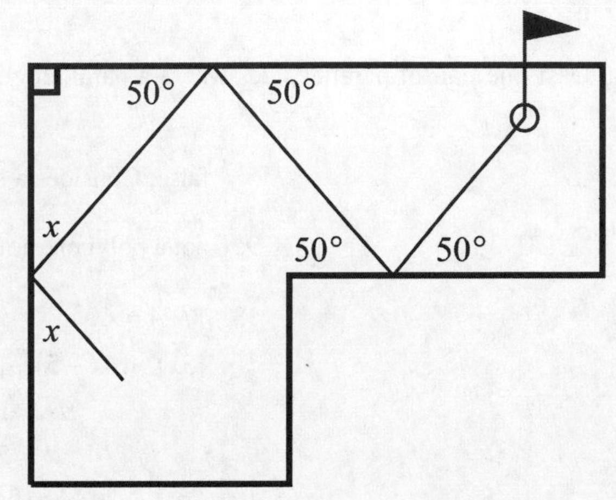

55. The sum of supplementary angles is 180°; the sum of complementary angles is 90°.

57. "Alternate" means that the angles are on opposite sides of the transversal. "Exterior" means that the angles are outside the parallel lines.

59. This is possible. If the two lines are perpendicular, then all four angles are right angles. Each angle measures 90°.

61. No; the angle opposite would have to be obtuse also.

63. It equals 90°.
$$\begin{array}{r} m\angle A+m\angle C=180^\circ \\ -(m\angle A+m\angle B=90^\circ) \\ \hline m\angle C-m\angle B=90^\circ \end{array}$$

65. The complement of $\angle A$ is 60°, and the supplement of 60° is 120°.

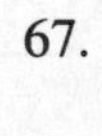
67.

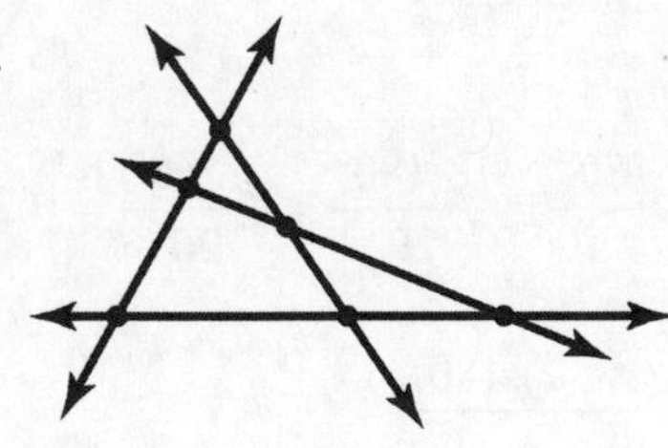

69.

Number of lines	Most number of angles formed
2	$4 = 4 \cdot 1$
3	$12 = 4 \cdot 3 = 4 \cdot (1+2)$
4	$24 = 4 \cdot 6 = 4 \cdot (1+2+3)$
5	$40 = 4 \cdot 10 = 4 \cdot (1+2+3+4)$
. .	. .
10	$4 \cdot (1+2+3+4+5+6+7+8+9) = 4 \cdot 45 = 180$

Section 9.2 Polygons

1. false; A trapezoid has at least one pair of parallel sides where a parallelogram must have two pairs of parallel sides.

3. true

5. false; Consider a rhombus and a square.

7. polygon

9. not a polygon; not made of line segments

11. $m\angle A = 22.5°$

$$\begin{aligned} x + 2x + 3x &= 180° \\ 8x &= 180° \\ x &= 22.5° \\ m\angle A = x &= 22.5° \end{aligned}$$

13. $m\angle A = 60°$

$$\begin{aligned} (3x) + (4x-5) + (2x+5) &= 180° \\ 9x &= 180° \\ x &= 20° \\ m\angle A = 3x &= 60° \end{aligned}$$

15. We can divide the hexagon into four triangles. The angle sum is $4 \cdot 180° = 720°$.

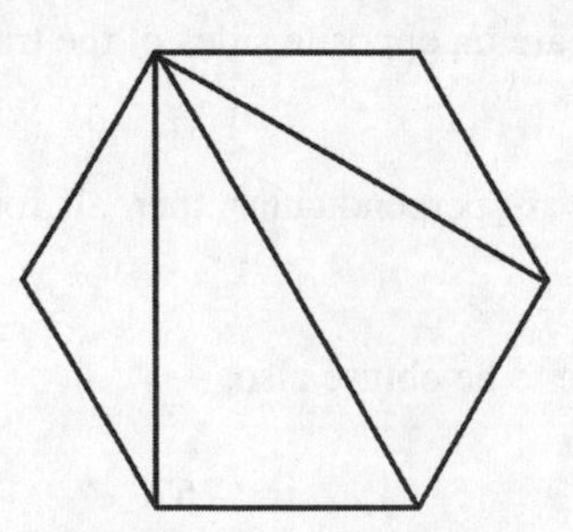

17. $\dfrac{(20-2) \cdot 180°}{20} = \dfrac{18 \cdot 180°}{20} = \dfrac{3,240°}{20} = 162°$

19. $160° = \dfrac{(n-2) \cdot 180°}{n}$

$$\begin{aligned} 160n &= (n-2) \cdot 180 \\ 160n &= 180n - 360 \\ -20n &= -360 \\ n &= \frac{-360}{-20} = 18 \end{aligned}$$

21. $\dfrac{x}{3} = \dfrac{10}{6}$

$$\begin{aligned} 6x &= 10 \cdot 3 \\ 6x &= 30 \\ x &= 5 \text{ in.} \end{aligned}$$

23. $m\angle E = 40°$

$$\frac{EH}{8} = \frac{21}{24}$$
$$24(EH) = 21 \cdot 8$$
$$24(EH) = 168$$
$$EH = \frac{168}{24} = 7$$

25. $m\angle I = 35°$

$$\frac{HI}{120} = \frac{105}{150}$$
$$150(HI) = 105 \cdot 120$$
$$150(HI) = 12,600$$
$$HI = \frac{12,600}{150} = 84$$

27. The angle of the triangle adjacent to $\angle A$ is a supplement to $\angle A$, so it is $180° - 138° = 42°$. $\angle B$ is a complement to $\angle A$, so it is $90° - 42° = 48°$.

29.

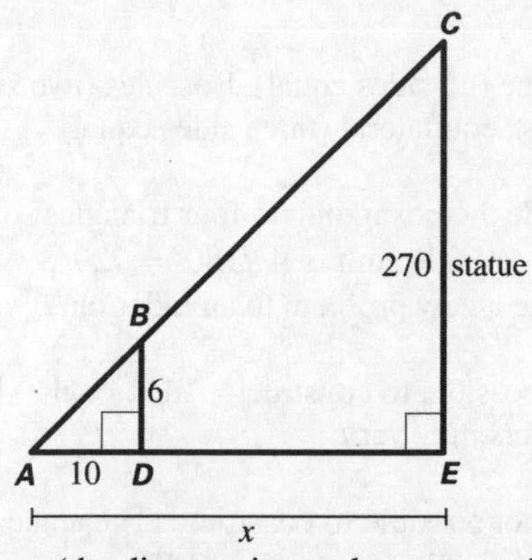

(the diagram is not drawn to scale)

Triangles ABD and ACE are similar. Therefore $\frac{BD}{AD} = \frac{CE}{AE}$, so $\frac{6}{10} = \frac{270}{x} \Rightarrow 6x = 2700 \Rightarrow x = 450$.

The length of the shadow of the statue is 450 ft.

31. Let x be the width of the river.

$$\frac{x}{200} = \frac{3}{8}$$
$$8x = 200 \cdot 3$$
$$8x = 600$$
$$x = 75 \text{ ft}$$

33. Let x be the length of the pond.

$$\frac{x}{270+60} = \frac{35}{60}$$
$$\frac{x}{330} = \frac{35}{60}$$
$$60x = 330 \cdot 35$$
$$60x = 11,550$$
$$x = \frac{11,550}{60} = 192.5 \text{ feet}$$

35. Since the triangles are isosceles, we have equal base angles.

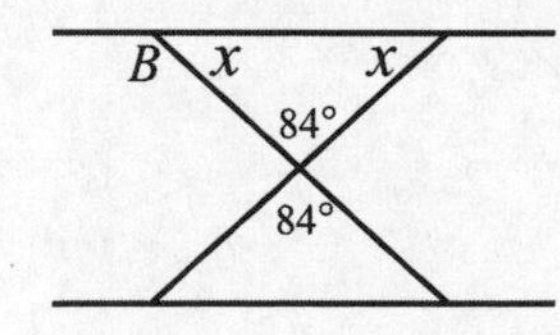

$$x + x + 84° = 180°$$
$$2x = 96°$$
$$x = 48°$$
$$B + 48° = 180°$$
$$B = 132°$$

37.

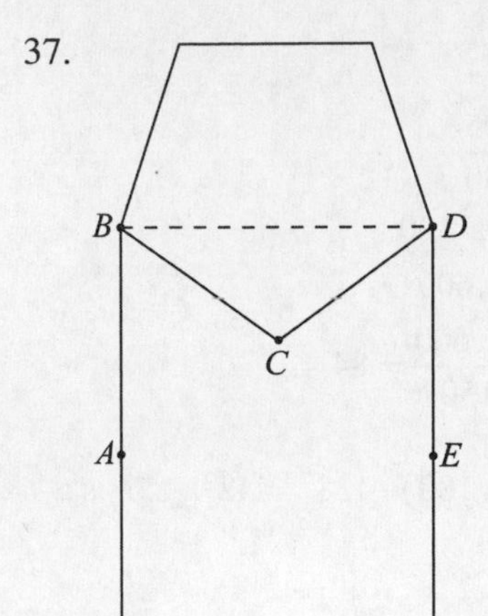

Label the points as shown. Draw a line connecting B to D. Angles BDE and ABD will be right angles. Find the measure of angle BCD.

$m\angle BCD = \frac{(5-2)\cdot 180°}{5} = \frac{3\cdot 180°}{5} = \frac{540°}{5} = 108°$

Since triangle BCD is isosceles, $m\angle BDC = \frac{180° - 108°}{2} = \frac{72°}{2} = 36°$. Also since $\angle BDC$ and $\angle CDE$ are complementary, the desired angles each measure $90° - 36° = 54°$.

In Exercise 39, the triangles formed by the sides and supports of the smaller and larger gazebos are similar, so the corresponding sides are proportional.

39. Let x be the length of the long support.

$\frac{x}{12} = \frac{17.4}{10}$
$10x = 12\cdot 17.4$
$10x = 208.8$
$x = \frac{208.8}{10} = 20.88$ ft

41. scalene (no sides equal), isosceles (two sides equal), equilateral (three sides equal)

43. Divide the hexagon into four triangles so the interior angle sum is $4\times 180° = 720°$. Relate a new problem to an older one.

45. It is possible to construct. Additional answers may vary.

47. It is not possible to construct. The angle sum would be greater than 180°.

49. No solution provided.

51. The bottom angle of the blue triangle is $360° - 108° - 60° - 60° = 132°$ since interior angles of a regular pentagon, and an equilateral triangle are $108°$ and $60°$, respectively. Since the red triangles are equilateral that share sides with the regular pentagon, the blue triangle is isosceles, which means the base angles must equally share the remaining $180° - 132° = 48°$ degrees of the triangle, so they are each $24°$.

53. $x + 130° = 180° \Rightarrow x = 50°$
$m\angle A + 50° + 105° = 180° \Rightarrow m\angle A = 25°$

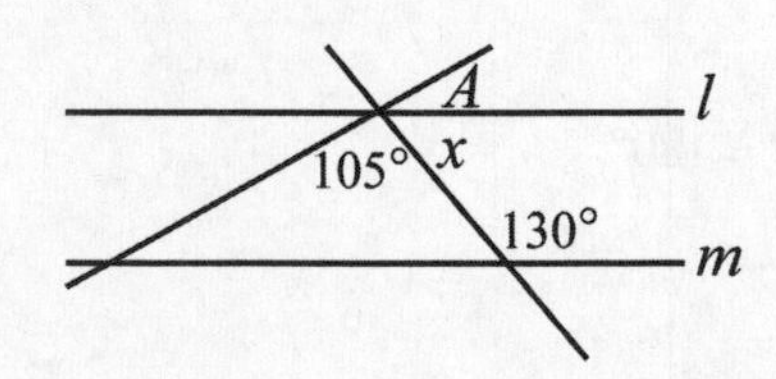

55. The measure of the interior angles gets larger. For very large n, the interior angles have measures close to 180°. For example, evaluate the expression $\frac{(n-2)\cdot 180°}{n}$ with n=10, 100, 1000, 10000, etc.

n	$\frac{[(n-2)\cdot 180°]}{n}$	simplified
10	$\frac{(10\text{-}2)\cdot 180°}{10}$	144°
100	$\frac{(100-2)\cdot 180°}{100}$	176.4°
1,000	$\frac{(1{,}000-2)\cdot 180°}{1{,}000}$	179.64°
10,000	$\frac{(10{,}000-2)\cdot 180°}{10{,}000}$	179.964°
⋮	⋮	⋮

57. The interior angle of the regular n-gon is $\frac{(n-2)\cdot 180°}{n}$.

Since the triangle formed on the bottom of the n-gon is isosceles, the desired angles each measure

$$90° - \frac{180° - \frac{(n-2)\cdot 180°}{n}}{2} = 90° - 90° + \frac{(n-2)\cdot 180°}{2n} = \frac{(n-2)\cdot 90°}{n}.$$

Section 9.3 Perimeter and Area

1. $A = l\cdot w = 16\cdot 10 = 160\ \text{ft}^2$

3. $A = h\cdot b = 7\cdot 20 = 140\ \text{in}^2$

5. $A = \frac{1}{2}\cdot(b_1 + b_2)\cdot h = \frac{1}{2}\cdot(22+14)\cdot 6 = \frac{1}{2}\cdot(36)\cdot 6 = 18\cdot 6 = 108\ \text{cm}^2$

7. $A = \frac{1}{2}\cdot h\cdot b = \frac{1}{2}\cdot 6\cdot 24 = 3\cdot 24 = 72\ \text{yd}^2$

9. $A = \pi\, r^2 \approx 3.14\cdot 5^2 = 3.14\cdot 25 = 78.5\ \text{cm}^2$

11. $r = \frac{d}{2} = \frac{8}{2} = 4\ \text{m}$

$A = \pi\, r^2 \approx 3.14\cdot 4^2 = 3.14\cdot 16 = 50.24\ \text{m}^2$

13. Entire area: $A = l \cdot w = 18 \cdot 8 = 144 \text{ yd}^2$

Unshaded area:

$$A = \frac{1}{2} \cdot h \cdot b = \frac{1}{2} \cdot 4 \cdot 18 = 2 \cdot 18 = 36 \text{ yd}^2$$

$$A = \frac{1}{2} \cdot h \cdot b = \frac{1}{2} \cdot 4 \cdot 9 = 2 \cdot 9 = 18 \text{ yd}^2$$

$$A = \frac{1}{2} \cdot h \cdot b = \frac{1}{2} \cdot 4 \cdot 9 = 2 \cdot 9 = 18 \text{ yd}^2$$

Shaded area: $144 - 36 - 18 - 18 = 72 \text{ yd}^2$

15. Entire area:

$$A = l \cdot w = 8 \cdot (2+2) = 8 \cdot 4 = 32 \text{ m}^2$$

Unshaded area:

$$A = \pi r^2 \approx 3.14 \cdot 2^2 = 3.14 \cdot 4 = 12.56 \text{ m}^2$$

$$A = \pi r^2 \approx 3.14 \cdot 2^2 = 3.14 \cdot 4 = 12.56 \text{ m}^2$$

Shaded area: $32 - 12.56 - 12.56 = 6.88 \text{ m}^2$

17. Entire area: $A = l \cdot w = 2 \cdot 2 = 4 \text{ m}^2$

Unshaded area:

$$A = \pi r^2 \approx 3.14 \cdot \left(\frac{2}{2}\right)^2 = 3.14 \cdot 1 = 3.14 \text{ m}^2$$

Shaded area: $4 - 3.14 = 0.86 \text{ m}^2$

19. Trapezoid $ABCD$

$$A = \frac{1}{2} \cdot (b_1 + b_2) \cdot h$$

$$54 = \frac{1}{2} \cdot (8+10) \cdot h$$

$$9h = 54$$

$$h = 6 \text{ ft}$$

Triangle ABC

$$A = \frac{1}{2} \cdot b \cdot h$$

$$A = \frac{1}{2} \cdot 6 \cdot 6$$

$$A = 18 \text{ ft}^2$$

21. $s = \frac{1}{2}(a+b+c) = \frac{1}{2}(15+6+18) = \frac{39}{2}$

$$A = \sqrt{s(s-a)(s-b)(s-c)} = \sqrt{\frac{39}{2}\left(\frac{39}{2}-15\right)\left(\frac{39}{2}-6\right)\left(\frac{39}{2}-18\right)} = \sqrt{\frac{39}{2} \cdot \frac{9}{2} \cdot \frac{27}{2} \cdot \frac{3}{2}} = \sqrt{\frac{28,431}{16}}$$

$A \approx 42.15 \text{ cm}^2$

23. $s = \frac{1}{2}(a+b+c) = \frac{1}{2}(10+15+12) = \frac{37}{2}$

$$A = \sqrt{s(s-a)(s-b)(s-c)} = \sqrt{\frac{37}{2}\left(\frac{37}{2}-10\right)\left(\frac{37}{2}-15\right)\left(\frac{37}{2}-12\right)} = \sqrt{\frac{37}{2} \cdot \frac{17}{2} \cdot \frac{7}{2} \cdot \frac{13}{2}} = \sqrt{\frac{57,239}{16}}$$

$A \approx 59.81 \text{ m}^2$

25. $s = \frac{1}{2}(a+b+c) = \frac{1}{2}(19+7+18) = \frac{44}{2} = 22$

$$A = \sqrt{s(s-a)(s-b)(s-c)} = \sqrt{22(22-19)(22-7)(22-18)} = \sqrt{22 \cdot 3 \cdot 15 \cdot 4} = \sqrt{3,960} \approx 62.93 \text{ cm}^2$$

$$A = \frac{1}{2} \cdot h \cdot b$$

$$\sqrt{3,960} = \frac{1}{2} \cdot h \cdot 19$$

$$h = \frac{2\sqrt{3,960}}{19} \approx 6.62 \text{ cm}$$

27. $a^2 + b^2 = c^2$

$$5^2 + 12^2 = x^2$$
$$25 + 144 = x^2$$
$$169 = x^2$$
$$x = \sqrt{169} = 13 \text{ m}$$

29. $a^2 + b^2 = c^2$

$$x^2 + 11^2 = 13^2$$
$$x^2 + 121 = 169$$
$$x^2 = 48$$
$$x = \sqrt{48} \approx 6.93 \text{ yd}$$

31. We need to first find the height, h, of the triangle using the Pythagorean theorem.

$$8^2 + h^2 = 10^2$$
$$64 + h^2 = 100$$
$$h^2 = 36$$
$$h = \sqrt{36} = 6 \text{ m}$$

$$A = \frac{1}{2} \cdot h \cdot b$$
$$A = \frac{1}{2} \cdot 6 \cdot (8 + 15)$$
$$A = \frac{1}{2} \cdot 6 \cdot 23 = 3 \cdot 23 = 69 \text{ m}^2$$

33. 7 in^2

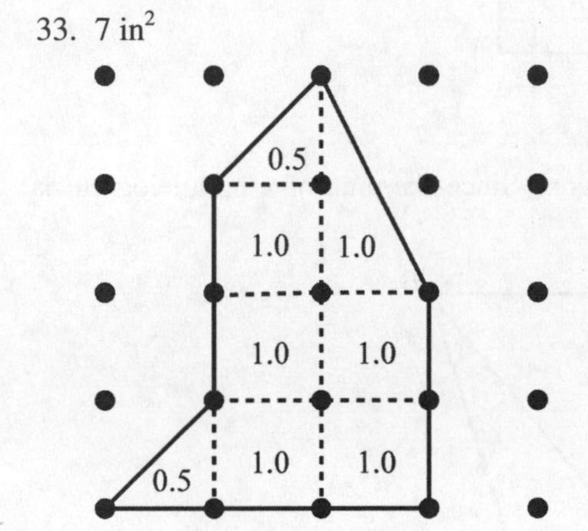

35. 3.5 in^2

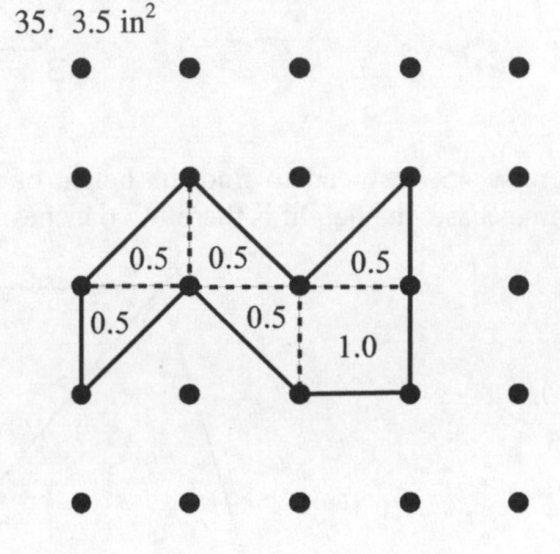

37. area

39. perimeter

41. The distance from home plate to second base represents the hypotenuse of an isosceles right triangle. Let d represent this distance.

$$90^2 + 90^2 = d^2$$
$$8{,}100 + 8{,}100 = d^2$$
$$16{,}200 = d^2$$
$$d = \sqrt{16{,}200} \approx 127.28 \text{ ft}$$

43. AB is 2 m.

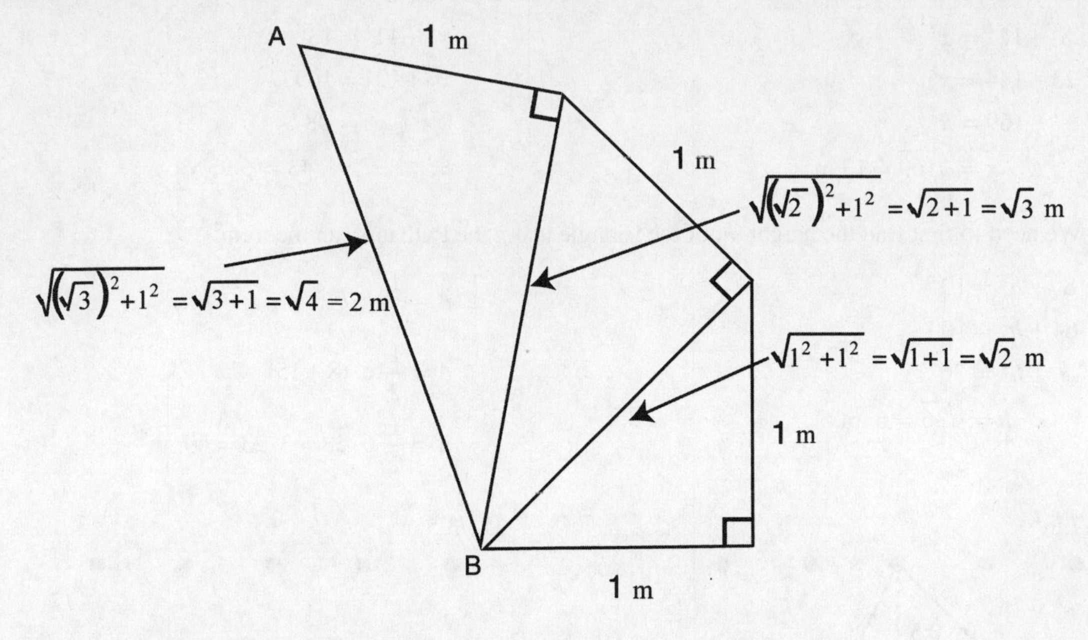

For Exercise 45, we need to find the height of the parallelogram. Since the area of a parallelogram is height times base, the height is therefore 6 inches.

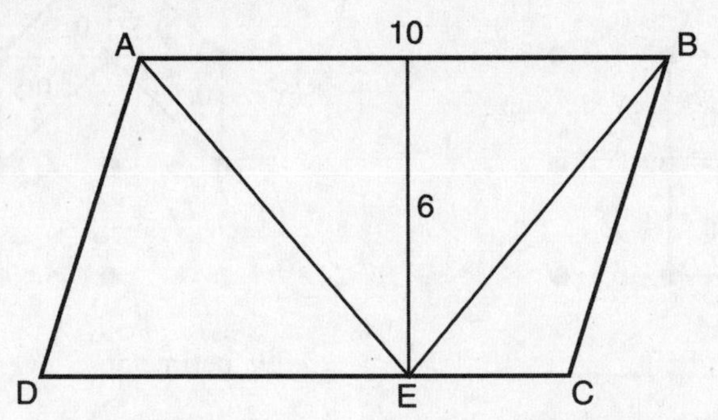

45. It is given that the area of ΔBEC is 6 square inches. The area of ΔABE is $A=\frac{1}{2}\cdot h\cdot b=\frac{1}{2}\cdot 6\cdot 10=30$ square inches. Thus, the area of ΔADE is $60-6-30=24$ square inches.

For Exercise 47, we need to find the areas ΔEDB and ΔAEF. Since the area of ΔBAE is 30 square yards, ΔEDB also has an area of 30 square yards. This is a result of the right angles that yields these two triangles congruent. Since the area of trapezoid $ABDF$ is 66 square yards, the area of ΔAEF is $66-30-30=6$ square yards. Also, lengths AB and DE are 10 yards each because they must satisfy the relation $30=\frac{1}{2}\cdot 6\cdot b \Rightarrow 30=3b \Rightarrow b=10$.

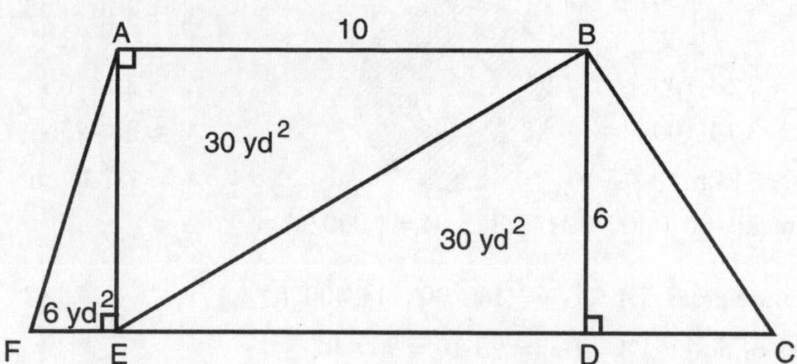

47. Since the area of ΔBDC is twice ΔAEF, the area of ΔBDC is $2\cdot 6=12$ square yards.

49. We need to sum the areas of a rectangle and a semi-circle (half of a circle).

    Area of rectangle: $A=h\cdot b=12\cdot 6=72\text{ ft}^2$

    Area of semi-circle: $A=\frac{1}{2}\pi\ r^2 \approx \frac{1}{2}\cdot 3.14\cdot\left(\frac{6}{2}\right)^2=\frac{1}{2}\cdot 3.14\cdot 3^2=\frac{1}{2}\cdot 3.14\cdot 9=14.13\text{ ft}^2$

    Total area: $72 + 14.13 = 86.13\text{ ft}^2$

51. We need to determine the cost per square inch in order to compare the value of the two pizzas.

    12-inch pizza: $A=\pi\ r^2$

    $$A\approx 3.14\cdot\left(\frac{12}{2}\right)^2=3.14\cdot 6^2=3.14\cdot 36=113.04$$

    The cost per square inch is approximately $\frac{5.99}{113.04}\approx 0.053$ or 5.3 cents.

    16-inch pizza: $A=\pi\ r^2$

    $$A\approx 3.14\cdot\left(\frac{16}{2}\right)^2=3.14\cdot 8^2=3.14\cdot 64=200.96$$

    The cost per square inch is approximately $\frac{8.99}{200.96}\approx 0.045$ or 4.5 cents.

    The 16-inch pizza is a better deal. The cost per square inch is less.

53. There are two rectangular pieces. Each of these has an area of $4 \cdot 100 = 400 \text{ m}^2$.
The two ends combined would form a washer-like figure in which we need to find the area of the shaded region.

Outer Area: $A_0 = \pi r^2$

$$A_0 \approx 3.14 \cdot \left(\frac{20}{2}\right)^2$$
$$A_0 \approx 3.14 \cdot 10^2$$
$$A_0 \approx 3.14 \cdot 100$$
$$A_0 \approx 314 \text{ m}^2$$

Inner Area: $A_I = \pi r^2$

$$A_I \approx 3.14 \left(\frac{20-4-4}{2}\right)^2$$
$$A_I \approx 3.14 \cdot 6^2$$
$$A_I \approx 3.14 \cdot 36$$
$$A_I \approx 113.04 \text{ m}^2$$

Surface area of track: $400 + 400 + 314 - 113.04 = 1{,}000.96 \text{ m}^2$

55.

Entire area: $A = l \cdot w = 160 \cdot 90 = 14{,}400 \text{ ft}^2$

Area of desk: $A = l \cdot w = 90 \cdot 10 = 900 \text{ ft}^2$

Area of left part of pool: $A = l \cdot w = (70-30) \cdot 10 = 40 \cdot 10 = 400 \text{ ft}^2$

Area of right part of pool: $A = l \cdot w = 30 \cdot 16 = 480 \text{ ft}^2$

Area to be tiled: $14{,}400 - 900 - 400 - 480 = 12{,}620 \text{ ft}^2$

57. By the Pythagorean theorem:

$$x = \sqrt{219^2 - 115^2}$$
$$x = \sqrt{34{,}736} \approx 186.38 \text{ m}$$

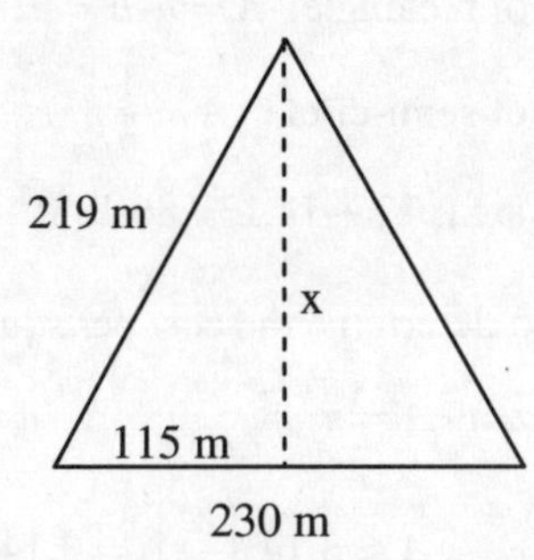

59. a) The circumference of a circle with an 18 foot radius is $C = 2 \cdot \pi r \approx 2 \cdot 3.14 \cdot 18 = 113.04$ ft. The amount of fencing we need would be $\frac{72}{360} = \frac{1}{5}$ of this or approximately 22.61 ft.

b) The area of a circle with an 18-foot radius is $A = \pi r^2 \approx 3.14 \cdot 18^2 = 3.14 \cdot 324 = 1{,}017.36 \text{ ft}^2$. One fifth of this is about 203.47 ft^2

61. We need to first determine the area of the rectangular and the circle.

Rectangle area: $V = 130 \cdot 60 = 7800 \text{ yd}^2$

Circle area: $A = \pi r^2 \approx 3.14 \cdot \left(\frac{60}{2}\right)^2 = 3.14 \cdot 30^2 = 3.14 \cdot 900 = 2826 \text{ yd}^2$

The area covered with stones will be $7800 - 2826 = 4974 \text{ yd}^2$.

63. rectangle, parallelogram, triangle, trapezoid

65. A triangle is one-half of a parallelogram.

67.

Circle

$$C = 2\pi r$$
$$6 = 2\pi r$$
$$r = \frac{6}{2\pi} = \frac{3}{\pi}$$
$$A = \pi r^2$$
$$A = \pi\left(\frac{3}{\pi}\right)^2 = \pi\left(\frac{9}{\pi^2}\right) = \frac{9}{\pi} \approx \frac{9}{3.14} = 2.866$$

Square

$$P = 4s$$
$$6 = 4s$$
$$s = \frac{6}{4} = \frac{3}{2}$$
$$A = s^2$$
$$A = \left(\frac{3}{2}\right)^2 = \frac{9}{4} = 2.25$$

The are of the circle is $\frac{2.866}{2.25} \approx 1.27$ times larger than the area of the square.

69. To find the area of the side of the museum, we must find the area of the trapezoid and the equilateral triangles.

The area of trapezoid is $A = \frac{1}{2}\cdot(b_1 + b_2)\cdot h = \frac{1}{2}\cdot[100 + 60]\cdot 40 = \frac{1}{2}\cdot(160)\cdot 40 = 80\cdot 40 = 3,200 \text{ ft}^2$.

square yards.

To find the area of each of these triangles, we will use Heron's formula.

$$s = \frac{1}{2}(a+b+c) = \frac{1}{2}(18+18+18) = \frac{54}{2} = 27$$
$$A = \sqrt{s(s-a)(s-b)(s-c)} = \sqrt{27(27-18)(27-18)(27-18)}$$
$$= \sqrt{27\cdot 9\cdot 9\cdot 9} = \sqrt{19,683} = 81\sqrt{3} \approx 140.3 \text{ ft}^2.$$

Therefore, the area of the side of the museum is $3,200 - 3(140.3) = 3,200 - 420.9 = 2,779.1 \text{ ft}^2$.

71. false; it quadruples

Area of original circle: $A_0 = \pi\ r^2$

Area of circle with double radius: $A_d = \pi\ (2r)^2 = \pi\cdot 4\cdot\ r^2 = 4\cdot\pi\ r^2 = 4A_0$

73. a) $A = \pi r^2 = \pi\left(\frac{d}{2}\right)^2 \approx 3.14\left(\frac{14}{2}\right)^2 = 3.14(7)^2 = 49\cdot 3.14 = 153.86 \text{ in}^2$

b) $100 \text{ acre}\left(\frac{43,560 \text{ ft}^2}{1 \text{ acre}}\right)\left(\frac{144 \text{ in}^2}{1 \text{ ft}^2}\right) = 627,264,000 \text{ in}^2$

c) $\frac{627,264,000 \text{ in}^2/\text{day}}{153.86 \text{ in}^2/\text{pizza}} \approx 4,076,849 \text{ pizzas/day}$

d) $\frac{4,076,849 \text{ pizzas}}{1 \text{ day}}\left(\frac{8 \text{ slices}}{1 \text{ pizza}}\right)\left(\frac{1 \text{ day}}{24 \text{ hr}}\right)\left(\frac{1 \text{ hr}}{60 \text{ min}}\right)\left(\frac{1 \text{ min}}{60 \text{ sec}}\right) \approx 377 \text{ slices/sec}$

75. The area of quadrilateral (parallelogram) *WXYZ* is one-half of the area of rectangle *ABCD*. Explanations may vary.

Let b be the base of rectangle *ABCD* and h the height. Although it is more common to use length and width, we will use base and height in order to mark the triangles clearly. The area of rectangle *ABCD* will be $A = h \cdot b$. The sides of the four congruent triangles will be as marked.

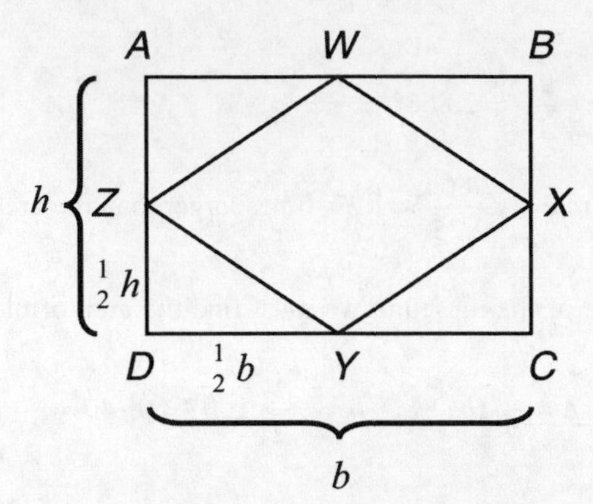

The area of each of the four congruent triangles will be $\frac{1}{2} \cdot \left(\frac{1}{2}h\right) \cdot \left(\frac{1}{2}b\right) = \frac{1}{8}hb$. The area of the four congruent triangles together will be $4 \cdot \left(\frac{1}{8}hb\right) = \frac{1}{2}hb$. Thus the area of quadrilateral (parallelogram) *WXYZ* will be $h \cdot b - \frac{1}{2}hb = \frac{1}{2}hb$ or one-half the area of rectangle *ABCD*.

77. The figure that will have the maximum area will be a square with dimensions 50 feet by 50 feet.

79. We need to first find the approximate radius of the open area.

$$A = \pi r^2$$
$$11{,}304 \approx 3.14 \cdot r^2$$
$$\frac{11{,}304}{3.14} \approx r^2$$
$$r \approx \sqrt{\frac{11{,}304}{3.14}} = 60 \text{ ft}$$

We are next concerned with a washer-like figure in which we need to find the area of the shaded region.

Outer Area:
$$A_0 = \pi r^2$$
$$A_0 \approx 3.14 \cdot (180 + 60)^2$$
$$A_0 \approx 3.14 \cdot 240^2$$
$$A_0 \approx 3.14 \cdot 57{,}600$$
$$A_0 \approx 180{,}864 \text{ ft}^2$$

Inner Area: 11,304 ft^2

Outer Area – Inner Area = 180,864 – 11,304 = 169,560 ft^2

Each of the lots will be one-eighth of this area or about 21,195 ft^2

Section 9.4 Volume and Surface Area

1. a) Surface area: $S = 2lw + 2lh + 2wh = 2\cdot6\cdot4 + 2\cdot6\cdot5 + 2\cdot4\cdot5 = 48 + 60 + 40 = 148 \text{ cm}^2$

   b) Volume: $V = lwh = 6\cdot4\cdot5 = 120 \text{ cm}^3$

3. a) Surface area: $S = \pi r\sqrt{r^2 + h^2} \approx 3.14\cdot3\cdot\sqrt{3^2 + 8^2} = 9.42\sqrt{9 + 64} = 9.42\sqrt{73} \approx 80.48 \text{ in}^2$

   b) Volume: $V = \frac{1}{3}\pi r^2 h \approx \frac{1}{3}\cdot3.14\cdot3^2\cdot8 = \frac{3.14\cdot9\cdot8}{3} = \frac{226.08}{3} = 75.36 \text{ in}^3$

5. a) Surface area: $S = 2\pi rh + 2\pi r^2 \approx 2\cdot3.14\cdot5\cdot8 + 2\cdot3.14\cdot5^2 = 251.2 + 157 = 408.2 \text{ ft}^2$

   b) Volume: $V = \pi r^2 h \approx 3.14\cdot5^2\cdot8 = 3.14\cdot25\cdot8 = 628 \text{ ft}^3$

7. a) Surface area: $S = 4\pi r^2 \approx 4\cdot3.14\cdot20^2 = 12.56\cdot400 = 5{,}024 \text{ cm}^2$

   b) Volume: $V = \frac{4}{3}\pi r^3 \approx \frac{4}{3}\cdot3.14\cdot20^3 = \frac{4\cdot3.14\cdot8{,}000}{3} = \frac{100{,}480}{3} \approx 33{,}493.33 \text{ cm}^3$

9. Volume = area of the base times height = $25\cdot3 = 75 \text{ ft}^3$

11. Area of base = $A = \frac{1}{2}\cdot(b_1 + b_2)\cdot h = \frac{1}{2}\cdot(11 + 7)\cdot4 = \frac{1}{2}\cdot(18)\cdot4 = 9\cdot4 = 36 \text{ in}^2$

    Volume = area of the base times height = $36\cdot5 = 180 \text{ in}^3$

13. $s = \frac{1}{2}(a + b + c) = \frac{1}{2}(5 + 5 + 4) = \frac{14}{2} = 7$

    Area of base applying Heron's formula is

    $A = \sqrt{s(s-a)(s-b)(s-c)} = \sqrt{7(7-5)(7-5)(7-4)} = \sqrt{7\cdot2\cdot2\cdot3} = \sqrt{84} \approx 9.17 \text{ m}^2$

    Volume = area of the base times height = $\sqrt{84}\cdot2 \approx 18.33 \text{ m}^3$

15. $12^3 = 1{,}728$

17. We need to first determine the volume of the punch bowl and the ladle.

    Punch bowl volume: $V = \frac{1}{2}\cdot\left(\frac{4}{3}\pi r^3\right) \approx \frac{4}{6}\cdot3.14\cdot9^3 = \frac{4\cdot3.14\cdot729}{6} = \frac{9{,}156.24}{6} = 1{,}526.04 \text{ in}^3$

    Ladle volume: $V = \frac{1}{2}\cdot\left(\frac{4}{3}\pi r^3\right) \approx \frac{4}{6}\cdot3.14\cdot2^3 = \frac{4\cdot3.14\cdot8}{6} = \frac{100.48}{6} \approx 16.75 \text{ in}^3$

    We then divide the volume of the punch bowl by the volume of the ladle. $\frac{1{,}526.04}{16.75} \approx 91.11$

    There are approximately 91 ladles of punch.

19. We need to first determine the volume of the cylindrical pitcher and the conical glass.

Pitcher volume: $V = \pi r^2 h \approx 3.14 \cdot (2.5)^2 \cdot 8 = 3.14 \cdot 6.25 \cdot 8 = 157 \text{ in}^3$

Glass volume: $V = \frac{1}{3}\pi r^2 h \approx \frac{1}{3} \cdot 3.14 \cdot (2)^2 \cdot 1.5 = \frac{3.14 \cdot 4 \cdot 1.5}{3} = \frac{18.84}{3} = 6.28 \text{ in}^3$

We then divide the volume of the pitcher by the volume of the glass. $\frac{157}{6.28} = 25$

There are approximately 25 glasses of punch.

21. We need to first determine the volume of the trapezoidal wheelbarrow.

Area of base = $A = \frac{1}{2} \cdot (b_1 + b_2) \cdot h = \frac{1}{2} \cdot (2+3) \cdot 1 = \frac{1}{2} \cdot (5) \cdot 1 = \frac{5}{2} \text{ ft}^2$

Volume = area of the base times height = $\frac{5}{2} \cdot \frac{5}{2} = \frac{25}{4} = 6.25 \text{ ft}^3$

The volume of the wheelbarrow is then multiplied by 16. $\frac{25}{4} \cdot 16 = 100 \text{ ft}^3$

She needs 100 ft^3 of stone and this is approximately $\frac{100}{27} \approx 3.7 \text{ yd}^3$ of stone.

23. Volume of the rectangular cake: $V = lwh = 14 \cdot 9 \cdot 3 = 378 \text{ in}^3$

Volume of the round cake: $V = \pi r^2 h \approx 3.14 \cdot 5^2 \cdot 4 = 3.14 \cdot 25 \cdot 4 = 314 \text{ in}^3$

The rectangular cake has more volume.

25. We need to first determine the volume of the rectangular block and the cylinder.

Block volume: $V = l \cdot w \cdot h = 10 \cdot 6 \cdot 7 = 420 \text{ ft}^3$

Cylinder volume: $V = \pi r^2 h \approx 3.14 \cdot \left(\frac{1}{2}\right)^2 \cdot 6 = 3.14 \cdot \frac{1}{4} \cdot 6 = 3.14 \cdot \frac{3}{2} = 4.71 \text{ ft}^3$

The actual volume of the shape is $420 - 4.71 = 415.29 \text{ ft}^3$.

27. We need to first determine the volume of the rectangular block and the half-cylinder.

Block volume: $V = l \cdot w \cdot h = 4 \cdot 5 \cdot 3 = 60 \text{ yd}^3$

Half-cylinder volume: $V = \frac{1}{2}\pi r^2 h \approx \frac{1}{2} \cdot 3.14 \cdot \left(\frac{2}{2}\right)^2 \cdot 4 = \frac{1}{2} \cdot 3.14 \cdot 1 \cdot 4 = 3.14 \cdot 2 = 6.28 \text{ yd}^3$

The actual volume of the shape is $60 - 6.29 = 53.72 \text{ yd}^3$.

29. a) $V = \pi r^2 h \approx 3.14 \cdot 12^2 \cdot 5 = 3.14 \cdot 144 \cdot 5 = 2,260.8 \text{ in}^3$

b) $V = \pi r^2 h \approx 3.14 \cdot 10^2 \cdot 7 = 3.14 \cdot 100 \cdot 7 = 2,198 \text{ in}^3$

We get a larger increase by increasing the radius because in the formula for the volume of a cylinder, we square the radius. This is not the case in general, however.

31. a) $V = \frac{1}{3}\pi r^2 h \approx \frac{1}{3}\cdot 3.14\cdot 8^2\cdot 3 = \frac{3.14\cdot 64\cdot 3}{3} = \frac{602.88}{3} = 200.96 \text{ in}^3$

b) $V = \frac{1}{3}\pi r^2 h \approx \frac{1}{3}\cdot 3.14\cdot 6^2\cdot 5 = \frac{3.14\cdot 36\cdot 5}{3} = \frac{565.2}{3} = 188.4 \text{ in}^3$

We get a larger increase by increasing the radius because in the formula for the volume of a cylinder, we square the radius. This is not the case in general, however.

33. We must first find the volume of the concrete floor in cubic feet. To find the volume, we need to find the base area; that is, the area of the regular hexagon. The regular hexagon can be divided up into 6 equilateral triangles. To find the area of each of these triangles, we can either apply Heron's formula or use the Pythagorean theorem to find the height and then apply the area formula for a triangle. In this solution, we will use Heron's formula.

$s = \frac{1}{2}(a+b+c) = \frac{1}{2}(8+8+8) = \frac{24}{2} = 12$

Area of base applying Heron's formula is

$A = \sqrt{s(s-a)(s-b)(s-c)} = \sqrt{12(12-8)(12-8)(12-8)} = \sqrt{12\cdot 4\cdot 4\cdot 4} = \sqrt{768} = 16\sqrt{3} \approx 27.71 \text{ ft}^2$

Since there are 6 of these triangles, we have the area of the hexagon to be $16\sqrt{3}\cdot 6 = 96\sqrt{3}$ $\approx 166.28 \text{ ft}^2$.

Since the height of 6 inches is $\frac{1}{2}$ foot, we have

Volume = area of the base times height = $96\sqrt{3}\cdot\frac{1}{2} = 48\sqrt{3} \approx 83.14 \text{ ft}^3$. He must purchase 84 cubic feet of concrete.

35. The volume of the earth is approximately:

$V = \frac{4}{3}\pi r^3 \approx \frac{4}{3}\cdot 3.14\cdot\left(\frac{7{,}920}{2}\right)^3 = \frac{4}{3}\cdot 3.14\cdot 3{,}960^3 = \frac{4}{3}\cdot 3.14\cdot 62{,}099{,}136{,}000 = 259{,}988{,}382{,}720 \text{ mi}^3$.

The volume of the moon is therefore approximately $\frac{259{,}988{,}382{,}720}{49} \approx 5{,}305{,}885{,}361.63$.

$$\begin{aligned} V &= \frac{4}{3}\pi r^3 \\ 5{,}305{,}885{,}361.63 &\approx \frac{4}{3}\cdot 3.14\cdot r^3 \\ 1{,}267{,}329{,}306.12 &\approx r^3 \\ r &\approx \sqrt[3]{1{,}267{,}329{,}306.12} \\ r &\approx 1{,}082.17 \text{ mi} \end{aligned}$$

The diameter is therefore approximately $2\cdot 1{,}082.17 \approx 2{,}164$ miles.

37. The volume of a cone is less than half the volume of a cylinder with the same height and radius, so it is less than half of $\pi r^2 h$, namely $\frac{1}{3}\pi r^2 h$.

39. The volume is four times as large because in the formula of the volume of a cone, the radius is squared.

$$V=\frac{1}{3}\pi r^2h=\frac{1}{3}\cdot\pi\cdot(2r)^2\cdot h=\frac{1}{3}\pi\cdot4\cdot r^2h=4\cdot\left(\frac{1}{3}\pi r^2h\right)$$

41. Let d be the diameter of the can. The height of the can is $3d$; the circumference of the can is $\pi d\approx 3.14d$, which is larger.

43. The smaller cubes have more surface area. For example, compare the surface area of 1 3-inch cube with the surface of 27 1-inch cubes.

45. The volume of the $4\times4\times¼$ hamburger is $4\cdot4\cdot\frac{1}{4}=4\text{ in}^3$. The volume of the "tri-burger" will also be 4 in^3. This volume can be found from base area times height, where the height is ¼ in and the base is an equilateral triangle. Since the height is ¼ inch, the base area is 16 in^2. Since each side of an equilateral triangle has the same measure, let this measure be a. The perimeter of this triangle will be $3a$.

Applying Heron's formula we have $s=\frac{1}{2}(3a)=\frac{3a}{2}$ and

$$16=\sqrt{\frac{3a}{2}(\frac{3a}{2}-a)(\frac{3a}{2}-a)(\frac{3a}{2}-a)}$$

$$16=\sqrt{\frac{3a}{2}\cdot\frac{a}{2}\cdot\frac{a}{2}\cdot\frac{a}{2}}$$

$$16=\sqrt{\frac{3a^4}{16}}$$

$$256=\frac{3a^4}{16}$$

$$4{,}096=3a^4$$

$$\frac{4{,}096}{3}=a^4$$

$$a=\left(\frac{4{,}096}{3}\right)^{1/4}\approx 6.08\text{ in.}$$

47. $B=s^2=3^2=9\text{ in}^2$, so $V=\frac{1}{3}\cdot B\cdot h=\frac{1}{3}\cdot9\cdot5=15\text{ in}^3$.

49. To find the area of the triangle, we can apply Heron's formula.

$$s=\frac{1}{2}(a+b+c)=\frac{1}{2}(5+7+8)=\frac{20}{2}=10$$

Area of base applying Heron's formula is

$$A=\sqrt{s(s-a)(s-b)(s-c)}=\sqrt{10(10-5)(10-7)(10-8)}=\sqrt{10\cdot5\cdot3\cdot2}=\sqrt{300}=10\sqrt{3}\text{ ft}^2$$

$$V=\frac{1}{3}B\cdot h=\frac{1}{3}\cdot10\sqrt{3}\cdot6=20\sqrt{3}\approx34.64\text{ ft}^3.$$

51. $B = s^2 = 3^2 = 9$ ft

$$V = \frac{1}{3} \cdot B \cdot h$$

$$10 = \frac{1}{3} \cdot 9 \cdot h \qquad \text{in}^3.$$

$$3h = 10$$

$$h = \frac{10}{3} \approx 3.33 \text{ ft}$$

Section 9.5 The Metric System and Dimensional Analysis

1. 24,000 dl

3. 2,800 mm

5. 350 dL

7. h; $\frac{1}{4}\text{pound} \times \frac{454 \text{ grams}}{1 \text{ pound}} = 113.5 \text{ grams} = 1{,}135 \text{dg}$

9. e; $15 \text{ feet } = 15 \text{feet} \times \frac{1 \text{ yard}}{3 \text{ feet}} = 5 \text{ yards } \times \frac{1 \text{ meter}}{1.0936 \text{ yards}} \approx 4.572 \text{ meters} = 4{,}572 \text{ mm}$ or

$15 \text{ feet } = 15 \text{feet} \times \frac{12 \text{ inches}}{1 \text{ foot}} = 180 \text{ inches } \times \frac{1 \text{ meter}}{39.37 \text{ inches}} \approx 4.572 \text{ meters} = 4{,}572 \text{ mm}$

11. g; $6 \text{ inches } \times \frac{1 \text{ meter}}{39.37 \text{ inches}} \approx 0.1524 \text{ meters} = 15.24 \text{ cm}$

13. a; $8.5 \text{ inches } \times \frac{1 \text{ meter}}{39.37 \text{ inches}} \approx 0.2159 \text{ meters} = 0.02159 \text{ dam}$

15. d; $6 \text{ ounces } = 6 \text{ ounces} \times \frac{1 \text{ quart}}{32 \text{ ounces}} = 0.1875 \text{ quarts} \times \frac{1 \text{ liter}}{1.0567 \text{ quarts}} \approx 0.177 \text{ liters} = 1.77 \text{ dl}$

The explanations to Exercises 17 – 21 may vary.

17. a; A juice glass is about one eighth of a quart or about one-eighth of a liter.

19. c; A nose is about 2 inches. Since 39.37 inches is about one meter or 100 centimeters, one inch is about $\frac{100}{39.37} \approx 2.5$ cm. Five centimeters would correspond to a two-inch nose.

21. b; The dog might be about two feet tall or about $\frac{2}{3}$ of a meter.

23. $18 \text{ meters} \times \frac{1.0936 \text{ yards}}{1 \text{ meter}} = 19.6848 \text{ yards} \times \frac{3 \text{ feet}}{1 \text{ yard}} \approx 59.05 \text{ feet}$

25. $3 \text{ kilograms} = 3000 \text{ grams} \times \frac{1 \text{ pound}}{454 \text{ grams}} \approx 6.608 \text{ pounds} \times \frac{16 \text{ ounces}}{1 \text{ pound}} \approx 105.73 \text{ ounces}$

27. $2.1 \text{ kiloliters} = 2{,}100 \text{ liters} \times \frac{1.0567 \text{ quarts}}{1 \text{ liter}} = 2{,}219.07 \text{ quarts} \times \frac{1 \text{ gallon}}{4 \text{ quarts}} \approx 554.77 \text{ gallons}$

29. $10{,}000 \text{ deciliters} = 1{,}000 \text{ liters} \times \frac{1.0567 \text{ quarts}}{1 \text{ liter}} = 1{,}056.7 \text{ quarts}$

31. $176 \text{ centimeters} = 1.76 \text{ meters} \times \frac{1.0936 \text{ yards}}{1 \text{ meter}} \approx 1.9247 \text{ yards} \times \frac{3 \text{ feet}}{1 \text{ yard}} = 5.7741 \text{ feet} \times \frac{12 \text{ inches}}{1 \text{ foot}} \approx$

69.29 inches or $176 \text{ centimeters} = 1.76 \text{ meters} \times \frac{39.37 \text{ inches}}{1 \text{ meter}} \approx 69.29 \text{ inches}$

33. $45{,}000 \text{ kg} = 45{,}000{,}000 \text{ grams} \times \frac{1 \text{ pound}}{454 \text{ grams}} \approx 99{,}118.94 \text{ pounds} \times \frac{1 \text{ ton}}{2{,}000 \text{ pounds}} \approx 49.56 \text{ tons}$

35. $2.6 \text{ feet} \times \frac{1 \text{ yard}}{3 \text{ feet}} \approx 0.86667 \text{ yards} \times \frac{1 \text{ meter}}{1.0936 \text{ yards}} \approx 0.792 \text{ meters} = 7.92 \text{ decimeters}$ or

$2.6 \text{ feet} \times \frac{12 \text{ inches}}{1 \text{ foot}} = 31.2 \text{ inches} \times \frac{1 \text{ meter}}{39.37 \text{ inches}} \approx 0.792 \text{ meters} = 7.92 \text{ decimeters}$

37. Don't give him an inch.

39. It is first down and ten yards to go.

41. $8850 \text{ meters} \times \frac{1.0936 \text{ yards}}{1 \text{ meter}} \times \frac{3 \text{ feet}}{1 \text{ yard}} \approx 29{,}035.08 \text{ feet}$

43. a) $1 \text{ meter}^2 \times \frac{1.0936 \text{ yards}}{1 \text{ meter}} \times \frac{1.0936 \text{ yards}}{1 \text{ meter}} \times \frac{3 \text{ feet}}{1 \text{ yard}} \times \frac{3 \text{ feet}}{1 \text{ yard}} \approx 10.76 \text{ feet}^2$

b) $3 \text{ meter} \times 4 \text{ meter} = 12 \text{ meter}^2 \times \frac{10.76 \text{ feet}^2}{1 \text{ meter}^2} \approx 129.12 \text{ feet}^2$

45. $11 \text{ feet} \times 8 \text{ feet} \times 3 \text{ feet} = 264 \text{ feet}^3$

$264 \text{ feet}^3 \times \frac{1 \text{ yard}}{3 \text{ feet}} \times \frac{1 \text{ yard}}{3 \text{ feet}} \times \frac{1 \text{ yard}}{3 \text{ feet}} \approx 9.7778 \text{ yard}^3$

$9.7778 \text{ yard}^3 \times \frac{1 \text{ meter}}{1.0936 \text{ yards}} \times \frac{1 \text{ meter}}{1.0936 \text{ yards}} \times \frac{1 \text{ meter}}{1.0936 \text{ yards}} \approx 7.476 \text{ meter}^3$

$7.476 \text{ meter}^3 \times \frac{1000 \text{ liter}}{1 \text{ meter}^3} \approx 7476 \text{ liter}$

47. $\frac{80\,km}{1\,hr} \times \frac{1\,mi}{1.609\,km} \approx 49.72\,mph$

49. $V = lwh = 40 \cdot 20 \cdot 6 = 4{,}800 \text{ cubic feet} \times \dfrac{1 \text{ cubic yard}}{3 \times 3 \times 3 \text{ cubic feet}} \approx 177.7778 \text{ cubic yards}$

We now convert to cubic meters.

$$177.7778 \text{ cubic yards} \times \frac{1 \text{ cubic meter}}{1.0936 \times 1.0936 \times 1.0936 \text{ cubic yards}} \approx 135.9258 \text{ cubic meters}$$

We now convert to kiloliters.

$$135.9258 \text{ cubic meters} \times \frac{1 \text{ kiloliter}}{1 \text{ cubic meter}} \approx 135.93 \text{ kiloliters}$$

51. a) A diameter of 10 centimeters implies we have a radius of 5 centimeters. Thus the volume of the can in cubic centimeters would be $V = \pi r^2 h \approx 3.14 \cdot 5^2 \cdot 14 = 3.14 \cdot 25 \cdot 14 = 1{,}099 \text{ cm}^3$. Since 1,000 cubic centimeters is equal to one liter, we have a volume of 1.099 liters.

b) $1.099 \text{ liters} \times \dfrac{1.0567 \text{ quarts}}{1 \text{ liter}} = 1.1613133 \text{quarts} \times \dfrac{32 \text{ ounces}}{1 \text{ quarts}} \approx 37.16 \text{ ounces}$

53. $\dfrac{\$2.75}{1 \text{ kilogram}} \times \dfrac{1 \text{ kilogram}}{1{,}000 \text{ grams}} = \dfrac{\$2.75}{1{,}000 \text{ grams}} \times \dfrac{454 \text{ grams}}{1 \text{ pound}} = \dfrac{\$1{,}248.5}{1{,}000 \text{ pounds}} \approx \$1.25 \text{ per pound}$

55. $\dfrac{\$2.18}{1 L} \times \dfrac{0.9464 L}{1 \, quart} \times \dfrac{4 \, quart}{1 \, gallon} \approx \$8.25 \text{ per gallon}$

57. Since $P = 2l + 2w$, we have $2 \cdot 62 + 2 \cdot 35 = 124 + 70 = 194$ feet of fencing. We need to convert this to meters. $194 \text{ feet} \times \dfrac{1 \text{ yard}}{3 \text{ feet}} \approx 64.6667 \text{ yards} \times \dfrac{1 \text{ meter}}{1.0936 \text{ yards}} \approx 59.13 \text{ meters}$. She will need to buy 60 meters.

59. $\dfrac{30 \text{ miles}}{1 \text{ gallon}} \times \dfrac{1 \text{ gallon}}{4 \text{ quarts}} \times \dfrac{1.0567 \text{ quarts}}{1 \text{ liter}} \times \dfrac{1.609 \text{ km}}{1 \text{ mile}} \approx \dfrac{51.0069 \text{ km}}{4 \text{ liters}} \approx 12.75 \text{ kilometers per liter}$

61. $\dfrac{\$8.00}{1 ft^2} \times \dfrac{9 \, ft^2}{1 \, yd^2} \times \dfrac{1 \, yd}{0.9144 \, m} \times \dfrac{1 \, yd}{0.9144 \, m} \approx \$86.11 \text{ per square meter}$

63. 65° C

$149 = \frac{9}{5}C + 32$

$117 = \frac{9}{5}C$

$585 = 9C$

$65 = C$

65. 140° F

$F = \frac{9}{5} \cdot 60 + 32$

$F = 9 \cdot 12 + 32$

$F = 108 + 32$

$F = 140$

67. 58° C

$$136 = \frac{9}{5}C + 32$$
$$105 = \frac{9}{5}C$$
$$520 = 9C$$
$$58 \approx C$$

69. 104.4° C

$$59 = \frac{9}{5}C + 32 \qquad -129 = \frac{9}{5}C + 32$$
$$27 = \frac{9}{5}C \qquad -161 = \frac{9}{5}C$$
$$135 = 9C \qquad -805 = 9C$$
$$15 = C \qquad -89.4 \approx C$$
$$15° - (-89.4°) = 104.4°$$

71. One hectare $= 100 \text{ m} \cdot 100 \text{ m} = 10{,}000$ square meters

73. The land that Thiep purchased is $0.75 \text{ km} \cdot 1.2 \text{ km} = 0.9 \text{ km}^2$. We know from Exercise 71 that one square kilometer is equal to 100 hectares. So, Thiep purchased 90 hectares.

75. More decimeters because decimeters are smaller than hectometers.

77. Kiloliters are larger than dekaliters, therefore we need fewer kiloliters. So, b is larger than a.

79. $A = (0.05)(4)(12) = 2.4$ oz

$$BAC = \frac{A \times 5.14}{W \times r} - 0.015 \times H = \frac{2.4 \times 5.14}{160 \times 0.73} - 0.015 \times 2 = \frac{12.336}{116.8} - 0.03 \approx 0.1056 - 0.03 = 0.08$$

81. $\frac{1.585 \text{ euro}}{1 \text{ liter}} \times \frac{\$1}{0.7564 \text{ euro}} \times \frac{1 \text{ liter}}{1.0567 \text{ quarts}} \times \frac{4 \text{ quarts}}{1 \text{ gallon}} \approx \$7.93/\text{gallon}$

83. $\frac{78 \text{ baht}}{2 \text{ kg}} \times \frac{1 \text{ kg}}{2.2 \text{ lb}} \times \frac{\$1}{31 \text{ baht}} \approx \$0.57/\text{pound}$

85. No solution provided.

Section 9.6 Geometric Symmetry and Tessellations

1. – 3.

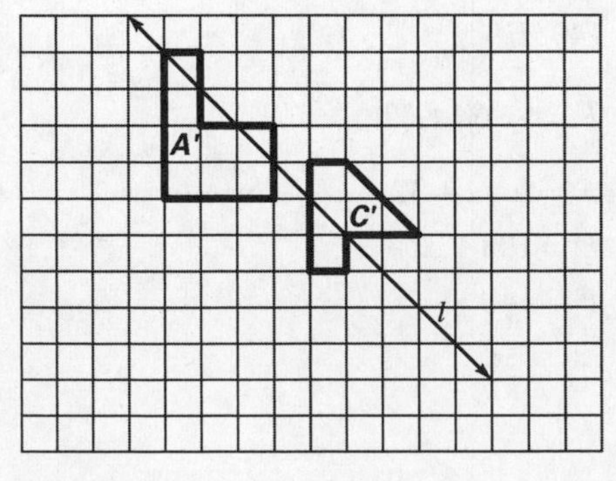

5.

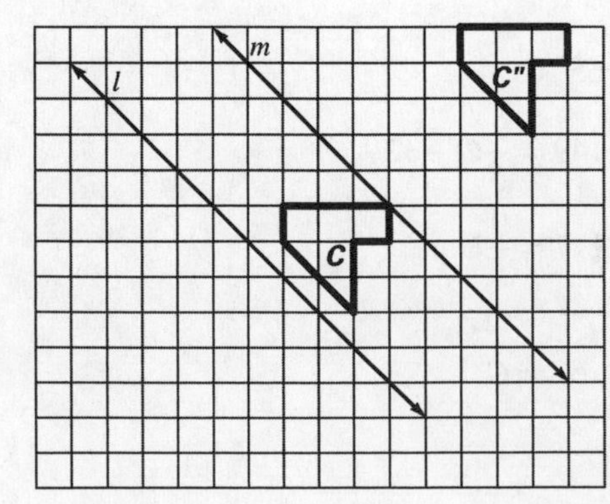

7.

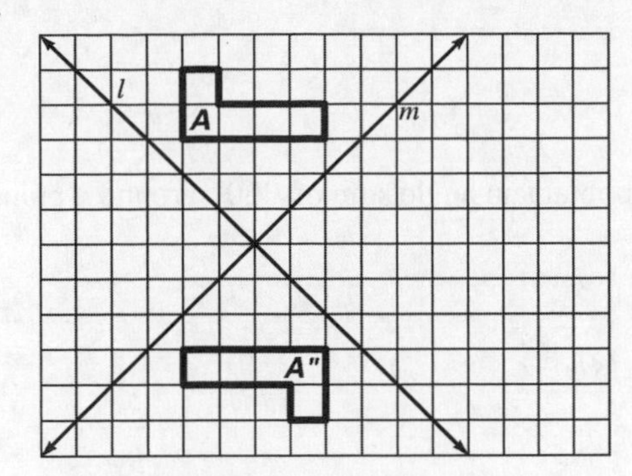

9. a) yes
   b) The effect is the same as if we performed a translation.

11.

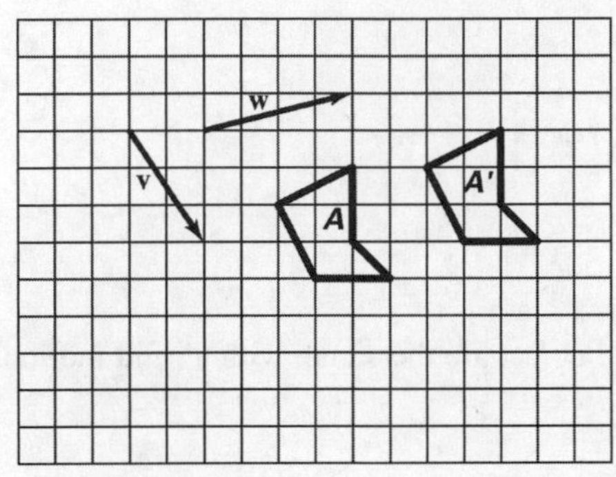

13.

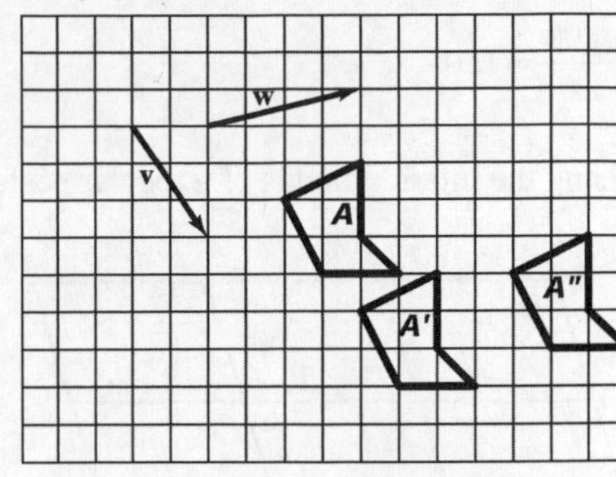

15.

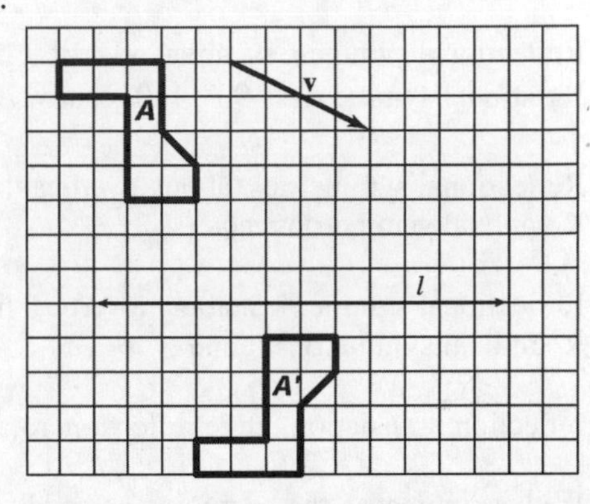

17.

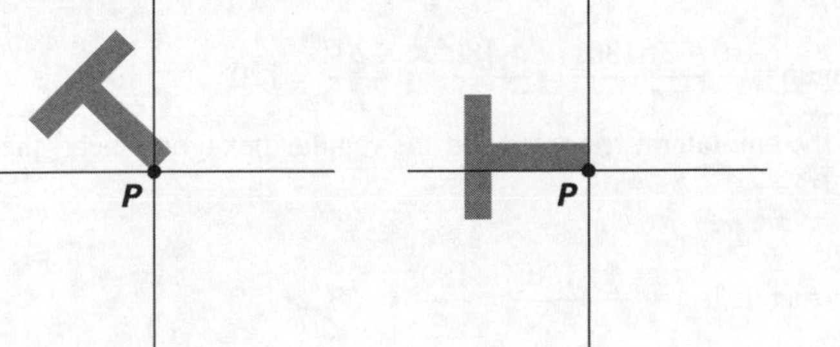

19.

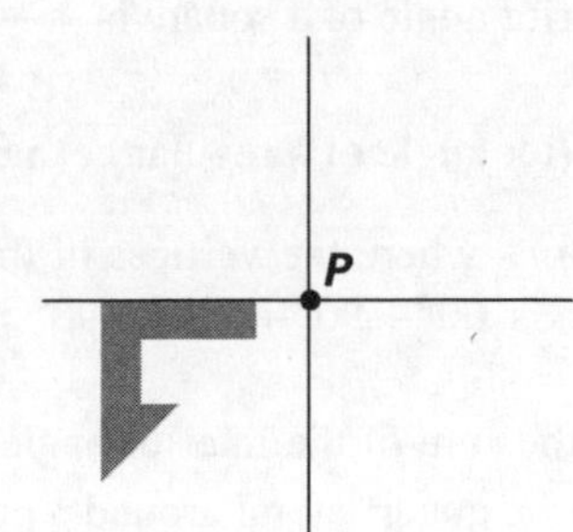

21. $(12-2)\cdot 180 = 10\cdot 180 = 1{,}800°$

$\dfrac{1{,}800}{12} = 150°$

23. Using the interior angles of a regular pentagon, we cannot obtain an angle sum of 360° around a point.

25.

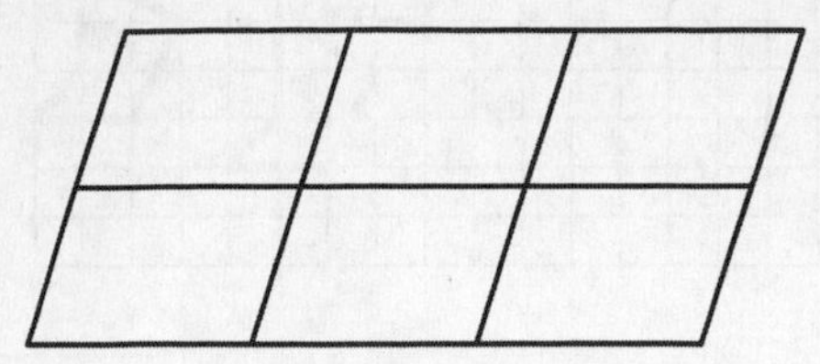

27. (c), (d)

29. (a), (e)

31. Reflectional symmetries: about a vertical line, about a horizontal line, and two diagonal lines
Rotational symmetries: 90°, 180°, and 270°

33. Reflectional symmetries: about a vertical line
Rotational symmetries: none

35. Reflectional symmetries: about a vertical line
Rotational symmetries: none

37. reflection, translation, glide reflection, rotation

39. With a symmetry, the beginning and ending positions of the object are the same; with a rigid motion, this does not have to be the case.

41. Each interior angle of an equilateral triangle is $\dfrac{(3-2)\cdot 180°}{3} = \dfrac{180°}{3} = 60°$.

Each interior angle of a regular hexagon is $\dfrac{(6-2)\cdot 180°}{6} = \dfrac{4\cdot 180°}{6} = \dfrac{720°}{6} = 120°$.

At the points where the vertices of the equilateral triangles and the regular hexagons meet, the angle sum is $120° + 60° + 120° + 60° = 360°$.

43. Each interior angle of an equilateral triangle is $\dfrac{(3-2)\cdot 180°}{3} = \dfrac{180°}{3} = 60°$.

Each interior angle of a square is $\dfrac{(4-2)\cdot 180°}{4} = \dfrac{360°}{4} = 90°$.

Each interior angle of a regular hexagon is $\dfrac{(6-2)\cdot 180°}{6} = \dfrac{4\cdot 180°}{6} = \dfrac{720°}{6} = 120°$.

At the points where the vertices of the equilateral triangle, squares, and the regular hexagon meet, the angle sum is $60° + 90° + 120° + 90° = 360°$.

45. Because the sum of the interior angles of a convex quadrilateral is 360°, you can always arrange four copies of the quadrilateral around a point to make an angle sum of 360°.

47. In the figure, we constructed the perpendicular bisectors of segments AA' and CC'. The point where these perpendicular bisectors meet is the center of rotation.

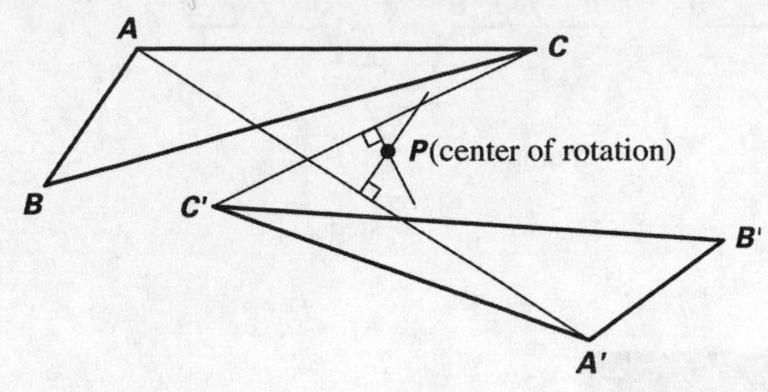

49. No solution provided.

Section 9.7 Looking Deeper: Fractals

1. $4^5 = 1,024$

3. $\left(\frac{4}{3}\right)^{10} = \frac{1,048,576}{59,049} \approx 17.76$

5. a)

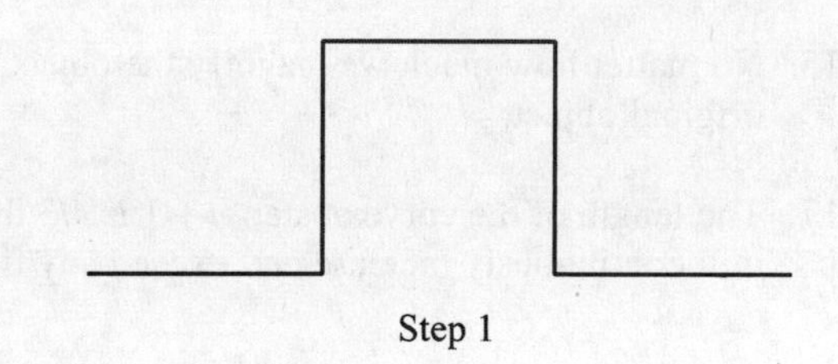

Step 0 Step 1

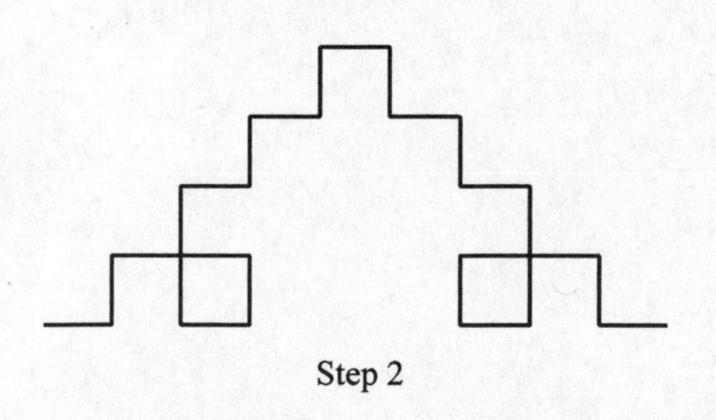

Step 2

b) Length in step 5: $\left(\frac{5}{3}\right)^5 = \frac{3,125}{243} \approx 12.86$

7.
$$4^D = 6$$
$$\log 4^D = \log 6$$
$$D \log 4 = \log 6$$
$$D = \frac{\log 6}{\log 4} \approx 1.29$$

9.
$$3^D = 5$$
$$\log 3^D = \log 5$$
$$D \log 3 = \log 5$$
$$D = \frac{\log 5}{\log 3} \approx 1.46$$

11.

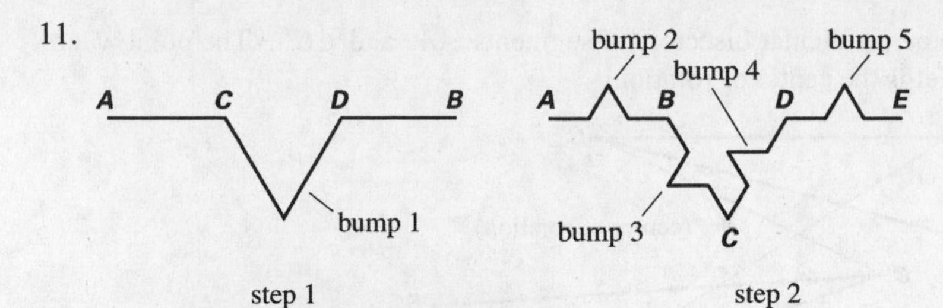

13.

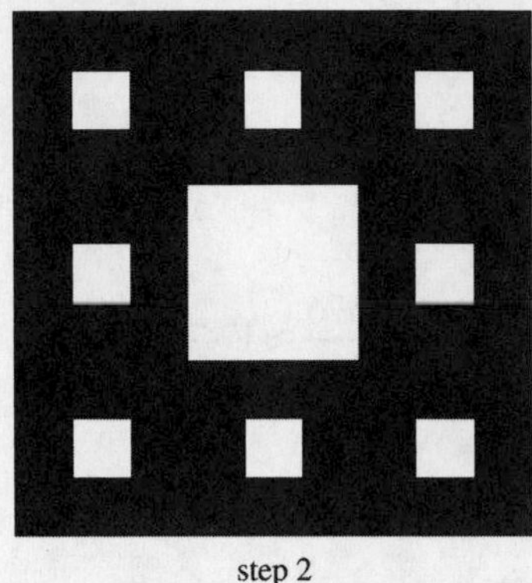

step 2

15. No matter how much we magnify the object, we see patterns that are similar to the patterns seen in the original object.

17. The length of the curve at step $n + 1$ is 4/3 the length at step n, so as n increases, the length of the curve will continuously increase and exceed any fixed number.

19. 5^n

21. At step 10: $\left(\frac{3}{4}\right)^{10} = \frac{59,049}{1,048,576} \approx 0.06$

At step n: $\left(\frac{3}{4}\right)^n$

23. – 25 No solution provided.

Chapter Review Exercises

1. a) b, g
   b) d, e

2. $m\angle a = 42°$; $m\angle b = 180° - 42° = 138°$;
   $m\angle c = 180° - 90° - 42° = 48°$

3.
$$\frac{\text{measure of central angle}}{360°} = \frac{\text{arc length}}{\text{circumference}}$$
$$\frac{m}{360°} = \frac{12-3}{24 \text{ in}}$$
$$\frac{m}{360°} = \frac{9}{24 \text{ in}}$$
$$24a = 360 \cdot 9$$
$$24a = 3,240$$
$$a = 135°$$

4.
$$x + 2x = 90°$$
$$3x = 90°$$
$$x = 30°$$

5. an arc of a great circle

6. $\frac{(18-2) \cdot 180°}{18} = \frac{16 \cdot 180°}{18} = \frac{2,880°}{18} = 160°$

7. $m\angle E = 45°$

$$\frac{EH}{8} = \frac{10}{12}$$
$$12(EH) = 10 \cdot 8$$
$$12(EH) = 80$$
$$EH = \frac{80}{12} = \frac{20}{3}$$

$$\frac{FG}{8} = \frac{6}{12}$$
$$12(FG) = 6 \cdot 8$$
$$12(FG) = 48$$
$$FG = \frac{48}{12} = 4$$

8. $$\frac{x}{4} = \frac{15}{6}$$
$$6x = 4 \cdot 15$$
$$6x = 60$$
$$x = 10 \text{ in}$$

9. a) $A = \frac{1}{2} \cdot (b_1 + b_2) \cdot h = \frac{1}{2} \cdot (20 + 12) \cdot 5 = \frac{1}{2} \cdot (32) \cdot 5 = 16 \cdot 5 = 80 \text{ cm}^2$

b) $A = \frac{1}{2} \cdot h \cdot b = \frac{1}{2} \cdot 4 \cdot 10 = 2 \cdot 10 = 20 \text{ in}^2$

10. a) Entire area: $A = \frac{1}{2} \cdot (b_1 + b_2) \cdot h = \frac{1}{2} \cdot (10 + 6) \cdot 4 = \frac{1}{2} \cdot (16) \cdot 4 = 8 \cdot 4 = 32 \text{ m}^2$

Unshaded area: $A = \frac{1}{2} \cdot h \cdot b = \frac{1}{2} \cdot 4 \cdot 10 = 2 \cdot 10 = 20 \text{ m}^2$

Shaded area: $32 - 20 = 12 \text{ m}^2$

b) Semi-circle's area: $A = \frac{1}{2}\pi \ r^2 \approx \frac{1}{2} \cdot 3.14 \cdot \left(\frac{8}{2}\right)^2 = \frac{1}{2} \cdot 3.14 \cdot 4^2 = \frac{1}{2} \cdot 3.14 \cdot 16 = 25.12 \text{ ft}^2$

Triangle's area: $A = \frac{1}{2} \cdot h \cdot b = \frac{1}{2} \cdot 4 \cdot 8 = 2 \cdot 8 = 16 \text{ ft}^2$

Shaded area: $25.12 + 16 = 41.12 \text{ ft}^2$

11. a) $s = \frac{1}{2}(a + b + c) = \frac{1}{2}(10 + 8 + 4) = \frac{22}{2} = 11$

$$A = \sqrt{s(s-a)(s-b)(s-c)} = \sqrt{11(11-10)(11-8)(11-4)} = \sqrt{11 \cdot 1 \cdot 3 \cdot 7} = \sqrt{231} \approx 15.2 \text{ cm}^2$$

b) $$A = \frac{1}{2} \cdot h \cdot b$$
$$\sqrt{231} = \frac{1}{2} \cdot h \cdot 10$$
$$\sqrt{231} = 5h$$
$$h = \frac{\sqrt{231}}{5} \approx 3.04 \text{ cm}$$

12. There are two rectangular pieces. Each of these has an area of $5 \cdot 100 = 500 \text{ m}^2$.
The two ends combined would form a washer-like figure in which we need to find the area of the shaded region.

Outer Area: $A_0 = \pi r^2$

$A_0 \approx 3.14 \cdot \left(\frac{20}{2}\right)^2$

$A_0 \approx 3.14 \cdot 10^2$

$A_0 \approx 3.14 \cdot 100$

$A_0 \approx 314 \text{ m}^2$

Inner Area: $A_I = \pi r^2$

$A_I \approx 3.14 \left(\frac{20-5-5}{2}\right)^2$

$A_I \approx 3.14 \cdot 5^2$

$A_I \approx 3.14 \cdot 25$

$A_I \approx 78.5 \text{ m}^2$

Surface area of track: 500 + 500 + 314 – 78.5 = 1,235.5 m^2

13. a) Area of base = $\frac{60}{360} \pi r^2 \approx \frac{1}{6} \cdot 3.14 \cdot 6^2 = \frac{3.14 \cdot 36}{6} = \frac{113.04}{6} = 18.84 \text{ in}^2$
Volume = area of the base times height $\approx 18.84 \cdot 2 = 37.68 \text{ in}^3$

b) Area of base = $A = \frac{1}{2} \cdot (b_1 + b_2) \cdot h = \frac{1}{2} \cdot (12+10) \cdot 5 = \frac{1}{2} \cdot (22) \cdot 5 = 11 \cdot 5 = 55 \text{ cm}^2$
Volume = area of the base times height = $55 \cdot 6 = 330 \text{ cm}^3$

14. We need to first determine the volume of the punch bowl and the cylindrical glass.
Punch bowl volume: $V = \frac{1}{2} \cdot \left(\frac{4}{3} \pi r^3\right) \approx \frac{4}{6} \cdot 3.14 \cdot 9^3 = \frac{4 \cdot 3.14 \cdot 729}{6} = \frac{9{,}156.24}{6} = 1{,}526.04 \text{ in}^3$

Glass volume: $V = \pi r^2 h \approx 3.14 \cdot \left(\frac{3}{2}\right)^2 \cdot 3 = 3.14 \cdot \frac{9}{4} \cdot 3 = 21.195 \text{ in}^3$

We then divide the volume of the punch bowl by the volume of the glass. $\frac{1{,}526.04}{21.195} = 72$

There are approximately 72 glasses of punch.

15. The volume is four times as large, because in the formula of the volume of a cone, the radius is squared.

$$V = \frac{1}{3} \pi r^2 h = \frac{1}{3} \cdot \pi \cdot (2r)^2 \cdot h = \frac{1}{3} \pi \cdot 4 \cdot r^2 h = 4 \cdot \left(\frac{1}{3} \pi r^2 h\right)$$

16. a) 3.5 meters
b) 43,150 centigrams
c) 38,600 deciliters

17. $514 \text{ decimeters} = 51.4 \text{ meters} \times \frac{1.0936 \text{ yards}}{1 \text{ meter}} \approx 56.21 \text{ yards}$

18. $2.1 \text{ kiloliters} = 2{,}100 \text{ liters} \times \frac{1.0567 \text{ quarts}}{1 \text{ liter}} = 2{,}219.07 \text{ quarts}$

19. $\frac{\$11.00}{1 ft^2} \times \frac{9 ft^2}{1 yd^2} \times \frac{1 yd}{0.9144 m} \times \frac{1 yd}{0.9144 m} \approx \$118.40$ per square meter

20.

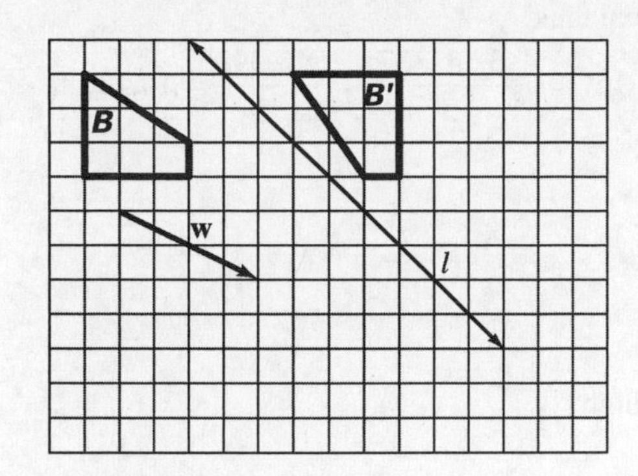

21. a) c, e b) d

22. Reflectional symmetries: about a vertical line and about a horizontal line
Rotational symmetries: 180°

23.

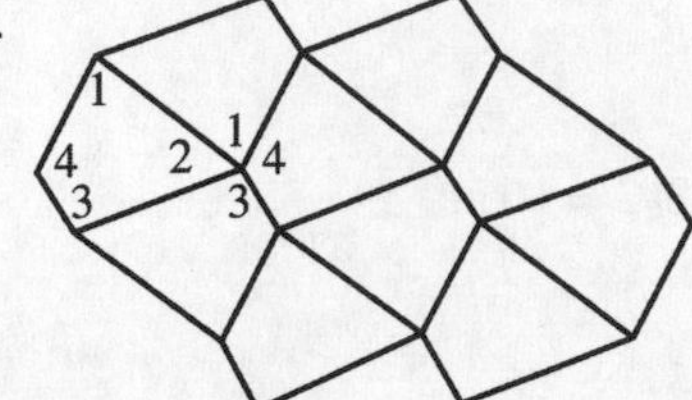

24. At each stage, the length increases by a factor of $\frac{4}{3}$. Therefore the length is increasing without bound.

25. zero

26. Assuming that the line segment in step 0 has length 1, in step 8 the length will be $\left(\frac{5}{3}\right)^8 \approx 59.5374$.

Chapter Test

1. a) vertical angles
   b) corresponding angles
   c) alternate exterior angles
   d) alternate interior angles

2. $12\cdot\left(\frac{(12-2)\cdot 180°}{12}\right) = \cancel{12}\cdot\left(\frac{(12-2)\cdot 180°}{\cancel{12}}\right) = (10)\cdot 180° = 1800°$

3. $m\angle c = 40°$; $m\angle b = 180° - 40° = 140°$;
   $m\angle a = 180° - (180° - 40° - 90°) = 180° - 50° = 130°$

4. We need to first determine the volume of the spherical tank.

   Spherical tank volume: $V = \left(\frac{4}{3}\pi r^3\right) \approx \frac{4}{3} \cdot 3.14 \cdot 15^3 = \frac{4 \cdot 3.14 \cdot 3375}{3} = \frac{42,390}{3} = 14,130 \text{ ft}^3$

   Let h be the height of the cylindrical tank.

   $V = \pi r^2 h \Rightarrow 14,130 \approx 3.14 \cdot (15)^2 h \Rightarrow$

   $h = 3.14 \cdot 225h \Rightarrow 706.5h = 14,130 \Rightarrow$

   $h = \frac{14,130}{706.5} = 20 \text{ ft}$

   The cylindrical tank should be approximately 20 ft high.

5. a) Area of base: $s = \frac{1}{2}(a+b+c) = \frac{1}{2}(6+6+8) = \frac{20}{2} = 10$

   $A = \sqrt{s(s-a)(s-b)(s-c)} = \sqrt{10(10-6)(10-6)(10-8)} =$

   $\sqrt{10 \cdot 4 \cdot 4 \cdot 2} = \sqrt{320} \approx 17.9 \text{ cm}^2$

   Volume = area of the base times height $\approx 17.9 \cdot 4 = 71.6 \text{ cm}^3$

   b) Area of base = $A = \frac{1}{2} \cdot (b_1 + b_2) \cdot h = \frac{1}{2} \cdot (10+6) \cdot 2 = \frac{1}{2} \cdot (16) \cdot 2 = 8 \cdot 2 = 16 \text{ in}^2$

   Volume = area of the base times height = $16 \cdot 3 = 48 \text{ in}^3$

6. a) *d*, *e*, and *f*
   b) *b* and *c*

7. $\frac{\text{measure of central angle}}{360°} = \frac{\text{arc length}}{\text{circumference}}$

   $\frac{180° - 150°}{360°} = \frac{a}{36 \text{ in}}$

   $\frac{30°}{360°} = \frac{a}{36 \text{ in}}$

   $360a = 30 \cdot 36$

   $360a = 1,080$

   $a = 3$

   The arc length is 3 inches.

8. Assuming that the line segment in step 0 has length 1, in step 5 the length will be $\left(\frac{4}{3}\right)^5 \approx 4.21$ units

9. a) $A = \frac{1}{2} \cdot (b_1 + b_2) \cdot h = \frac{1}{2} \cdot (15+12) \cdot 2 = \frac{1}{2} \cdot (27) \cdot 2 = \frac{1}{2} \cdot 2 \cdot (27) = 27 \text{ in}^2$

   b) $A = \frac{1}{2} \cdot h \cdot b = \frac{1}{2} \cdot 4 \cdot 16 = 2 \cdot 16 = 32 \text{ cm}^2$

10. a) Entire area: $A = \frac{1}{2} \cdot (b_1 + b_2) \cdot h = \frac{1}{2} \cdot (20+13) \cdot 4 = \frac{1}{2} \cdot (33) \cdot 4 = \frac{1}{2} \cdot 4 \cdot (33) = 2 \cdot 33 = 66 \text{ m}^2$

Unshaded area: $A = \frac{1}{2} \cdot h \cdot b = \frac{1}{2} \cdot 4 \cdot 20 = 2 \cdot 20 = 40 \text{ m}^2$

Shaded area: $66 - 40 = 26 \text{ m}^2$

b) Semi-circle's area: $A = \frac{1}{2}\pi r^2 \approx \frac{1}{2} \cdot 3.14 \cdot 3^2 = \frac{1}{2} \cdot 3.14 \cdot 9 = 14.13 \text{ ft}^2$

Triangle's area: $A = \frac{1}{2} \cdot h \cdot b = \frac{1}{2} \cdot 3 \cdot 6 = \frac{1}{2} \cdot 18 = 9 \text{ ft}^2$

Shaded area: $14.13 - 9 = 5.13 \text{ ft}^2$

11. $m\angle F = 35°$

$$\frac{FG}{12} = \frac{4}{6} \qquad\qquad \frac{GH}{10} = \frac{4}{6}$$

$$6(FG) = 4 \cdot 12 \qquad\qquad 6(GH) = 10 \cdot 4$$

$$6(FG) = 48 \qquad\qquad 6(GH) = 40$$

$$FG = \frac{48}{6} = 8 \qquad\qquad GH = \frac{40}{6} = \frac{20}{3}$$

12. a) The pool consists of a rectangle with area $(16-4-4) \cdot 20 = 160 \text{ ft}^2$. The two ends combine to form a circle with area $\pi(4)^2 = \pi(16) = 16\pi \text{ ft}^2$. So, the surface area of the pool is : $160 \text{ ft}^2 + 16\pi \text{ ft}^2 =$ or approximately 210.24 ft^2.

b) Volume = area of the base times height = $210.24 \cdot 3 = 630.72 \text{ ft}^2$.

c) There are two rectangular pieces. Each of these has an area of $4 \cdot 20 = 80 \text{ ft}^2$.
The two ends combined would form a washer-like figure in which we need to find the area of the shaded region.

Outer Area: $A_0 = \pi\ r^2$

$A_0 \approx 3.14 \cdot \left(\frac{16}{2}\right)^2$

$A_0 \approx 3.14 \cdot 8^2$

$A_0 \approx 3.14 \cdot 64$

$A_0 \approx 200.96 \text{ ft}^2$

Inner Area: $A_I = \pi\ r^2$

$A_I \approx 3.14 \left(\frac{16-4-4}{2}\right)^2$

$A_I \approx 3.14 \cdot 4^2$

$A_I \approx 3.14 \cdot 16$

$A_I \approx 50.24 \text{ ft}^2$

Surface area of track: $80 + 80 + 200.96 - 50.24 = 310.72 \text{ ft}^2$

13. $4x + 5x = 90°$

$9x = 90°$

$x = 10°$

14.

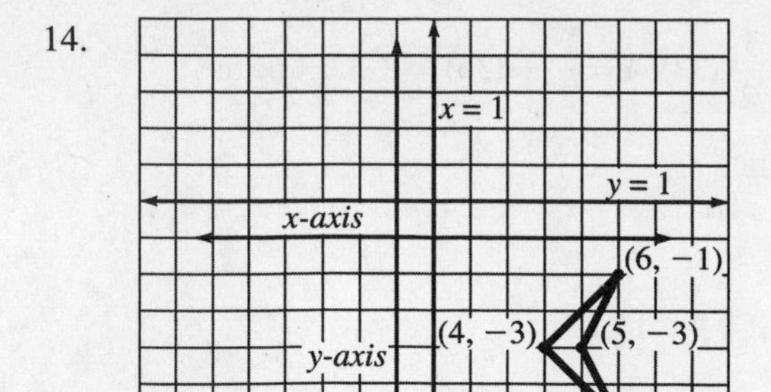

15. a) $s=\frac{1}{2}(a+b+c)=\frac{1}{2}(6+9+12)=\frac{27}{2}=13.5$

$$A=\sqrt{s(s-a)(s-b)(s-c)}=\sqrt{13.5(13.5-6)(13.5-9)(13.5-12)}$$
$$=\sqrt{(13.5)\cdot(7.5)\cdot(4.5)\cdot(1.5)}=\sqrt{683.4375}\approx 26.1 \text{ ft}^2$$

b) $A=\frac{1}{2}\cdot h\cdot b$

$$\sqrt{683.4375}=\frac{1}{2}\cdot h\cdot 12$$
$$\sqrt{683.4375}=6h$$
$$h=\frac{\sqrt{683.4375}}{6}\approx 4.4 \text{ ft}$$

16. zero

17. a) 24 meters; b) 3,460,000 milligrams; c) 2,140 centiliters

18. The surface area is four times as large, because in the formula for the surface area of a sphere, $A=4\pi r^2$, the radius is squared.

If the radius is doubled: $A=4\pi r^2=4\pi(2r)^2=4\pi(4r^2)=4\pi(4r^2)=4(4\pi r^2)$

19. 18 yards $\times\frac{1 \text{ meter}}{1.0936 \text{ yards}}\approx 16.459$ meters = 164.59 decimeters

20. 2,614.35 quarts$\times\frac{1 \text{ liter}}{1.0567 \text{ quarts}}\approx 2{,}474$ liters = 2.474 kiloliters

21. Reflectional symmetries about lines through *AE, BF, CG, DH*; rotational symmetries through 0°,90°,180°, and 270°.

22.

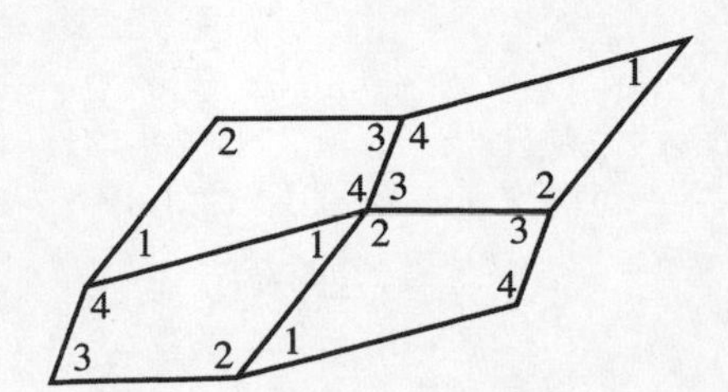

Chapter 10
Apportionment: How Do We Measure Fairness?

Section 10.1 Understanding Apportionment

1.

State	Number of customer %	Step 1: Determine exact number of seats deserved	Step 2: Assign integer parts	Examine fractional parts	Step 3: Assign 1 additional seat
California	56,000 (37.6%)	$0.376 \times 11 = 4.136$	4	0.136	4
Arizona	52,000 (34.9%)	$0.349 \times 11 = 3.839$	3	0.839	4
Nevada	41,000 (27.5%)	$0.275 \times 11 = 3.025$	3	0.025	3
Total	149,000 (100%)	11	10		11

3.

Guild	Number of members %	Step 1: Determine exact number of booths	Step 2: Assign integer parts	Examine fractional parts	Step 3: Assign 2 additional booths
Painters	87 (46.8%)	$0.468 \times 31 = 14.508$	14	0.508	14
Sculptures	46 (24.7%)	$0.247 \times 31 = 7.657$	7	0.657	8
Weavers	53 (28.5%)	$0.285 \times 31 = 8.835$	8	0.835	9
Total	186 (100%)	31	29		31

5.

State	Number of residents %	Step 1: Determine exact number of Representative	Step 2: Assign integer parts	Examine fractional parts	Step 3: Assign 1 additional Representative
Alabama	4,803 (38.9%)	$0.389 \times 17 = 6.613$	6	0.613	7
Mississippi	2,978 (24.2%)	$0.242 \times 17 = 4.114$	4	0.114	4
Louisiana	4,554 (36.9%)	$0.369 \times 17 = 6.273$	6	0.273	6
Total	12,335 (100%)	17	16		17

7. $\text{average constituency} = \dfrac{\text{population}}{\text{number of representatives}} = \dfrac{177,408}{3} = 59,136$

9. A is more poorly represented.

$$\text{average constituency of A} = \frac{\text{population}}{\text{number of representatives}} = \frac{27,600}{16} = 1,725$$

$$\text{average constituency of B} = \frac{\text{population}}{\text{number of representatives}} = \frac{23,100}{14} = 1,650$$

The absolute unfairness is $1,725 - 1,650 = 75$

$$\text{relative unfairness} = \frac{\text{absolute unfairness}}{\text{smaller average consituency}} = \frac{75}{1,650} \approx 0.045$$

11. $\text{average constituency of Naxxon} = \dfrac{\text{stockholders}}{\text{number of members}} = \dfrac{4,700}{5} = 940$

$\text{average constituency of Aroco} = \dfrac{\text{stockholders}}{\text{number of members}} = \dfrac{3,700}{4} = 925$

The absolute unfairness is 940 – 925 = 15.

$\text{relative unfairness} = \dfrac{\text{absolute unfairness}}{\text{smaller average consituency}} = \dfrac{15}{925} \approx 0.016$

13. $\text{average constituency of Texas} = \dfrac{\text{population}}{\text{number of representatives}} = \dfrac{25,268}{36} \approx 701.889$

$\text{average constituency of Georgia} = \dfrac{\text{population}}{\text{number of representatives}} = \dfrac{9,728}{14} \approx 694.857$

The absolute unfairness is 701.889 – 694.857 = 7.032.

$\text{relative unfairness} = \dfrac{\text{absolute unfairness}}{\text{smaller average consituency}} = \dfrac{7.032}{694.857} \approx 0.010$

15. a)

Division	Number of employees %	Step 1: Determine exact number of council members	Step 2: Assign integer parts	Examine fractional parts	Step 3: Assign 2 additional members
(D)igital	140 (54.9%)	$0.549 \times 12 = 6.588$	6	0.588	7
(C)omputers	85 (33.3%)	$0.333 \times 12 = 3.996$	3	0.996	4
(B)usiness Products	30 (11.8%)	$0.118 \times 12 = 1.416$	1	0.416	1
Total	255 (100%)	12	10		12

b)

Division	Number of employees %	Step 1: Determine exact number of council members	Step 2: Assign integer parts	Examine fractional parts	Step 3: Assign 1 additional member
(D)igital	140 (54.9%)	$0.549 \times 13 = 7.137$	7	0.137	7
(C)omputers	85 (33.3%)	$0.333 \times 13 = 4.329$	4	0.329	4
(B)usiness Products	30 (11.8%)	$0.118 \times 13 = 1.534$	1	0.534	2
Total	255 (100%)	13	12		13

Division	Number of employees %	Step 1: Determine exact number of council members	Step 2: Assign integer parts	Examine fractional parts	Step 3: Assign 2 additional members
(D)igital	140 (54.9%)	$0.549 \times 14 = 7.686$	7	0.686	8
(C)omputers	85 (33.3%)	$0.333 \times 14 = 4.662$	4	0.662	5
(B)usiness Products	30 (11.8%)	$0.118 \times 14 = 1.652$	1	0.652	1
Total	255 (100%)	14	12		14

c) Business Products loses a seat when the council size is increased from 13 to 14.

17. a)

Division	Number of patients %	Step 1: Determine exact number of doctors	Step 2: Assign integer parts	Examine fractional parts	Step 3: Assign 1 additional doctor
Center City	213 (37%)	$0.37 \times 23 = 8.51$	8	0.51	8
South Street	65 (11%)	$0.11 \times 23 = 2.53$	2	0.53	3
West Side	300 (52%)	$0.52 \times 23 = 11.96$	11	0.96	12
Total	100 (100%)	23	21		23

b) In going from 31 to 32, South Street loses a physician.

Division	Number of patients %	Step 1: Determine exact number of doctors	Step 2: Assign integer parts	Examine fractional parts	Step 3: Assign 1 additional doctor
Center City	213 (37%)	$0.37 \times 31 = 11.42$	11	0.42	11
South Street	65 (11%)	$0.11 \times 31 = 3.49$	3	0.49	4
West Side	300 (52%)	$0.52 \times 31 = 16.09$	16	0.09	16
Total	100 (100%)	31	30		31

Division	Number of patients %	Step 1: Determine exact number of doctors	Step 2: Assign integer parts	Examine fractional parts	Step 3: Assign 1 additional doctor
Center City	213 (37%)	$0.37 \times 32 = 11.79$	11	0.79	12
South Street	65 (11%)	$0.11 \times 32 = 3.60$	3	0.60	3
West Side	300 (52%)	$0.52 \times 32 = 16.61$	16	0.61	17
Total	100 (100%)	32	30		32

19. An Alabama paradox can occur when, as a legislature increases in size, we reassign representatives allocated earlier. Use an apportionment system that only assigns new representatives, and never reassigns representatives assigned earlier.

21. States will small populations generally will lose a representative at some point.

23. a) $\text{average constituency of Naxxon} = \dfrac{\text{stockholders}}{\text{number of members}} = \dfrac{4,700}{6} \approx 783$

$$\text{average constituency of Aroco} = \frac{\text{stockholders}}{\text{number of members}} = \frac{3,700}{4} = 925$$

The absolute unfairness is 925 – 783 = 142.

$$\text{relative unfairness} = \frac{\text{absolute unfairness}}{\text{smaller average consituency}} = \frac{142}{783} \approx 0.181$$

b) $\text{average constituency of Naxxon} = \dfrac{\text{stockholders}}{\text{number of members}} = \dfrac{4,700}{5} = 940$

$$\text{average constituency of Aroco} = \frac{\text{stockholders}}{\text{number of members}} = \frac{3,700}{5} = 740$$

The absolute unfairness is 940 – 740 = 200.

$$\text{relative unfairness} = \frac{\text{absolute unfairness}}{\text{smaller average consituency}} = \frac{200}{740} \approx 0.270$$

23. (continued)

c) Giving the extra representative to Naxxon will result in a smaller unfairness.

25. New York deserves the additional representative.

If New York receives the additional representative.

$$\text{average constituency of New York} = \frac{\text{population}}{\text{number of representatives}} = \frac{19{,}421{,}000}{28} \approx 693{,}607$$

$$\text{average constituency of New Jersey} = \frac{\text{population}}{\text{number of representatives}} = \frac{8{,}808{,}000}{12} = 734{,}000$$

The absolute unfairness is 734,000 – 693,607 = 40,393

$$\text{relative unfairness} = \frac{\text{absolute unfairness}}{\text{smaller average consituency}} = \frac{40{,}393}{693{,}607} \approx 0.058$$

If New Jersey receives the additional representative.

$$\text{average constituency of New York} = \frac{\text{population}}{\text{number of representatives}} = \frac{19{,}421{,}000}{27} \approx 719{,}296$$

$$\text{average constituency of New Jersey} = \frac{\text{population}}{\text{number of representatives}} = \frac{8{,}808{,}000}{13} \approx 677{,}538$$

The absolute unfairness is 719,296 – 677,538 = 41,758

$$\text{relative unfairness} = \frac{\text{absolute unfairness}}{\text{smaller average consituency}} = \frac{41{,}758}{677{,}538} \approx 0.062$$

27. Answers will vary.

Section 10.2 The Huntington–Hill Apportionment Principle

1. a) $$\text{average constituency of Musicians' Guild} = \frac{\text{members}}{\text{number of representatives}} = \frac{908}{5} = 181.6$$

$$\text{average constituency of Artists' Alliance} = \frac{\text{members}}{\text{number of representatives}} = \frac{633}{3} = 211$$

$$\text{relative unfairness} = \frac{\text{larger average consituency} - \text{smaller average consituency}}{\text{smaller average consituency}}$$

$$= \frac{211 - 181.6}{181.6} \approx 0.162$$

Note: If you round the average constituency of the Musicians' Guild to the nearest whole person, it would be approximately 182. This implies the relative unfairness would be approximately 0.159.

1. (continued)

b) average constituency of Musicians' Guild $= \dfrac{\text{members}}{\text{number of representatives}} = \dfrac{908}{4} = 227$

$$\text{average constituency of Artists' Alliance} = \frac{\text{members}}{\text{number of representatives}} = \frac{633}{4} \approx 158$$

$$\text{relative unfairness} = \frac{\text{larger average consituency} - \text{smaller average consituency}}{\text{smaller average consituency}}$$

$$= \frac{227 - 158}{158} \approx 0.437$$

Note: If you don't round the average constituency of the Artists' Alliance to the nearest whole person, it would be 156.25. The relative unfairness would be approximately 0.434.

c) Giving the additional representative to Musicians' Guild will result in a smaller unfairness.

3. a) average number of passengers on red line $= \dfrac{\text{passengers}}{\text{number of cars}} = \dfrac{405}{10} = 40.5$

$$\text{average number of passengers on blue line} = \frac{\text{passengers}}{\text{number of cars}} = \frac{287}{7} = 41$$

$$\text{relative unfairness} = \frac{\text{larger average} - \text{smaller average}}{\text{smaller average}} = \frac{41 - 40.5}{40.5} \approx 0.012$$

b) average number of passengers on red line $= \dfrac{\text{passengers}}{\text{number of cars}} = \dfrac{405}{9} = 45$

$$\text{average number of passengers on blue line} = \frac{\text{passengers}}{\text{number of cars}} = \frac{287}{8} = 35.875$$

$$\text{relative unfairness} = \frac{\text{larger average} - \text{smaller average}}{\text{smaller average}} = \frac{45 - 35.875}{35.875} \approx 0.254$$

c) Giving the additional car to red line will result in a smaller unfairness.

5. If New York receives the additional representative

$$\text{average constituency of New York} = \frac{\text{population}}{\text{number of representatives}} = \frac{19{,}400{,}000}{28} \approx 692{,}857$$

$$\text{average constituency of Pennsylvania} = \frac{\text{population}}{\text{number of representatives}} = \frac{12{,}700{,}000}{18} \approx 705{,}556$$

$$\text{relative unfairness} = \frac{\text{larger average consituency} - \text{smaller average consituency}}{\text{smaller average consituency}}$$

$$= \frac{705{,}556 - 692{,}857}{692{,}857} \approx 0.018$$

5. (continued)

If Pennsylvania receives the additional representative

$$\text{average constituency of New York} = \frac{\text{population}}{\text{number of representatives}} = \frac{19{,}400{,}000}{27} \approx 718{,}519$$

$$\text{average constituency of Pennsylvania} = \frac{\text{population}}{\text{number of representatives}} = \frac{12{,}700{,}000}{19} \approx 668{,}421$$

$$\begin{aligned}\text{relative unfairness} &= \frac{\text{larger average consituency} - \text{smaller average consituency}}{\text{smaller average consituency}}\\ &= \frac{718{,}519 - 668{,}421}{668{,}421} \approx 0.075\end{aligned}$$

Giving the additional representative to New York will result in a smaller unfairness.

7. If New Jersey receives the additional representative

$$\text{average constituency of New Jersey} = \frac{\text{population}}{\text{number of representatives}} = \frac{8{,}800{,}000}{13} \approx 679{,}923$$

$$\text{average constituency of Virgina} = \frac{\text{population}}{\text{number of representatives}} = \frac{8{,}000{,}000}{11} = 727{,}273$$

$$\begin{aligned}\text{relative unfairness} &= \frac{\text{larger average consituency} - \text{smaller average consituency}}{\text{smaller average consituency}}\\ &= \frac{727{,}273 - 679{,}923}{679{,}923} \approx 0.070\end{aligned}$$

If Virginia receives the additional representative

$$\text{average constituency of New Jersey} = \frac{\text{population}}{\text{number of representatives}} = \frac{8{,}800{,}000}{12} \approx 733{,}333$$

$$\text{average constituency of Virgina} = \frac{\text{population}}{\text{number of representatives}} = \frac{8{,}000{,}000}{12} = 666{,}667$$

$$\begin{aligned}\text{relative unfairness} &= \frac{\text{larger average consituency} - \text{smaller average consituency}}{\text{smaller average consituency}}\\ &= \frac{733{,}333 - 666{,}667}{666{,}667} \approx 0.100\end{aligned}$$

Giving the additional representative to New Jersey will result in a smaller unfairness.

9. $$\text{Huntington-Hill number} = \frac{(\text{population } X)^2}{x(x+1)} = \frac{908^2}{4 \cdot 5} = 41{,}223.2$$

11. $$\text{Huntington-Hill number} = \frac{(\text{population } X)^2}{x(x+1)} = \frac{287^2}{7 \cdot 8} \approx 1{,}470.9$$

13. $$\text{Huntington-Hill number} = \frac{(\text{population } X)^2}{x(x+1)} = \frac{6.5^2}{9 \cdot 10} \approx 0.469$$

15. Huntington-Hill number $= \dfrac{(\text{population } X)^2}{x(x+1)} = \dfrac{1.4^2}{2\cdot 3} \approx 0.327$

17. The electricians, plumbers, and carpenters each get one representative in no particular order. The remaining four assignments in order would be C, E, P, C. So the electricians would have 2 representatives, the plumbers would have 2, and the carpenters would have 3 according to the Huntington–Hill method

19. Theater, music and dance each get one representative in no particular order. The remaining eight assignments in order would be M, T, M, T, M, T, D, M. So theater would have 4 representatives, music would have 5, and dance would have 2 according to the Huntington–Hill method

21. a)

Boroughs	Number of residents %	Step 1: Determine exact number of seats	Step 2: Assign integer parts	Examine fractional parts	Step 3: Assign 2 additional seats
Bronx	1.4 (26.9%)	$0.269 \times 10 = 2.69$	2	0.69	3
Manhattan	1.6 (30.8%)	$0.308 \times 10 = 3.08$	3	0.08	3
Queens	2.2 (42.3%)	$0.423 \times 10 = 4.23$	4	0.23	4
Total	5.2 (100%)	10	9		10

b) Make the following table.

Current Representation	(B)ronx	(M)anhattan	(Q)ueens
1	$\dfrac{1.4^2}{1\cdot 2} = 0.98$	$\dfrac{1.6^2}{1\cdot 2} = 1.28$	$\dfrac{2.2^2}{1\cdot 2} = 2.42$
2	$\dfrac{1.4^2}{2\cdot 3} \approx 0.33$	$\dfrac{1.6^2}{2\cdot 3} \approx 0.43$	$\dfrac{2.2^2}{2\cdot 3} \approx 0.81$
3	$\dfrac{1.4^2}{3\cdot 4} \approx 0.16$	$\dfrac{1.6^2}{3\cdot 4} \approx 0.21$	$\dfrac{2.2^2}{3\cdot 4} \approx 0.40$
4	$\dfrac{1.4^2}{4\cdot 5} \approx 0.10$	$\dfrac{1.6^2}{4\cdot 5} = 0.13$	$\dfrac{2.2^2}{4\cdot 5} \approx 0.24$

Bronx, Manhattan, and Queens each get one representative in no particular order. The remaining seven assignments in order would be QMBQMQB. So Bronx should have 3 representatives, Manhattan should have 3, and Queens should have 4, according to the Huntington–Hill apportionment principle.

23. Make the following table.

Current Representation	Utah	Idaho	Oregon
1	$\dfrac{2.8^2}{1\cdot 2} = 3.92$	$\dfrac{1.6^2}{1\cdot 2} = 1.28$	$\dfrac{3.8^2}{1\cdot 2} = 7.22$
2	$\dfrac{2.8^2}{2\cdot 3} \approx 1.307$	$\dfrac{1.6^2}{2\cdot 3} \approx 0.428$	$\dfrac{3.8^2}{2\cdot 3} = 2.407$
3	$\dfrac{2.8^2}{3\cdot 4} \approx 0.653$	$\dfrac{1.6^2}{3\cdot 4} \approx 0.213$	$\dfrac{3.8^2}{3\cdot 4} \approx 1.203$
4	$\dfrac{2.8^2}{4\cdot 5} = 0.392$	$\dfrac{1.6^2}{4\cdot 5} = 0.128$	$\dfrac{3.8^2}{4\cdot 5} = 0.722$

23. (continued)

Utah, Idaho, and Oregon each get one representative in no particular order. If we assume "U" represents Utah, "I" represents Idaho, and "O" represents Oregon then the remaining seven assignments in order would be O, U, O, U, I, O, O, U. So Utah would have 4 representatives, Idaho would have 2, and Oregon would have 5 according to the Huntington–Hill method.

25. Make the following table.

Current Representation	Texas	Michigan	Florida
1	$\frac{25.3^2}{1\cdot 2} = 320.045$	$\frac{9.9^2}{1\cdot 2} = 49.005$	$\frac{18.9^2}{1\cdot 2} = 178.605$
2	$\frac{25.3^2}{2\cdot 3} \approx 106.682$	$\frac{9.9^2}{2\cdot 3} = 16.335$	$\frac{18.9^2}{2\cdot 3} = 59.535$
3	$\frac{25.3^2}{3\cdot 4} \approx 53.341$	$\frac{9.9^2}{3\cdot 4} \approx 8.168$	$\frac{18.9^2}{3\cdot 4} \approx 29.768$
4	$\frac{25.3^2}{4\cdot 5} \approx 32.005$	$\frac{9.9^2}{4\cdot 5} \approx 4.901$	$\frac{18.9^2}{4\cdot 5} \approx 17.861$
5	$\frac{25.3^2}{5\cdot 6} \approx 21.336$	$\frac{9.9^2}{5\cdot 6} \approx 3.267$	$\frac{18.9^2}{5\cdot 6} \approx 11.907$

Texas, Michigan, and Florida each get one representative in no particular order. If we assume "T" represents Texas, "M" represents Michigan, and "F" represents Florida then the remaining nine assignments in order would be T, F, T, F, T, M, T, F, T. So Texas would have 6 representatives, Michigan would have 2, and Florida would have 45 according to the Huntington–Hill method.

27. Make the following table.

Current Representation	Center 1	Center 2	Center 3
1	$\frac{98^2}{1\cdot 2} = 4802$	$\frac{34^2}{1\cdot 2} = 578$	$\frac{57^2}{1\cdot 2} \approx 1625$
2	$\frac{98^2}{2\cdot 3} \approx 1601$	$\frac{34^2}{2\cdot 3} \approx 193$	$\frac{57^2}{2\cdot 3} \approx 542$
3	$\frac{98^2}{3\cdot 4} \approx 800$	$\frac{34^2}{3\cdot 4} \approx 96$	$\frac{57^2}{3\cdot 4} \approx 271$

Center 1, Center 2, and Center 3 each get one representative in no particular order. If we assume "A" represents Center 1, "B" represents Center 2, and "C" represents Center 3, then the remaining four assignments in order would be ACAA. So Center 1 would have 4 representatives, Center 2 would have 1, and Center 3 would have 2, according to the Huntington–Hill method.

29. It minimizes relative unfairness.

31. This would be similar to the situation that Michigan has 15 representatives and Wisconsin has 9 and we want to assign one more representative.

$$\text{Huntington-Hill number for Michigan} = \frac{(\text{population } X)^2}{x(x+1)} = \frac{9.3^2}{15\cdot 16} \approx 0.360$$

$$\text{Huntington-Hill number for Wisconsin} = \frac{(\text{population } X)^2}{x(x+1)} = \frac{X^2}{9\cdot 10} = \frac{X^2}{90}$$

31. (continued)

$0.360 < \frac{X^2}{90} \Rightarrow 32.4 < X^2 \Rightarrow 5.6921 < X$ Therefore, Wisconsin's population needs to be approximately 5.7 million in order to take a seat away from Michigan.

33. 8 representatives are needed. In order to do this, A would have 2 representatives and C would have 3.

Section 10.3 Other Paradoxes and Apportionment Methods

1. a) $\text{Standard Divisor} = \frac{\text{Total Population}}{\text{Number of Representatives Allocated}} = \frac{512{,}000}{16} = 32{,}000$

$\text{Standard Quota} = \frac{\text{State's Population}}{\text{Standard Divisor}} = \frac{84{,}160}{32{,}000} = 2.63$

b) 2.63 rounded down is 2.

c) 2.63 rounded up is 3.

d) 2.63 rounded as usual is 3.

3. a) $\text{Standard Divisor} = \frac{\text{Total Population}}{\text{Number of Representatives Allocated}} = \frac{102{,}000}{11} = 9{,}272.73$

$\text{Standard Quota} = \frac{\text{State's Population}}{\text{Standard Divisor}} = \frac{19{,}975}{9{,}272.73} = 2.15$

b) 2.15 rounded down is 2.

c) 2.15 rounded up is 3.

d) 2.15 rounded as usual is 2.

5. The total population considered is $56{,}000 + 52{,}000 + 41{,}000 = 149{,}000$.

$\text{Standard Divisor} = \frac{149{,}000}{11} \approx 13{,}545.45$; California's Standard Quota $= \frac{56{,}000}{13{,}545.45} \approx 4.134$;

Nevada's Standard Quota $= \frac{41{,}000}{13{,}545.45} \approx 3.027$; Arizona's Standard Quota $= \frac{52{,}000}{13{,}545.45} \approx 3.839$

7. With the standard divisor of 13,545.45, a modified divisor is 14,000.

	California	Nevada	Arizona
Number of Customers	56,000	41,000	52,000
Standard Quota	4.134	3.027	3.839
Modified Quota	$\frac{56{,}000}{14{,}000} = 4$	$\frac{41{,}000}{14{,}000} \approx 2.93$	$\frac{52{,}000}{14{,}000} \approx 3.71$
Round Modified Quota Up	**4**	**3**	**4**
			Total = 11

9. The total number of workers considered is $213+273+178=664$.

Standard Divisor $= \frac{664}{20} = 33.2$; Performers' Standard Quota $= \frac{213}{33.2} \approx 6.416$;

Food Workers' Standard Quota $= \frac{273}{33.2} \approx 8.223$;

Maintenance Workers' Standard Quota $= \frac{178}{33.2} \approx 5.361$

11. With the standard divisor of 33.2, a modified divisor is 35.

	Performers	Food Workers	Maintenance
Number of Workers	213	273	178
Standard Quota	6.416	8.223	5.361
Modified Quota	$\frac{213}{35} \approx 6.09$	$\frac{273}{35} = 7.8$	$\frac{178}{35} \approx 5.09$
Round Modified Quota Up	7	8	6
			Total = 21

This total was too high, so try a modified divisor of 36.

	Performers	Food Workers	Maintenance
Number of Workers	213	273	178
Modified Quota	$\frac{213}{36} \approx 5.92$	$\frac{273}{36} \approx 7.58$	$\frac{178}{36} \approx 4.94$
Round Modified Quota Up	6	8	5
			Total = 19

This total was too low, so try a modified divisor of 35.5.

	Performers	Food Workers	Maintenance
Number of Workers	213	273	178
Modified Quota	$\frac{213}{35.5} = 6$	$\frac{273}{35.5} \approx 7.69$	$\frac{178}{35.5} \approx 5.01$
Round Modified Quota Up	**6**	**8**	**6**
			Total = 20

13. The total number of members considered is $25+18+29+31=103$.

Standard Divisor $= \frac{103}{20} = 5.15$; Electricians Union's Standard Quota $= \frac{25}{5.15} \approx 4.854$;

Plumber Union's Standard Quota $= \frac{18}{5.15} \approx 3.495$; Painters Union's Standard Quota $= \frac{29}{5.15} \approx 5.631$;

Carpenters Union's Standard Quota $= \frac{31}{5.15} \approx 6.019$

15. With the standard divisor of 5.15, a modified divisor is 5.5.

	Electricians	Plumbers	Painters	Carpenters
Number of Members	25	18	29	31
Standard Quota	4.85	3.50	5.63	6.02
Modified Quota	$\frac{25}{5.5} \approx 4.55$	$\frac{18}{5.5} \approx 3.27$	$\frac{29}{5.5} \approx 5.27$	$\frac{31}{5.5} \approx 5.64$
Round Modified Quota Up	5	4	6	6
				Total = 21

This total was too high, so try a modified divisor of 6.

	Electricians	Plumbers	Painters	Carpenters
Number of Members	25	18	29	31
Modified Quota	$\frac{25}{6} \approx 4.17$	$\frac{18}{6} = 3$	$\frac{29}{6} \approx 4.83$	$\frac{31}{6} \approx 5.17$
Round Modified Quota Up	5	3	5	6
				Total = 19

This total was too low, so try a modified divisor of 5.8.

	Electricians	Plumbers	Painters	Carpenters
Number of Members	25	18	29	31
Modified Quota	$\frac{25}{5.8} \approx 4.31$	$\frac{18}{5.8} \approx 3.10$	$\frac{29}{5.8} = 5$	$\frac{31}{5.8} \approx 5.34$
Round Modified Quota Up	**5**	**4**	**5**	**6**
				Total = 20

17. The total number of graduate students considered is $30 + 20 + 17 + 14 = 81$.

Standard Divisor $= \frac{81}{19} \approx 4.263$; Fiction Majors' Standard Quota $= \frac{30}{4.263} \approx 7.037$;

Poetry Majors' Standard Quota $= \frac{20}{4.263} \approx 4.692$;

Technical Writing Majors' Standard Quota $= \frac{17}{4.263} \approx 3.988$;

Media Writers Majors' Standard Quota $= \frac{14}{4.26} \approx 3.284$

19. With the standard divisor of 4.263 a modified divisor is 4.3.

	Fiction	Poetry	Technical	Media
Number of Students	30	20	17	14
Standard Quota	7.037	4.692	3.988	3.284
Modified Quota	$\frac{30}{4.3} \approx 6.98$	$\frac{20}{4.3} \approx 4.65$	$\frac{17}{4.3} \approx 3.95$	$\frac{14}{4.3} \approx 3.26$
Round Modified Quota Up	7	5	4	4
				Total = 20

19. (continued)

This total was too high, so try a modified divisor of 5.

	Fiction	Poetry	Technical	Media
Number of Students	30	20	17	14
Modified Quota	$\frac{30}{5} = 6$	$\frac{20}{5} = 4$	$\frac{17}{5} = 3.4$	$\frac{14}{5} = 2.8$
Round Modified Quota Up	6	4	4	3
				Total = 17

This total was too low, so try a modified divisor of 4.8.

	Fiction	Poetry	Technical	Media
Number of Students	30	20	17	14
Modified Quota	$\frac{30}{4.8} = 6.25$	$\frac{20}{4.8} \approx 4.17$	$\frac{17}{4.8} \approx 3.54$	$\frac{14}{4.8} \approx 2.92$
Round Modified Quota Up	**7**	**5**	**4**	**3**
				Total = 19

21. The total number of students considered is $56 + 29 + 11 + 4 = 100$.

Standard Divisor $= \frac{100}{6} \approx 16.667$; Pilates' Standard Quota $= \frac{56}{16.667} \approx 3.36$;

Kick Boxing's Standard Quota $= \frac{29}{16.667} \approx 1.74$;

Yoga's Standard Quota $= \frac{11}{16.667} \approx 0.66$;

Spinning's Standard Quota $= \frac{4}{16.667} \approx 0.24$

23. With the standard divisor of 16.67 a modified divisor of 27.5.

	Pilates	kick boxing	yoga	spinning
Number of Students	56	29	11	4
Standard Quota	3.36	1.74	0.66	0.24
Modified Quota	$\frac{56}{27.5} \approx 2.04$	$\frac{29}{27.5} \approx 1.05$	$\frac{11}{27.5} = 0.4$	$\frac{4}{27.5} \approx 0.15$
Round Modified Quota Up	3	2	1	1
				Total = 7

This total was too high, so try a modified divisor of 29.

	Pilates	kick boxing	yoga	spinning
Number of Students	56	29	11	4
Modified Quota	$\frac{56}{29} \approx 1.93$	$\frac{29}{29} = 1$	$\frac{11}{29} \approx 0.38$	$\frac{4}{29} \approx 0.14$
Round Modified Quota Up	2	1	1	1
				Total = 5

23. (continued)

This total was too low, so try a modified divisor of 28

	Pilates	kick boxing	yoga	spinning
Number of Students	56	29	11	4
Modified Quota	$\frac{56}{28} = 2$	$\frac{29}{28} \approx 1.04$	$\frac{11}{28} \approx 0.39$	$\frac{4}{28} \approx 0.14$
Round Modified Quota Up	**2**	**2**	**1**	**1**
				Total = 6

25. We must first find how the ten representatives were originally apportioned according to the Hamilton method. The standard divisor $= \frac{10,000}{10} = 1,000$.

State	Number of residents	Standard Quota	Integer parts	Fractional Parts	Assign 2 additional representatives
A	570	$\frac{570}{1,000} = 0.570$	0	0.570	1
B	2,557	$\frac{2,557}{1,000} = 2.557$	2	0.557	2
C	6,873	$\frac{6,873}{1,000} = 6.873$	6	0.873	7
Total	10,000		8		10

Determine the apportionment ten years later. The standard divisor $= \frac{10,080}{10} = 1,008$.

State	Number of residents	Standard Quota	Integer parts	Fractional Parts	Assign 2 additional representatives
A	590	$\frac{590}{1,008} \approx 0.585$	0	0.585	0
B	2,617	$\frac{2,617}{1,008} \approx 2.596$	2	0.596	3
C	6,873	$\frac{6,873}{1,008} \approx 6.818$	6	0.818	7
Total	10,080		8		10

Rate of increase in A's population $= \frac{20}{570} \approx 3.5\%$;

Rate of increase in B's population $= \frac{60}{2,557} \approx 2.3\%$

Yes, because A's population has increased faster than B's, but A loses a representative to B in the reapportionment.

27. We must first find how the 150 security personnel were originally apportioned according to the Hamilton method. The standard divisor $= \frac{150,000}{150} = 1,000$.

Airport	Number of passengers	Standard Quota	Integer parts	Fractional Parts	Assign 2 security personnel
Allenport	120,920	$\frac{120,920}{1,000} = 120.920$	120	0.920	121
Bakerstown	5,550	$\frac{5,550}{1,000} = 5.550$	5	0.550	6
Columbia City	23,530	$\frac{23,530}{1,000} = 23.530$	23	0.530	23
Total	150,000		148		150

Determine the apportionment one year later. The standard divisor $= \frac{151,000}{150} \approx 1006.67$.

Airport	Number of passengers	Standard Quota	Integer parts	Fractional Parts	Assign 2 security personnel
Allenport	121,420	$\frac{121,420}{1,006.67} \approx 120.615$	120	0.615	121
Bakerstown	5,650	$\frac{5,650}{1,006.67} \approx 5.613$	5	0.613	5
Columbia City	23,930	$\frac{23,930}{1,006.67} \approx 23.771$	23	0.771	24
Total	151,000		148		150

Rate of increase at Columbia City $= \frac{23,930 - 23,530}{23,530} = \frac{400}{23,530} \approx 1.7\%$

Rate of increase at Bakerstown $= \frac{5,650 - 5,550}{5,550} = \frac{100}{5,550} \approx 1.8\%$

Yes, because the number of passengers per week at Bakerstown has increased faster than at Columbia City, but Bakerstown loses a security guard to Columbia City.

29. Consider the apportionment among only states A and B. The standard divisor $= \frac{9,770}{100} = 97.7$.

State	Number of residents	Standard Quota	Integer parts	Fractional Parts	Assign 1 additional computer labs
A	2,000	$\frac{2,000}{97.7} \approx 20.471$	20	0.471	20
B	7,770	$\frac{7,770}{97.7} \approx 79.529$	79	0.529	**80**
Total	9,770		99		100

Now consider approximately the number of representatives C would get with the standard divisor of 97.7. C would get $\frac{600}{97.7} \approx 6.141$ representatives.

Determine the apportionment with 6 additional committee members.

The standard divisor $= \frac{10,370}{106} \approx 97.83$.

State	Number of residents	Standard Quota	Integer parts	Fractional Parts	Assign 1 additional computer labs
A	2,000	$\frac{2,000}{97.83} \approx 20.444$	20	0.444	21
B	7,770	$\frac{7,770}{97.83} \approx 79.423$	79	0.423	**79**
C	600	$\frac{600}{97.83} \approx 6.133$	6	0.133	6
Total	10,370		105		106

Yes, when C's representatives are added to the commission, B loses a representative to A.

31. Consider the apportionment among only states A, B, and C. The standard divisor $= \frac{16,200}{56} \approx 289.29$.

State	Number of residents	Standard Quota	Integer parts	Fractional Parts	Assign 1 additional computer labs
A	8,700	$\frac{8,700}{289.29} \approx 30.074$	30	0.074	30
B	4,300	$\frac{4,300}{289.29} \approx 14.864$	14	0.864	15
C	3,200	$\frac{3,200}{289.29} \approx 11.062$	11	0.062	11
Total	16,200		55		56

Now consider approximately the number of representatives D would get with the standard divisor of 289.29. D would get $\frac{6,500}{289.29} \approx 22.469$ representatives.

Determine the apportionment with 22 additional committee members.

The standard divisor $= \frac{22,700}{78} \approx 291.03$.

State	Number of residents	Standard Quota	Integer parts	Fractional Parts	Assign 3 additional computer labs
A	8,700	$\frac{8,700}{291.03} \approx 29.894$	29	0.894	30
B	4,300	$\frac{4,300}{291.03} \approx 14.775$	14	0.775	15
C	3,200	$\frac{3,200}{291.03} \approx 10.995$	10	0.995	11
D	6,500	$\frac{6,500}{291.03} \approx 22.334$	22	0.334	22
Total	22,700		75		78

No, when D's representatives are added to the commission, there is no change in A's, B's, or C's apportionment.

33. Total Population $= 700+1{,}500+820+4{,}530+2{,}200+550=10{,}300$

$$\text{Standard Divisor} = \frac{\text{Total Population}}{\text{Number of Representatives Allocated}} = \frac{10{,}300}{200} = 51.5$$

State	Number of residents	Standard Quota	Lower Quota	Upper Quota
A	700	$\frac{700}{51.5} \approx 13.592$	13	14
B	1,500	$\frac{1{,}500}{51.5} \approx 29.126$	29	30
C	820	$\frac{820}{51.5} \approx 15.922$	15	16
D	4,530	$\frac{4{,}530}{51.5} \approx 87.961$	87	**88**
E	2,200	$\frac{2{,}200}{51.5} \approx 42.718$	42	43
F	550	$\frac{550}{51.5} \approx 10.680$	10	11

Jefferson apportionment method: With the standard divisor of 51.5 a modified divisor is 50.5.

State	Population	Modified Quota	Round Modified Quota Down
A	700	$\frac{700}{50.5} \approx 13.861$	13
B	1,500	$\frac{1{,}500}{50.5} \approx 29.703$	29
C	820	$\frac{820}{50.5} \approx 16.238$	16
D	4,530	$\frac{4{,}530}{50.5} \approx 89.703$	**89**
E	2,200	$\frac{2{,}200}{50.5} \approx 43.564$	43
F	550	$\frac{550}{50.5} \approx 10.891$	10
Total	10,300		200

Adams apportionment method: With the standard divisor of 51.5, a modified divisor is 52.1.

State	Population	Modified Quota	Round Modified Quota Up
A	700	$\frac{700}{52.1} \approx 13.436$	14
B	1,500	$\frac{1{,}500}{52.1} \approx 28.791$	29
C	820	$\frac{820}{52.1} \approx 15.739$	16
D	4,530	$\frac{4{,}530}{52.1} \approx 86.948$	87
E	2,200	$\frac{2{,}200}{52.1} \approx 42.226$	43
F	550	$\frac{550}{52.1} \approx 10.557$	11
Total	10,300		200

Using the Jefferson method, D would receive 89 representatives, which is greater than its upper quota.

35. Total Population $= 1,400 + 2,000 + 1,500 + 9,000 + 4,300 + 500 = 18,700$

$$\text{Standard Divisor} = \frac{\text{Total Population}}{\text{Number of Representatives Allocated}} = \frac{18,700}{500} = 37.4$$

State	Number of residents	Standard Quota	Lower Quota	Upper Quota
A	1,400	$\frac{1,400}{37.4} \approx 37.433$	37	38
B	2,000	$\frac{2,000}{37.4} \approx 53.476$	53	54
C	1,500	$\frac{1,500}{37.4} \approx 40.107$	40	41
D	9,000	$\frac{9,000}{37.4} \approx 240.642$	**240**	**241**
E	4,300	$\frac{4,300}{37.4} \approx 114.973$	114	115
F	500	$\frac{500}{37.4} \approx 13.369$	13	14

Jefferson apportionment method: With the standard divisor of 37.4 a modified divisor is 37.1.

State	Population	Modified Quota	Round Modified Quota Down
A	1,400	$\frac{1,400}{37.1} \approx 37.736$	37
B	2,000	$\frac{2,000}{37.1} \approx 53.908$	53
C	1,500	$\frac{1,500}{37.1} \approx 40.431$	40
D	9,000	$\frac{9,000}{37.1} \approx 242.588$	**242**
E	4,300	$\frac{4,300}{37.1} \approx 115.903$	115
F	500	$\frac{500}{37.1} \approx 13.477$	13
Total	18,700		500

Adams apportionment method: With the standard divisor of 37.4 a modified divisor is 37.7.

State	Population	Modified Quota	Round Modified Quota Up
A	1,400	$\frac{1,400}{37.7} \approx 37.135$	38
B	2,000	$\frac{2,000}{37.7} \approx 53.050$	54
C	1,500	$\frac{1,500}{37.7} \approx 39.788$	40
D	9,000	$\frac{9,000}{37.7} \approx 238.727$	**239**
E	4,300	$\frac{4,300}{37.7} \approx 114.058$	115
F	500	$\frac{500}{37.7} \approx 13.263$	14
Total	18,700		500

Using the Jefferson method, D would receive 242 representatives, which is greater than its upper quota. Using the Adams method, D would receive 239 representatives, which is less than its lower quota.

37. The standard divisor is the number of constituents that each representative has. The standard quota is the exact number of representatives due a state. We calculate the standard divisor once and calculate the standard quota for each state.

39. Because the Jefferson method rounds the modified quotas down, we often need larger modified quotas and hence smaller modified divisors.

41. Answers will vary.

43. She needs to try a larger modified divisor.

45. smallest 926; largest 940

47. smallest 9.64; largest 9.88

49. No solution provided.

Section 10.4 Looking Deeper: Fair Division

1. a) discrete
   b) continuous
   c) discrete

3.

	Darnell $\left(\frac{1}{2}\right)$	Joy $\left(\frac{1}{2}\right)$
Bid on car	\$110,000	\$85,000
Fair share of estate	\$110,000/2 = \$55,000	\$85,000/2 = \$42,500
Item obtained with the highest bid	Car	

5.

	Dennis (40%)	Zadie (60%)
Bid on copyright	\$20,000	\$28,000
Fair share of copyright	$\$20,000 \times 0.40 = \$8,000$	$\$28,000 \times 0.60 = \$16,800$
Item obtained with the highest bid		Copyright

7.

	Ed $\left(\frac{1}{3}\right)$	Al $\left(\frac{1}{3}\right)$	Jerry $\left(\frac{1}{3}\right)$
Bid on painting	\$13,000	\$ 8,000	\$ 9,000
Bid on statue	\$14,000	\$13,000	\$15,000
Total value	\$27,000	\$21,000	\$24,000
Fair share of estate	$\frac{1}{3} \cdot 27000 = \$9,000$	$\frac{1}{3} \cdot 21000 = \$7,000$	$\frac{1}{3} \cdot 24000 = \$8,000$

9.

	Darnell $\left(\frac{1}{2}\right)$	Joy $\left(\frac{1}{2}\right)$
Pays to estate $(+)$ or receives from estate $(-)$	\$55,000 $(+)$	\$42,500 $(-)$
Division of estate balance (\$12,500)	\$6,250 $(-)$	\$6,250 $(-)$
Summary of cash	Pays \$48,750	Receives \$48,750

11.

	Dennis (40%)	Zadie (60%)
Pays to pool $(+)$ or receives from pool $(-)$	\$8,000 $(-)$	\$11,200 $(+)$
Division of pool balance (\$3,200)	\$1,280 $(-)$	\$ 1,920 $(-)$
Summary of cash	Receives \$9,280	Pays \$9,280

13.

	Ed $\left(\frac{1}{3}\right)$	Al $\left(\frac{1}{3}\right)$	Jerry $\left(\frac{1}{3}\right)$
Items obtained with highest bid	Painting		Statue
Pays to estate $(+)$ or receives from estate $(-)$	$13,000-9,000$ \$4,000.00 $(+)$	\$7,000.00$(-)$	$15,000-8,000$ \$7,000.00 $(+)$
Division of estate balance (\$4,000)	\$1,333.33 $(-)$	\$1,333.33$(-)$	\$1,333.33 $(-)$
Summary of cash	Pays \$2,666.67	Receives \$8,333.33	Pays \$5,666.67

15.

	Betty $\left(\frac{1}{2}\right)$	Dennis $\left(\frac{1}{2}\right)$
Bid on ring	\$16,000	\$18,000
Bid on desk	\$ 4,500	\$ 5,000
Bid on books	\$ 4,000	\$ 3,000
Total value	\$24,500	\$26,000
Fair share of estate	\$12,250	\$13,000

	Betty $\left(\frac{1}{2}\right)$	Dennis $\left(\frac{1}{2}\right)$
Item with highest bid	Books	Ring and Desk
Pays to pool $(+)$ or receives from pool $(-)$	\$8,250 $(-)$	\$10,000 $(+)$
Division of estate balance (\$1,750)	\$ 875 $(-)$	\$ 875 $(-)$
Summary of Cash	Receives \$9,125	Pays \$9,125

Betty receives the books, Dennis receives the ring and desk; Dennis pays Betty \$9,125.

17. Each person has a different estimate of the value of the estate and does not know the others' estimates.

19. Those who think that items in the estate are worth more receive the items and then must contribute cash to the estate that is then (partially) distributed to those who made low estimates of the items in the estate.

21. – 27. No solution provided.

Chapter Review Exercises

1. When adding a member to the legislature and without changing populations, a state loses a representative.

2.

Division	Number of members %	Step 1:Determine exact number of seats	Step 2: Assign integer parts	Examine fractional parts	Step 3: Assign additional seats
Revolutionary War	560 (28.00%)	$0.2800 \times 11 = 3.0800$	3	0.0800	3
Civil War	524 (26.20%)	$0.2620 \times 11 = 2.8820$	2	0.8820	3
World War I	431 (21.55%)	$0.2155 \times 11 = 2.3705$	2	0.3705	2
World War II	485 (24.25%)	$0.2425 \times 11 = 2.6675$	2	0.6675	3
Total	2,000 (100%)	11	9		11

3. If the number of representatives increases, we reassign representatives that have been assigned earlier.

4. a) B is more poorly represented.

$$\text{average constituency of A} = \frac{\text{population}}{\text{number of representatives}} = \frac{935}{5} = 187 \text{ (thousand)}$$

$$\text{average constituency of B} = \frac{\text{population}}{\text{number of representatives}} = \frac{2,343}{11} = 213 \text{ (thousand)}$$

The absolute unfairness is $213 - 187 = 26$ (thousand).

b) $$\text{relative unfairness} = \frac{\text{absolute unfairness}}{\text{smaller average consituency}} = \frac{26}{187} \approx 0.139$$

5. a) $$\text{Huntington-Hill number for Florida} = \frac{(\text{population } X)^2}{x(x+1)} = \frac{16.03^2}{25 \cdot 26} \approx 0.395$$

$$\text{Huntington-Hill number for Texas} = \frac{(\text{population } X)^2}{x(x+1)} = \frac{21.49^2}{32 \cdot 33} \approx 0.437$$

b) Since $0.395 < 0.437$, Texas should receive the additional representative.

6. Once a representative is assigned, it is not reassigned at a later date.

7. Qinhuangdao, Beijing, Qingdao, and Tianjin routes each get one bus in no particular order. The remaining nine buses in order would be given to HBQHBHQBT. So the Beijing route should have 4 buses, the Qingdao route should have 3, the Tianjin route should have 2, and the Qinhuandao route should have 4 according to the Huntington–Hill apportionment principle.

8. Make the following table.

Current Representation	Tae Bo	Karate	Weight	Tai-Chi
1	$\frac{66^2}{1 \cdot 2} = 2178$	$\frac{39^2}{1 \cdot 2} \approx 761$	$\frac{18^2}{1 \cdot 2} \approx 162$	$\frac{23^2}{1 \cdot 2} = 265$
2	$\frac{66^2}{2 \cdot 3} = 726$	$\frac{39^2}{2 \cdot 3} \approx 254$	$\frac{18^2}{2 \cdot 3} = 54$	$\frac{23^2}{2 \cdot 3} \approx 88$
3	$\frac{66^2}{3 \cdot 4} = 363$	$\frac{39^2}{3 \cdot 4} \approx 127$	$\frac{18^2}{3 \cdot 4} = 27$	$\frac{23^2}{3 \cdot 4} \approx 44$
4	$\frac{66^2}{4 \cdot 5} \approx 218$	$\frac{39^2}{4 \cdot 5} \approx 76$	$\frac{18^2}{4 \cdot 5} \approx 16$	$\frac{23^2}{4 \cdot 5} \approx 26$

Tae-Bo, karate, weight training, and Tae-Chi each get one class in no particular order. The remaining 4 classes in order would be Tae-Bo, karate, Tae-Bo, and Tae-Bo, according to the Huntington–Hill method. So Tae-Bo would have 4 classes, karate would have 2, weight training would have 1, and Tae-Chi would have 1.

9. Total Responses $= 8 + 64 + 11 + 31 = 114$

$$\text{Standard Divisor} = \frac{\text{Total Responses}}{\text{Number of Classes}} = \frac{114}{8} = 14.25$$

Jefferson apportionment method: With the standard divisor of 14.25, a modified divisor is 11.

Class	Responses	Modified Quota	Round Modified Quota Down
Pilates	8	$\frac{8}{11} \approx 0.727$	0
Kick boxing	64	$\frac{64}{11} \approx 5.818$	5
Yoga	11	$\frac{11}{11} = 1$	1
Spinning	31	$\frac{31}{11} \approx 2.818$	2
Total	114		8

10. Total Responses $= 8 + 64 + 11 + 31 = 114$

$$\text{Standard Divisor} = \frac{\text{Total Responses}}{\text{Number of Classes}} = \frac{114}{8} = 14.25$$

Adams apportionment method: With the standard divisor of 14.25, a modified divisor is 16.

Class	Responses	Modified Quota	Round Modified Quota Up
Pilates	8	$\frac{8}{16} = 0.5$	1
Kick boxing	64	$\frac{64}{16} = 4$	4
Yoga	11	$\frac{11}{16} \approx 0.688$	1
Spinning	31	$\frac{31}{16} \approx 1.938$	2
Total	114		8

11. Total Responses $= 8 + 64 + 11 + 31 = 114$

$$\text{Standard Divisor} = \frac{\text{Total Responses}}{\text{Number of Classes}} = \frac{114}{8} = 14.25$$

Webster apportionment method: With the standard divisor of 14.25, a modified divisor is 15.

Class	Responses	Modified Quota	Round Modified Quota
Pilates	8	$\frac{8}{15} \approx 0.533$	1
Kick boxing	64	$\frac{64}{15} \approx 4.267$	4
Yoga	11	$\frac{11}{15} \approx 0.733$	1
Spinning	31	$\frac{31}{15} \approx 2.067$	2
Total	114		8

12. We must first find how the 150 security personnel were originally apportioned according to the Hamilton method. The standard divisor $= \frac{180,600}{150} = 1,204$.

Airport	Number of passengers	Standard Quota	Integer parts	Fractional Parts	Assign 2 security personnel
A	80,500	$\frac{80,500}{1,204} \approx 66.860$	66	0.860	67
B	6,800	$\frac{6,800}{1,204} \approx 5.648$	5	0.648	6
C	93,300	$\frac{93,300}{1,204} \approx 77.492$	77	0.492	77
Total	180,600		148		150

Determine the apportionment one year later. The standard divisor $= \frac{196,410}{150} \approx 1,309.4$.

Airport	Number of passengers	Standard Quota	Integer parts	Fractional Parts	Assign 2 security personnel
A	87,700	$\frac{87,700}{1,309.4} \approx 66.977$	66	0.977	67
B	7,410	$\frac{7410}{1,309.4} \approx 5.659$	5	0.659	6
C	101,300	$\frac{101,300}{1,309.4} \approx 77.364$	77	0.364	77
Total	196,410		148		150

No, the apportionment stayed the same.

13. The new states paradox occurs when a new state is added, and its share of seats is added to the legislature causing a change in the allocation of seats previously given to another state.

14. Jefferson's, Adams', and Webster's

15.

	Tito $\left(\frac{1}{3}\right)$	Omarosa $\left(\frac{1}{3}\right)$	Piers $\left(\frac{1}{3}\right)$
Bid on rifle	\$12,000	\$17,000	\$10,000
Bid on sword	\$15,000	\$13,000	\$11,000
Total value	\$27,000	\$30,000	\$21,000
Fair share of estate	$\frac{1}{3} \cdot 27000 = \$9,000$	$\frac{1}{3} \cdot 30000 = \$10,000$	$\frac{1}{3} \cdot 21000 = \$7,000$

16.

	Tito $\left(\frac{1}{3}\right)$	Omarosa $\left(\frac{1}{3}\right)$	Piers $\left(\frac{1}{3}\right)$
Items obtained with highest bid	Sword	Rifle	
Pays to estate $(+)$ or receives from estate $(-)$	$15,000-9,000$ $6,000.00 $(+)$	$17,000-10,000$ $7,000.00 $(+)$	$7,000.00$(-)$
Division of estate balance ($6,000)	$2,000.00 $(-)$	$2,000.00$(-)$	$2,000.00 $(-)$
Summary of cash	Pays $4,000.00	Pays $5,000.00	Receives $9,000.00

Chapter Test

1. When adding a member to the legislature and without changing populations, a state loses a representative.

2. a) D is more poorly represented.

$$\text{average constituency of C} = \frac{\text{population}}{\text{number of representatives}} = \frac{1{,}640}{8} = 205 \text{ (thousand)}$$

$$\text{average constituency of D} = \frac{\text{population}}{\text{number of representatives}} = \frac{1{,}863}{9} = 207 \text{ (thousand)}$$

The absolute unfairness is $207-205=2$ (thousand).

b) $$\text{relative unfairness} = \frac{\text{absolute unfairness}}{\text{smaller average consituency}} = \frac{2}{205} \approx 0.0098$$

3.

Division	Number of members %	Step 1:Determine exact number of seats	Step 2: Assign integer parts	Examine fractional parts	Step 3: Assign additional seats
Art	47 (23.86%)	$0.2386\times 8 =$ 1.9088	1	0.9088	2
Music	111 (56.34%)	$0.5634\times 8 =$ 4.5072	4	0.5072	4
Theater	39 (19.80%)	$0.1980\times 8 =$ 1.5840	1	0.5840	2
Total	197 (100%)	8	6		8

4. If the number of representatives increases, representatives assigned earlier are reassigned.

5. a) $$\text{Huntington} - \text{Hill number for Oregon} = \frac{(\text{population } X)^2}{x(x+1)} = \frac{3.61^2}{5\cdot 6} \approx 0.434$$

$$\text{Huntington} - \text{Hill number for Arizona} = \frac{(\text{population } X)^2}{x(x+1)} = \frac{5.23^2}{8\cdot 9} \approx 0.380$$

b) Since $0.380 < 0.434$, Oregon should receive the additional representative.

6. Once a representative is assigned, it is not reassigned at a later date.

7. Mystery Mountain, City Sidewalks, Jungle Village, and Great Frontier routes each get one bus in no particular order. The remaining ten buses in order would be given to MCJMGCJMGC. So the Mystery Mountain route should have 4 buses, the City Sidewalks route should have 4, the Jungle Village route should have 3, and the Great Frontier route should have 3 according to the Huntington–Hill apportionment principle.

8. Make the following table.

Current Representation	Mathematics	Reading	Study Skills
1	$\frac{16^2}{1\cdot 2}=128$	$\frac{9^2}{1\cdot 2}\approx 41$	$\frac{6^2}{1\cdot 2}\approx 18$
2	$\frac{16^2}{2\cdot 3}\approx 43$	$\frac{9^2}{2\cdot 3}\approx 14$	$\frac{6^2}{2\cdot 3}=6$
3	$\frac{16^2}{3\cdot 4}\approx 21$	$\frac{9^2}{3\cdot 4}\approx 7$	$\frac{6^2}{3\cdot 4}=3$
4	$\frac{16^2}{4\cdot 5}\approx 13$	$\frac{9^2}{4\cdot 5}\approx 4$	$\frac{6^2}{4\cdot 5}\approx 2$
5	$\frac{16^2}{5\cdot 6}\approx 9$	$\frac{9^2}{5\cdot 6}\approx 3$	$\frac{6^2}{5\cdot 6}\approx 1$

Mathematics, Reading, and Study Skills each get one tutor in no particular order. The remaining 8 tutors in order would be, Mathematics, Mathematics, Reading, Mathematics, Study Skills, Reading, Mathematics, and Mathematics, according to the Huntington–Hill method. So Mathematics would have 6 tutors, Reading would have 3, and Study Skills would have 2.

9. Consider the apportionment among only airports A, B, and C.

The standard divisor $=\frac{180{,}600}{150}=1204.$

Town	Number of customers	Standard Quota	Integer parts	Fractional Parts	Assign 1 additional security officer
Allenwood	80,500	$\frac{80{,}500}{1204}\approx 66.86$	66	0.86	**67**
Black Hills	6,800	$\frac{6{,}800}{1204}\approx 5.65$	5	0.65	**6**
Colombia	93,300	$\frac{93{,}300}{1204}\approx 77.49$	77	0.49	77
Total	180,600		148		150

Now consider approximately the number of security personnel Devon would get with the standard divisor of 1204. Devon would get $\frac{41{,}400}{1204}\approx 34.39$ representatives.

Determine the apportionment with 34 additional committee members.

The standard divisor $=\frac{222{,}000}{184}\approx 1206.5.$

9. (continued)

Town	Number of customers	Standard Quota	Integer parts	Fractional Parts	Assign 1 additional security officer
Allenwood	80,500	$\frac{80,500}{1206.5} \approx 66.72$	66	0.72	**67**
Black Hills	6,800	$\frac{6,800}{1206.5} \approx 5.64$	5	0.64	**6**
Colombia	93,300	$\frac{93,300}{1206.5} \approx 77.33$	77	0.33	77
Devon	41,400	$\frac{41,400}{1206.5} \approx 34.31$	34	0.31	34
Total	222,000		182		184

No, because when Devon is added, no mall loses any security personnel.

10. The population paradox occurs when state A grows faster that state B but, with no change in the number of representatives, A loses a representative to B.

11. Total Responses $= 47 + 32 + 41 + 21 = 141$

$$\text{Standard Divisor} = \frac{\text{Total Responses}}{\text{Number of Classes}} = \frac{141}{9} = 15.67$$

Webster apportionment method: With the standard divisor of 15.67, a modified divisor is 15.

Class	Responses	Modified Quota	Round Modified Quota
massage	47	$\frac{47}{15} \approx 3.133$	3
aromatherapy	32	$\frac{32}{15} \approx 2.133$	2
yoga	41	$\frac{41}{15} \approx 2.733$	3
meditation	21	$\frac{21}{15} = 1.4$	1
Total	141		9

12. Total Responses $= 47 + 32 + 41 + 21 = 141$

Standard Divisor $= \dfrac{\text{Total Responses}}{\text{Number of Classes}} = \dfrac{141}{9} = 15.67$

Jefferson apportionment method: With the standard divisor of 15.67, a modified divisor is 12.

Class	Responses	Modified Quota	Round Modified Quota Down
massage	47	$\frac{47}{12} = 3.917$	3
aromatherapy	32	$\frac{32}{12} \approx 2.667$	2
yoga	41	$\frac{41}{12} = 3.417$	3
meditation	21	$\frac{21}{12} = 1.75$	1
Total	141		9

13. Total Responses $= 47 + 32 + 41 + 21 = 141$

Standard Divisor $= \dfrac{\text{Total Responses}}{\text{Number of Classes}} = \dfrac{141}{9} = 15.67$

Adams apportionment method: With the standard divisor of 15.67, a modified divisor is 20.8.

Class	Responses	Modified Quota	Round Modified Quota Up
massage	47	$\frac{47}{20.8} \approx 2.260$	3
aromatherapy	32	$\frac{32}{20.8} \approx 1.538$	2
yoga	41	$\frac{41}{20.8} \approx 1.971$	2
meditation	21	$\frac{21}{20.8} \approx 1.010$	2
Total	141		9

14. only Hamilton's

15.

	Larry $\left(\frac{1}{3}\right)$	Moe $\left(\frac{1}{3}\right)$	Curly $\left(\frac{1}{3}\right)$
Bid on Branch A	\$13 million	\$9 million	\$10 million
Bid on Branch B	\$17 million	\$18 million	\$14 million
Total value	\$30 million	\$27 million	\$24 million
Fair share of estate	$\frac{1}{3}\cdot 30 = \$10$ million	$\frac{1}{3}\cdot 27 = \$9$ million	$\frac{1}{3}\cdot 24 = \$8$ million

	Larry $\left(\frac{1}{3}\right)$	Moe $\left(\frac{1}{3}\right)$	Curly $\left(\frac{1}{3}\right)$
Items obtained with highest bid	Branch A	Branch B	
Pays to estate $(+)$ or receives from estate $(-)$	$13-10$ \$3 million $(+)$	$18-9$ \$9 million $(+)$	\$8 million $(-)$
Division of estate balance (\$4 million)	\$1,333,333.33 $(-)$	\$1,333,333.33 $(-)$	\$1,333,333.33 $(-)$
Summary of cash	Pays \$1,666,666.67	Pays \$7,666,666.67	Receives \$9,333,333.33

Chapter 11
Voting: Using Mathematics to Make Choices

Section 11.1 Voting Methods

1. a) No. The total number of votes is 2,156 + 1,462 + 986 + 428 = 5,032; $2156 \div 5032 \approx 0.428$, which is less than 50%

   b) Edelson

3. Athletic facilities with 33 votes

5. Since Student Union Building had the least votes, it is removed. The new table would be

Preference	15	30	18	17	10	2
1st	A	D	A	C	C	A
2nd	C	C	D	D	A	C
3rd	D	A	C	A	D	D

By combining identical columns, we have

Preference	17	30	18	17	10
1st	A	D	A	C	C
2nd	C	C	D	D	A
3rd	D	A	C	A	D

Since Campus Security (C) has the fewest first-place votes, it is then eliminated

Preference	17	30	18	17	10
1st	A	D	A	D	A
2nd	D	A	D	A	D

By combining identical columns, we have

Preference	45	47
1st	A	D
2nd	D	A

Since Dining Facilities (D) now has the most first-place votes, it wins the election.

7. Comedy with 23 votes

9. Since Greek Tragedy had the least votes, it is removed. The new table would be

Preference	10	15	13	12	5	7
1st	C	D	C	M	M	C
2nd	M	M	D	D	C	M
3rd	D	C	M	C	D	D

By combining identical columns, we have

Preference	17	15	13	12	5
1st	C	D	C	M	M
2nd	M	M	D	D	C
3rd	D	C	M	C	D

9. (continued)

Since Drama (D) has the fewest first-place votes, it is then eliminated.

Preference	17	15	13	12	5
1st	C	M	C	M	M
2nd	M	C	M	C	C

By combining identical columns, we have

Preference	30	32
1st	C	M
2nd	M	C

Since Mystery (M) now has the most first-place votes, it wins the election.

11. Technology with 15 votes

13. Since Families (F) had the least votes, it is removed. The new table would be

Preference	15	7	13	5	2
1st	T	E	G	P	E
2nd	P	P	E	G	G
3rd	E	G	P	E	P
4th	G	T	T	T	T

Since Poverty in Third World Countries (P) has the fewest first-place votes, it is then eliminated.

Preference	15	7	13	5	2
1st	T	E	G	G	E
2nd	E	G	E	E	G
3rd	G	T	T	T	T

By combining identical columns we have

Preference	15	9	18
1st	T	E	G
2nd	E	G	E
3rd	G	T	T

Since Environment (E) has the fewest first-place votes, it is then eliminated.

Preference	15	9	18
1st	T	G	G
2nd	G	T	T

By combining identical columns, we have

Preference	15	27
1st	T	G
2nd	G	T

Since Gender Roles (G) now has the most first-place votes, it wins the election.

15. Londram with 17 votes

17. Since E-Mall (E) had the least votes, it is removed. The new table would be

Preference	15	11	9	10	2
1st	L	F	S	T	L
2nd	F	S	L	F	S
3rd	S	L	F	L	F
4th	T	T	T	S	T

17. (continued)

Since Securenet (S) has the fewest first-place votes, it is then eliminated.

Preference	15	11	9	10	2
1st	L	F	L	T	L
2nd	F	L	F	F	F
3rd	T	T	T	L	T

By combining identical columns, we have

Preference	26	11	10
1st	L	F	T
2nd	F	L	F
3rd	T	T	L

Since Techenium (T) has the fewest first-place votes, it is then eliminated.

Preference	26	11	10
1st	L	F	F
2nd	F	L	L

By combining identical columns, we have

Preference	26	21
1st	L	F
2nd	F	L

Since Londram (L) now has the most first-place votes, it wins the election.

19. The Borda count is being used. Three points for 1st, two points for 2nd, and one point for 3rd.

Player, Team	1st	2nd	3rd	Total	
Robert Griffin III, Baylor	405	**168**	136	1,687	$(1687-405\times3-136\times1)\div2=168$
Andrew Luck, Stanford	**247**	250	166	1,407	$(1407-250\times2-166\times1)\div3=247$
Trent Richardson, Alabama	138	207	**150**	978	$(978-138\times3-207\times2)\div1=150$
Montee Ball, Wisconsin	22	83	116	348	$22\times3+83\times2+116\times1=348$

21. Answers may vary.
The first one eliminated should be in last place. The second one eliminated should be in the second-to-last place. This process continues until the winner is determined.

23. 1st : Athletic Facilities, 2nd :Dining Facilities, 3rd :Campus Security, and 4th : Student Union Building

25. 1st :Dining Facilities, 2nd :Athletic Facilities, 3rd :Campus Security, and 4th : Student Union Building

27. 1st :Technology, 2nd :Gender Roles, 3rd :Environment, 4th : Poverty in Third World Countries, and 5th :Families

29. 1st :Gender Roles, 2nd:Technology, 3rd :Environment, 4th : Poverty in Third World Countries, and 5th :Families

31. The possible points awarded are 4, 3, 2, or 1. $(4+3+2+1)\cdot20=10\cdot20=200$

33. If the candidate is in first place in all selections, then $5\cdot22=110$ is the Borda count.

35. Assume the four candidates are A, B, C, and D. The comparisons you need to make are A to B, A to C, A to D, B to C, B to D, and C to D. Since 1 point is awarded to each of the winners of these comparisons, there would be 6 points total.

37. Assume the five candidates are A, B, C, D and E. The comparisons you need to make are A to B, A to C, A to D, A to E, B to C, B to D, B to E, C to D, C to E, and D to E. Since 1 point is awarded to each of the winners of these comparisons and any individual candidate appears in a comparison only 4 times, then the maximum that could be earned is 4 points.

39. Voters can state their second, third, and fourth preferences.

41. A candidate who wins an election should be able to beat the other candidates head to head.

43. Answers will vary.

45. S, with 92 votes.

Candidate	Votes
A	15 + 0 + 18 + 10 + 2 = 45
C	15 + 30 + 0 + 17 + 10 + 2 = 74
D	0 + 30 + 18 + 17 + 0 + 0 = 65
S	**15 + 30 + 18 + 17 + 10 + 2 = 92**

47. E, with 37 votes.

Candidate	Votes
E	15 + 7 + 13 + 0 + 2 = 37
F	0 + 0 + 13 + 5 + 2 = 20
G	0 + 7 + 13 + 5 + 2 = 27
P	15 + 7 + 0 + 5 + 0 = 27
T	15 + 0 + 0 + 0 + 0 = 15

49. Answers may vary.
If you have two candidates, say A and B, the two choices are A 1st and B 2nd or B 1st and A 2nd.

Preference	p votes	q votes
1st	A	B
2nd	B	A

Preference	1st place votes × 2 (points)	2nd place votes × 1 (points)	Total Points
A	$p\times 2$	$q\times 1$	$2p+q$
B	$q\times 2$	$p\times 1$	$2q+p$

Suppose A is declared the winner; that is; $2p+q>2q+p\Rightarrow p>q$; thus, A had the majority of the votes.

51. Answers may vary.
If you have two candidates, say A and B, the two choices are A 1st and B 2nd or B 1st and A 2nd.

Preference	p votes	q votes
1st	A	B
2nd	B	A

We must compare A with B. Thus, p prefer A over B and q prefer B over A. The higher of the two amounts, p or q, determines who is awarded the point and wins the election. If $p > q$, then A wins the election and had the majority of the votes.

Section 11.2 Defects in Voting Methods

1. Option B with a total of 149 points

Preference	1st place votes × 4 (points)	2nd place votes × 3 (points)	3rd place votes × 2 (points)	4th place votes × 1 (points)	Total Points
A	25×4	0×3	15×2	9×1	139
B	15×4	21×3	13×2	0×1	149
C	9×4	15×3	12×2	13×1	118
D	0×4	13×3	9×2	27×1	84

3.

Preference		1st place votes × 4 (points)	2nd place votes × 3 (points)	3rd place votes × 2 (points)	4th place votes × 1 (points)	Total Points
Lower the age to 18	(A)	3×4	18×3	10×2	14×1	100
Lower the age to 19	(B)	10×4	14×3	21×2	0×1	124
Lower the age to 20	(C)	22×4	10×3	0×2	13×1	131
Keep the age at 21	(D)	10×4	3×3	14×2	18×1	95

Option C is the winner by the Borda count. The Condorcet's criterion is not satisfied in this example because option B can beat all of the other options in a head-to-head competition, but C is the winner by the Borda count. 24 out of 45 prefer B to A, 23 out of 45 prefer B to C, and 32 out of 45 prefer B to D.

5.

Preference		1st place votes × 4 (points)	2nd place votes × 3 (points)	3rd place votes × 2 (points)	4th place votes × 1 (points)	Total Points
Atlanta	(A)	2×4	16×3	29×2	0×1	114
Boston	(B)	18×4	19×3	2×2	8×1	141
Chicago	(C)	23×4	2×3	8×2	14×1	128
Denver	(D)	4×4	10×3	8×2	25×1	87

Option B is the winner by the Borda count. The independence-of-irrelevant-alternatives criterion is not satisfied in this example because if option A were taken out, then the preference table would be

Preference	15	4	8	10	8	2
1st	C	D	C	B	B	C
2nd	B	B	D	D	C	B
3rd	D	C	B	C	D	D

5. (continued)

According to the Borda count, option C would be declared the winner.

Preference		1st place votes × 3 (points)	2nd place votes × 2 (points)	3rd place votes × 1 (points)	Total Points
Chicago	(C)	25×3	8×2	14×1	105
Boston	(B)	18×3	21×2	8×1	104
Denver	(D)	4×3	18×2	25×1	73

7. Since option B (reduce expenditures on art and music programs) had the least votes, it is removed.

The new table would be

Preference	9	12	4	5	4	1
1st	C	D	C	A	A	C
2nd	A	A	D	D	C	A
3rd	D	C	A	C	D	D

By combining identical columns, we have

Preference	10	12	4	5	4
1st	C	D	C	A	A
2nd	A	A	D	D	C
3rd	D	C	A	C	D

Since option A (reduce sports programs) has the fewest first-place votes, it is then eliminated.

Preference	10	12	4	5	4
1st	C	D	C	D	C
2nd	D	C	D	C	D

By combining identical columns, we have

Preference	18	17
1st	C	D
2nd	D	C

Since option C (increase class size) now has the most first-place votes, it wins the election. The Condorcet's criterion is not satisfied in this example because option A (reduce sport programs) can beat all of the other options in a head-to-head competition, but C (increase class size) is the winner by the plurality-with-elimination method. 30 out of 35 prefer A to B, 21 out of 35 prefer A to C, and 19 out of 35 prefer A to D.

9. We must compare a) A with B, b) A with C, and c) B with C.

a) In comparing A with B, we ignore C. Thus, 13 prefer A over B, and 10+5=15 prefer B over A. Thus, we award 1 point to option B.

d) In comparing A with C, we ignore B. Thus, 13 prefer A over C, and 10+5=15 prefer C over A. Thus, we award 1 point to option C.

e) In comparing B with C, we ignore A. Thus, 13+10=23 prefer B over C, and 5 prefer C over B. Thus, we award 1 point to option B.

Since B has 2 points and C has 1, option B is declared the winner.

This example doesn't violate the majority criterion because the criterion states that <u>if</u> a majority of the voters rank a candidate as their first choice, <u>then</u> that candidate should win the election. The hypothesis of this conditional criterion was not satisfied. The majority (over half) did not vote for any one candidate.

11. Answers may vary.

Preference	**5**	**4**
1st	B	A
2nd	A	C
3rd	C	B

Preference	1st place votes × 3 (points)	2nd place votes × 2 (points)	3rd place votes × 1 (points)	Total Points
A	4×3	5×2	0×1	22
B	5×3	0×2	4×1	19
C	0×3	4×2	5×1	13

Option A is the winner by the Borda count. Condorcet's criterion is not satisfied in this example because option B can beat all of the other options in a head-to-head competition, but A is the winner by Borda count. 5 out of 9 prefer B to A and 5 out of 9 prefer B to C.

13. Answers may vary.

Preference									
1st	A	A	A	A	B	B	B	C	C
2nd	B	B	B	B	A	A	A	B	B
3rd	C	C	C	C	C	C	C	A	A

Option A is the winner by the plurality method. In head-to-head competition, B defeats both A and C, so Condorcet's criterion is not satisfied.

15. Answers may vary.

Preference	20	**30**	8	8	12
1st	C	D	A	A	B
2nd	A	A	D	B	C
3rd	B	B	B	C	A
4th	D	C	C	D	D

Since option B had the least votes, it is removed. The new table would be

Preference	20	30	8	8	12
1st	C	D	A	A	C
2nd	A	A	D	C	A
3rd	D	C	C	D	D

By combining identical columns, we have

Preference	32	30	8	8
1st	C	D	A	A
2nd	A	A	D	C
3rd	D	C	C	D

Since option A had the least votes, it is removed. The new table would be

Preference	32	30	8	8
1st	C	D	D	C
2nd	D	C	C	D

By combining identical columns, we have

Preference	40	38
1st	C	D
2nd	D	C

15. (continued)

Since option C now has the most first-place votes, it wins the election. Condorcet's criterion is not satisfied in this example because option A can beat all of the other options in a head-to-head competition, but option C is the winner by the plurality-with-elimination method. 66 out of 78 prefer A to B, 46 out of 78 prefer A to C, and 48 out of 78 prefer A to D

17. Answers may vary.
In the plurality-with-elimination method, if an option has the majority (over 50%) of the votes, then that option will not be eliminated during this process. If anything, the option will gain strength as other options are eliminated and remain the winner of the election.

19. a) With only Bush, Gore and Nader remaining, the vote totals are:

$$\text{Bush: } 50,456,002 + 448,895 + 98,020 + 83,714 + \frac{51,186}{2} = 51,112,224$$

$$\text{Gore: } 50,999,897 + \frac{51,186}{2} = 51,025,490$$

$$\text{Nader: } 2,882,955 + 384,431 = 3,267,386$$

b) With only Bush and Gore remaining, the vote totals are:

$$\text{Bush: } 51,112,224 + 0.70 \times 2,882,955 = 51,977,110$$

$$\text{Gore: } 51,025,490 + 384,431 + 0.70 \times 2,882,955 = 53,427,989$$

c) Gore wins with $\frac{53,427,989}{105,405,100} \approx 50.69\%$ of the total vote.

21. E and K receive the two greatest numbers of first place votes, so the runoff will be held between them. The revised preference table below shows that E wins with 17 first place votes to 10 first place votes for K.

Preference	3	7	5	3	9
1st	E	K	E	K	E
2nd	K	E	K	E	K

23. majority, Condorcet's, independence-of-irrelevant-alternatives, monotonicity

25. Answers may vary.
If the majority (over 50%) of the voters rank an option first, then that option will be chosen over any other choice in a head-to-head competition and thus will win using the pairwise comparison method.

27. – 33. Answers will vary.

35. pairwise comparison; independence-of-irrelevant-alternatives

Section 11.3 Weighted Voting Systems

1. a) The quota is 5.
   b) There are 5 voters. Each voter has 1 vote.
   c) There are no dictators.
   d) The sum of the votes is the same as the quota. Each voter has veto power.

3. a) The quota is 11.
   b) There are 4 voters. Voter A has 10 votes, voter B has 3, voter C has 4, and voter D has 5.
   c) There are no dictators.
   d) No voter has veto power.

5. a) The quota is 15.
   b) There are 5 voters. Voter A has 1 vote, voter B has 2, two voters (C and D) have 3, and voter E has 4. Notice that the sum of the votes is less than the quota. No resolutions can be passed.
   c) There are no dictators.
   d) No voter has veto power, because no resolutions can be passed.

7. a) The quota is 12.
   b) There are 4 voters. Voter A has 1 vote, voter B has 3, voter C has 5, and voter D has 7.
   c) There are no dictators.
   d) Two voters have veto power. Voters C and D each have veto power.

9. a) The quota is 25.
   b) There are 5 voters. Two voters (A and B) have 4 votes, voter C has 6, voter D has 7, and voter E has 9.
   c) There are no dictators.
   d) Three voters have veto power. Voters C, D, and E each have veto power.

11. a) The quota is 51.
    b) There are 5 voters. Three voters (A, B, and C) have 20 votes and two (D and E) have 10.
    c) There are no dictators.
    d) No voter has veto power.

13.

Coalition	Weight	
{A}	1	
{B}	3	
{C}	5	
{D}	7	
{A,B}	4	
{A,C}	6	
{A,D}	8	
{B,C}	8	
{B,D}	10	
{C,D}	12	Winning
{A,B,C}	9	
{A,B,D}	11	
{A,C,D}	13	Winning
{B,C,D}	15	Winning
{A,B,C,D}	16	Winning

15.

Coalition	Weight	
{A}	4	
{B}	4	
{C}	6	
{D}	7	
{E}	9	
{A,B}	8	
{A,C}	10	
{A,D}	11	
{A,E}	13	
{B,C}	10	
{B,D}	11	
{B,E}	13	
{C,D}	13	
{C,E}	15	
{D,E}	16	
{A,B,C}	14	
{A,B,D}	15	
{A,B,E}	17	
{A,C,D}	17	
{A,C,E}	19	
{A,D,E}	20	
{B,C,D}	17	
{B,C,E}	19	
{B,D,E}	20	
{C,D,E}	22	
{A,B,C,D}	21	
{A,B,C,E}	23	
{A,B,D,E}	24	
{A,C,D,E}	26	Winning
{B,C,D,E}	26	Winning
{A,B,C,D,E}	30	Winning

17.

Coalition	Weight		Critical Voters
{C,D}	12	Winning	C,D
{A,C,D}	13	Winning	C,D
{B,C,D}	15	Winning	C,D
{A,B,C,D}	16	Winning	C,D

19.

Coalition	Weight		Critical Voters
{A,C,D,E}	26	Winning	A,C,D,E
{B,C,D,E}	26	Winning	B,C,D,E
{A,B,C,D,E}	30	Winning	C,D,E

25.

Coalition	Weight		Critical Voters
{P,T}	9	Winning	P,T
{P,S}	7	Winning	P,S
{T,S}	6	Winning	T,S
{P,S,T}	11	Winning	None

21. A simple majority is 6, since there are 11 total voters.

Coalition	Weight	
{P}	5	
{T}	4	
{S}	2	
{P,T}	9	Winning
{P,S}	7	Winning
{T,S}	6	Winning
{P,S,T}	11	Winning

23.

Coalition	Weight	
{A}	3	
{C}	4	
{T}	3	
{N}	2	
{A,C}	7	
{A,T}	6	
{A,N}	5	
{C,T}	7	
{C,N}	6	
{T,N}	5	
{A,C,T}	10	Winning
{A,C,N}	9	Winning
{A,T,N}	8	Winning
{C,T,N}	9	Winning
{A,C,T,N}	12	Winning

27.

Coalition	Weight		Critical Voters
{A,C,T}	10	Winning	A,C,T
{A,C,N}	9	Winning	A,C,N
{A,T,N}	8	Winning	A,T,N
{C,T,N}	9	Winning	C,T,N
{A,C,T,N}	12	Winning	None

29. The only winning coalition in Exercise 1 is when all five members vote for the issue. Each voter would therefore have a Banzhaf power index of $\frac{1}{5}$.

31.

Coalition	Weight		Critical Voters
{A}	4		
{B}	4		
{C}	6		
{D}	7		
{E}	9		
{A,B}	8		
{A,C}	10		
{A,D}	11		
{A,E}	13		
{B,C}	10		
{B,D}	11		
{B,E}	13		
{C,D}	13		
{C,E}	15		
{D,E}	16		
{A,B,C}	14		
{A,B,D}	15		
{A,B,E}	17		
{A,C,D}	17		
{A,C,E}	19		
{A,D,E}	20		
{B,C,D}	17		
{B,C,E}	19		
{B,D,E}	20		
{C,D,E}	22		
{A,B,C,D}	21		
{A,B,C,E}	23		
{A,B,D,E}	24		
{A,C,D,E}	26	Winning	A,C,D,E
{B,C,D,E}	26	Winning	B,C,D,E
{A,B,C,D,E}	30	Winning	C,D,E

31. (continued)

$$\frac{\text{The number of times A is critical in winning coalitions}}{\text{The total number of times voters are critical in winning coalitions}} = \frac{1}{11}$$

$$\frac{\text{The number of times B is critical in winning coalitions}}{\text{The total number of times voters are critical in winning coalitions}} = \frac{1}{11}$$

$$\frac{\text{The number of times C is critical in winning coalitions}}{\text{The total number of times voters are critical in winning coalitions}} = \frac{3}{11}$$

$$\frac{\text{The number of times D is critical in winning coalitions}}{\text{The total number of times voters are critical in winning coalitions}} = \frac{3}{11}$$

$$\frac{\text{The number of times E is critical in winning coalitions}}{\text{The total number of times voters are critical in winning coalitions}} = \frac{3}{11}$$

33.

Coalition	Weight		Critical Voters
{C,D}	12	Winning	C,D
{A,C,D}	13	Winning	C,D
{B,C,D}	15	Winning	C,D
{A,B,C,D}	16	Winning	C,D

$$\frac{\text{The number of times A is critical in winning coalitions}}{\text{The total number of times voters are critical in winning coalitions}} = \frac{0}{8} = 0$$

$$\frac{\text{The number of times B is critical in winning coalitions}}{\text{The total number of times voters are critical in winning coalitions}} = \frac{0}{8} = 0$$

$$\frac{\text{The number of times C is critical in winning coalitions}}{\text{The total number of times voters are critical in winning coalitions}} = \frac{4}{8} = \frac{1}{2}$$

$$\frac{\text{The number of times D is critical in winning coalitions}}{\text{The total number of times voters are critical in winning coalitions}} = \frac{4}{8} = \frac{1}{2}$$

35. a) Consider the votes to be labels A, B, C, D, E. There are $2^5 - 1 = 31$ different coalitions. Many of them have the same weight, however. The sixth line of Pascal's triangle (recall from earlier chapters) is

1 5 10 10 5 1.

This means there are 1 coalition with none (not counted), 5 with 2 (not a winning coalition because the quota is 3), 10 with 3, 10 with 4, 5 with 5, and 1 with 6. Of the 25 winning coalitions, only 10 of them will have critical voters.

Coalition	Critical Voters	Coalition	Critical Voters
{A,B,C}	A,B,C	{A,D,E}	A,D,E
{A,B,D}	A,B,D	{B,C,D}	B,C,D
{A,B,E}	A,B,E	{B,C,E}	B,C,E
{A,C,D}	A,C,D	{B,D,E}	B,D,E
{A,C,E}	A,C,E	{C,D,E}	C,D,E

Each voter would have the same Banzhaf power index of

$$\frac{\text{The number of times A is critical in winning coalitions}}{\text{The total number of times voters are critical in winning coalitions}} = \frac{6}{30} = \frac{1}{5}$$

b) No solution provided.

37. a) The only winning coalitions are those that have the dictator included. Let voter A be the voter with 15 votes, voter B be the voter with 2 votes, voters C and D each with 3 votes, and voter E with 5 votes. The winning coalitions are

Coalition	Critical Voters
{A}	A
{A,B}	A
{A,C}	A
{A,D}	A
{A,E}	A
{A,B,C}	A
{A,B,D}	A
{A,B,E}	A
{A,C,D}	A
{A,C,E}	A
{A,D,E}	A
{A,B,C,D}	A
{A,B,C,E}	A
{A,B,D,E}	A
{A,C,D,E}	A
{A,B,C,D,E}	A

$$\frac{\text{The number of times A is critical in winning coalitions}}{\text{The total number of times voters are critical in winning coalitions}} = \frac{16}{16} = 1$$

The Banzhaf power index of A is 1, the index of the other voters is 0.

b) No solution provided.

39.

Coalition	Weight		Critical Voters
{TX}	38		
{FL}	29		
{CA}	55		
{TX,FL}	67	Winning	TX,FL
{TX,CA}	93	Winning	TX,CA
{FL,CA}	84	Winning	FL,CA
{TX,FL,CA}	122	Winning	None

$$\frac{\text{The number of times TX is critical in winning coalitions}}{\text{The total number of times voters are critical in winning coalitions}} = \frac{2}{6} = \frac{1}{3}$$

$$\frac{\text{The number of times FL is critical in winning coalitions}}{\text{The total number of times voters are critical in winning coalitions}} = \frac{2}{6} = \frac{1}{3}$$

$$\frac{\text{The number of times CA is critical in winning coalitions}}{\text{The total number of times voters are critical in winning coalitions}} = \frac{2}{6} = \frac{1}{3}$$

41.

Coalition	Weight		Critical Voters
{PA}	20		
{NC}	15		
{NM}	5		
{PA,NC}	35	Winning	PA,NC
{PA,NM}	25	Winning	PA,NM
{NC,NM}	20		
{PA,NC,NM}	40	Winning	PA

$$\frac{\text{The number of times PA is critical in winning coalitions}}{\text{The total number of times voters are critical in winning coalitions}} = \frac{3}{5}$$

$$\frac{\text{The number of times NC is critical in winning coalitions}}{\text{The total number of times voters are critical in winning coalitions}} = \frac{1}{5}$$

$$\frac{\text{The number of times NM is critical in winning coalitions}}{\text{The total number of times voters are critical in winning coalitions}} = \frac{1}{5}$$

43. a) No solution provided.

b) Let the members of the firm be represented by {K,C,V,W,X,Y,Z}. The winning coalitions are given in the following table.

Coalition	Critical Voters	Coalition	Critical Voters
{K,C,V,W}	K,C,V,W	{K,C,V,X,Y}	K,C
{K,C,V,X}	K,C,V,X	{K,C,V,X,Z}	K,C
{K,C,V,Y}	K,C,V,Y	{K,C,V,Y,Z}	K,C
{K,C,V,Z}	K,C,V,Z	{K,C,W,X,Y}	K,C
{K,C,W,X}	K,C,W,X	{K,C,W,X,Z}	K,C
{K,C,W,Y}	K,C,W,Y	{K,C,W,Y,Z}	K,C
{K,C,W,Z}	K,C,W,Z	{K,C,X,Y,Z}	K,C
{K,C,X,Y}	K,C,X,Y	{K,C,V,W,X,Y}	K,C
{K,C,X,Z}	K,C,X,Z	{K,C,V,W,X,Z}	K,C
{K,C,Y,Z}	K,C,Y,Z	{K,C,V,W,Y,Z}	K,C
{K,C,V,W,X}	K,C	{K,C,V,X,Y,Z}	K,C
{K,C,V,W,Y}	K,C	{K,C,W,X,Y,Z}	K,C
{K,C,V,W,Z}	K,C	{K,C,V,W,X,Y,Z}	K,C

$$\frac{\text{The number of times K is critical in winning coalitions}}{\text{The total number of times voters are critical in winning coalitions}} = \frac{26}{72} = \frac{13}{36}$$

$$\frac{\text{The number of times C is critical in winning coalitions}}{\text{The total number of times voters are critical in winning coalitions}} = \frac{26}{72} = \frac{13}{36}$$

$$\frac{\text{The number of times V is critical in winning coalitions}}{\text{The total number of times voters are critical in winning coalitions}} = \frac{4}{72} = \frac{1}{18}$$

$$\frac{\text{The number of times W is critical in winning coalitions}}{\text{The total number of times voters are critical in winning coalitions}} = \frac{4}{72} = \frac{1}{18}$$

$$\frac{\text{The number of times X is critical in winning coalitions}}{\text{The total number of times voters are critical in winning coalitions}} = \frac{4}{72} = \frac{1}{18}$$

$$\frac{\text{The number of times Y is critical in winning coalitions}}{\text{The total number of times voters are critical in winning coalitions}} = \frac{4}{72} = \frac{1}{18}$$

43. (continued)

$$\frac{\text{The number of times Z is critical in winning coalitions}}{\text{The total number of times voters are critical in winning coalitions}} = \frac{4}{72} = \frac{1}{18}$$

45. a) No solution provided.

b) Let the members of the firm be represented by {K,C,H,X,Y,Z}. The winning coalitions are given in the following table.

Coalition	Critical Voters
{K,C,X,Y}	K,C,X,Y
{K,C,X,Z}	K,C,X,Z
{K,C,Y,Z}	K,C,Y,Z
{K,H,X,Y}	K,H,X,Y
{K,H,X,Z}	K,H,X,Z
{K,H,Y,Z}	K,H,Y,Z
{C,H,X,Y}	C,H,X,Y
{C,H,X,Z}	C,H,X,Z
{C,H,Y,Z}	C,H,Y,Z
{K,C,H,X,Y}	X,Y
{K,C,H,X,Z}	X,Z
{K,C,H,Y,Z}	Y,Z
{K,C,X,Y,Z}	K,C
{K,H,X,Y,Z}	K,H
{C,H,X,Y,Z}	C,H
{K,C,H,X,Y,Z}	None

$$\frac{\text{The number of times K is critical in winning coalitions}}{\text{The total number of times voters are critical in winning coalitions}} = \frac{8}{48} = \frac{1}{6}$$

$$\frac{\text{The number of times H is critical in winning coalitions}}{\text{The total number of times voters are critical in winning coalitions}} = \frac{8}{48} = \frac{1}{6}$$

$$\frac{\text{The number of times C is critical in winning coalitions}}{\text{The total number of times voters are critical in winning coalitions}} = \frac{8}{48} = \frac{1}{6}$$

$$\frac{\text{The number of times X is critical in winning coalitions}}{\text{The total number of times voters are critical in winning coalitions}} = \frac{8}{48} = \frac{1}{6}$$

$$\frac{\text{The number of times Y is critical in winning coalitions}}{\text{The total number of times voters are critical in winning coalitions}} = \frac{8}{48} = \frac{1}{6}$$

$$\frac{\text{The number of times Z is critical in winning coalitions}}{\text{The total number of times voters are critical in winning coalitions}} = \frac{8}{48} = \frac{1}{6}$$

47. No; for example, in the system [10 : 4, 3, 2, 1] voters have different weights but all have the same index, $\frac{1}{4}$.

49. False; see answer to Exercise 47.

51. 22 and 28

53. It is generally larger.

55. No solution provided.

Section 11.4: Looking Deeper: The Shapley–Shubik Index

1.

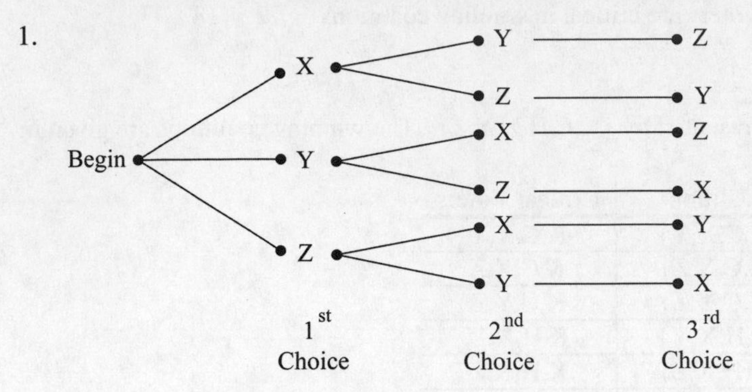

The permutations are (X,Y,Z), (X,Z,Y), (Y,X,Z), (Y,Z,X), (Z,X,Y), and (Z,Y,X).

3.

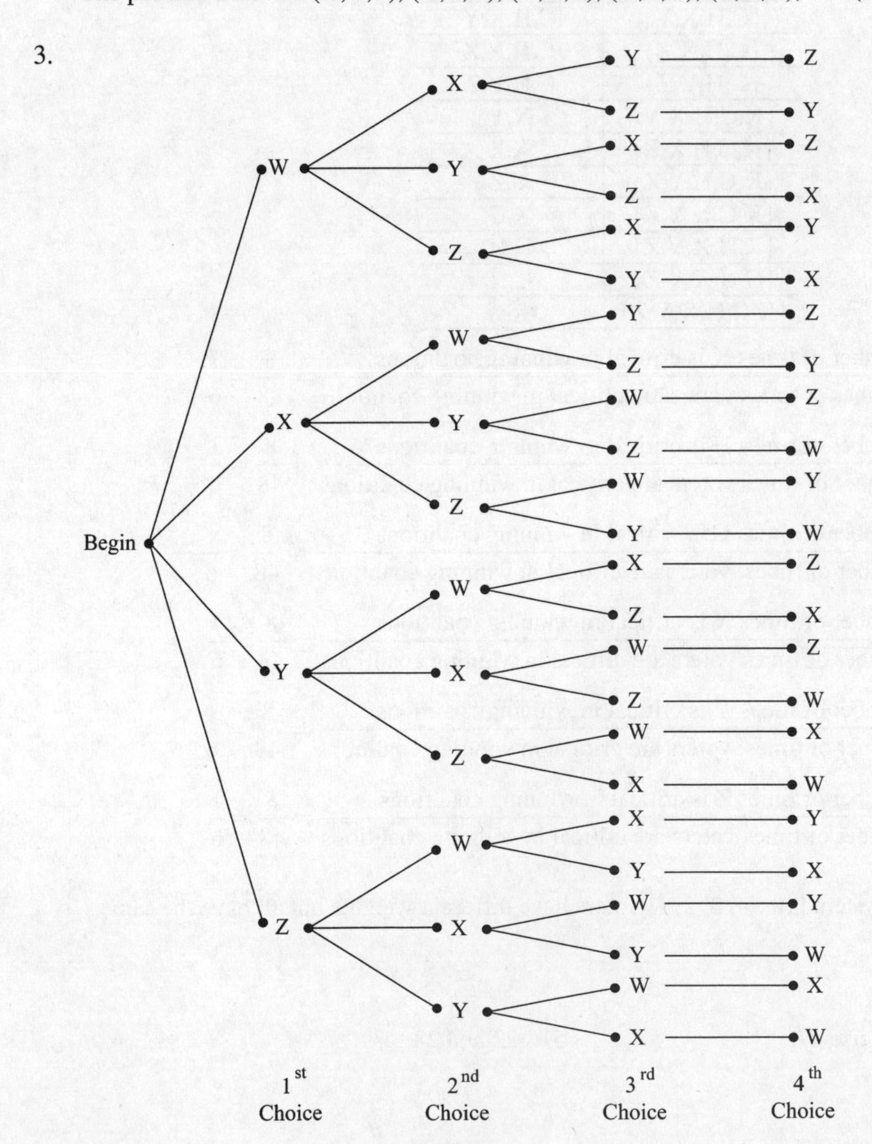

3. (continued)

The permutations are (W,X,Y,Z), (W,X,Z,Y), (W,Y,X,Z), (W,Y,Z,X), (W,Z,X,Y), (W,Z,Y,X), (X,W,Y,Z), (X,W,Z,Y), (X,Y,W,Z), (X,Y,Z,W), (X,Z,W,Y), (X,Z,Y,W), (Y,W,X,Z), (Y,W,Z,X), (Y,X,W,Z), (Y,X,Z,W), (Y,Z,W,X), (Y,Z,X,W), (Z,W,X,Y), (Z,W,Y,X), (Z,X,W,Y), (Z,X,Y,W), (Z,Y,W,X), and (Z,Y,X,W).

5. $6! = 6\times5\times4\times3\times2\times1 = 720$

7. $12! = 12\times11\times10\times9\times8\times7\times6\times5\times4\times3\times2\times1 = 479{,}001{,}600$

9.

Coalition	Weight after 1st, 2nd, and 3rd voter is added	Pivotal voter
a) (A,B,C)	3,5,7	B
b) (A,C,B)	3,5,7	C
c) (B,A,C)	2,5,7	A
d) (B,C,A)	2,4,7	A
e) (C,A,B)	2,5,7	A
f) (C,B,A)	2,4,7	A

11. Let the voters be labeled A, B, and C.

Coalition	Weight after 1st, 2nd, and 3rd voter is added	Pivotal voter
(A,B,C)	3,6,8	B
(A,C,B)	3,5,8	B
(B,A,C)	3,6,8	A
(B,C,A)	3,5,8	A
(C,A,B)	2,5,8	B
(C,B,A)	2,5,8	A

$$\frac{\text{The number of times A is pivotal in some permutation of voters}}{\text{The total number of permutations of voters}} = \frac{3}{6} = \frac{1}{2}$$

$$\frac{\text{The number of times B is pivotal in some permutation of voters}}{\text{The total number of permutations of voters}} = \frac{3}{6} = \frac{1}{2}$$

$$\frac{\text{The number of times C is pivotal in some permutation of voters}}{\text{The total number of permutations of voters}} = \frac{0}{6} = 0$$

13.

Coalition	Weight after 1st, 2nd, 3rd and 4th voter is added	Pivotal voter
(A,B,C,D)	3,6,8,10	C
(A,B,D,C)	3,6,8,10	D
(A,C,B,D)	3,5,8,10	B
(A,C,D,B)	3,5,7,10	B
(A,D,B,C)	3,5,8,10	B
(A,D,C,B)	3,5,7,10	B
(B,A,C,D)	3,6,8,10	C
(B,A,D,C)	3,6,8,10	D
(B,C,A,D)	3,5,8,10	A
(B,C,D,A)	3,5,7,10	A
(B,D,A,C)	3,5,8,10	A
(B,D,C,A)	3,5,7,10	A
(C,A,B,D)	2,5,8,10	B
(C,A,D,B)	2,5,7,10	B
(C,B,A,D)	2,5,8,10	A
(C,B,D,A)	2,5,7,10	A
(C,D,A,B)	2,4,7,10	B
(C,D,B,A)	2,4,7,10	A
(D,A,B,C)	2,5,8,10	B
(D,A,C,B)	2,5,7,10	B
(D,B,A,C)	2,5,8,10	A
(D,B,C,A)	2,5,7,10	A
(D,C,A,B)	2,4,7,10	B
(D,C,B,A)	2,4,7,10	A

$$\frac{\text{The number of times A is pivotal in some permutation of voters}}{\text{The total number of permutations of voters}} = \frac{10}{24} = \frac{5}{12}$$

$$\frac{\text{The number of times B is pivotal in some permutation of voters}}{\text{The total number of permutations of voters}} = \frac{10}{24} = \frac{5}{12}$$

$$\frac{\text{The number of times C is pivotal in some permutation of voters}}{\text{The total number of permutations of voters}} = \frac{2}{24} = \frac{1}{12}$$

$$\frac{\text{The number of times D is pivotal in some permutation of voters}}{\text{The total number of permutations of voters}} = \frac{2}{24} = \frac{1}{12}$$

15.

Coalition	Weight after 1st, 2nd, 3rd and 4th voter is added	Pivotal voter
(A,B,C,D)	2,3,4,5	C
(A,B,D,C)	2,3,4,5	D
(A,C,B,D)	2,3,4,5	B
(A,C,D,B)	2,3,4,5	D
(A,D,B,C)	2,3,4,5	B
(A,D,C,B)	2,3,4,5	C
(B,A,C,D)	1,3,4,5	C
(B,A,D,C)	1,3,4,5	D
(B,C,A,D)	1,2,4,5	A
(B,C,D,A)	1,2,3,5	A
(B,D,A,C)	1,2,4,5	A
(B,D,C,A)	1,2,3,5	A
(C,A,B,D)	1,3,4,5	B
(C,A,D,B)	1,3,4,5	D
(C,B,A,D)	1,2,4,5	A
(C,B,D,A)	1,2,3,5	A
(C,D,A,B)	1,2,4,5	A
(C,D,B,A)	1,2,3,5	A
(D,A,B,C)	1,3,4,5	B
(D,A,C,B)	1,3,4,5	C
(D,B,A,C)	1,2,4,5	A
(D,B,C,A)	1,2,3,5	A
(D,C,A,B)	1,2,4,5	A
(D,C,B,A)	1,2,3,5	A

$$\frac{\text{The number of times A is pivotal in some permutation of voters}}{\text{The total number of permutations of voters}} = \frac{12}{24} = \frac{1}{2}$$

$$\frac{\text{The number of times B is pivotal in some permutation of voters}}{\text{The total number of permutations of voters}} = \frac{4}{24} = \frac{1}{6}$$

$$\frac{\text{The number of times C is pivotal in some permutation of voters}}{\text{The total number of permutations of voters}} = \frac{4}{24} = \frac{1}{6}$$

$$\frac{\text{The number of times D is pivotal in some permutation of voters}}{\text{The total number of permutations of voters}} = \frac{4}{24} = \frac{1}{6}$$

17. a) The Shapley–Shubik index for each person will be $\frac{1}{5}$.

    b) No solution provided.

    c) No solution provided.

19. a) The Shapley–Shubik index for A is 1. It will be 0 for each of the others.

    b) No solution provided.

    c) No solution provided.

21.

Coalition	Pivotal voter
(A,B,C,D)	B
(A,B,D,C)	B
(A,C,B,D)	C
(A,C,D,B)	C
(A,D,B,C)	D
(A,D,C,B)	D
(B,A,C,D)	A
(B,A,D,C)	A
(B,C,A,D)	A
(B,C,D,A)	A
(B,D,A,C)	A
(B,D,C,A)	A
(C,A,B,D)	A
(C,A,D,B)	A
(C,B,A,D)	A
(C,B,D,A)	A
(C,D,A,B)	A
(C,D,B,A)	A
(D,A,B,C)	A
(D,A,C,B)	A
(D,B,A,C)	A
(D,B,C,A)	A
(D,C,A,B)	A
(D,C,B,A)	A

$$\frac{\text{The number of times A is pivotal in some permutation of voters}}{\text{The total number of permutations of voters}} = \frac{18}{24} = \frac{3}{4}$$

$$\frac{\text{The number of times B is pivotal in some permutation of voters}}{\text{The total number of permutations of voters}} = \frac{2}{24} = \frac{1}{12}$$

$$\frac{\text{The number of times C is pivotal in some permutation of voters}}{\text{The total number of permutations of voters}} = \frac{2}{24} = \frac{1}{12}$$

$$\frac{\text{The number of times D is pivotal in some permutation of voters}}{\text{The total number of permutations of voters}} = \frac{2}{24} = \frac{1}{12}$$

23. a) (A,C,T,N),(A,C,N,T),(A,T,C,N),(A,T,N,C),(A,N,C,T),(A,N,T,C),(C,A,T,N),(C,A,N,T),(C,T,A,N),(C,T,N,A),(C,N,T,A),(C,N,A,T),(T,A,C,N),(T,A,N,C),(T,C,A,N),(T,C,N,A),(T,N,A,C),(T,N,C,A),(N,A,B,T),(N,A,C,C),(N,C,A,T),(N,C,T,A),(N,T,A,C), and (N,T,C,A)

b)

Coalition	Weight after 1st, 2nd, 3rd and 4th voter is added	Pivotal group
(A,C,T,N)	3,7,10,12	T
(A,C,N,T)	3,7,9,12	N
(A,T,C,N)	3,6,10,12	C
(A,T,N,C)	3,6,8,12	N
(A,N,C,T)	3,5,9,12	C
(A,N,T,C)	3,5,8,12	T
(C,A,T,N)	4,7,10,12	T
(C,A,N,T)	4,7,9,12	N
(C,T,A,N)	4,7,10,12	A
(C,T,N,A)	4,7,9,12	N
(C,N,A,T)	4,6,9,12	A
(C,N,T,A)	4,6,9,12	T
(T,A,C,N)	3,6,10,12	C
(T,A,N,C)	3,6,8,12	N
(T,C,A,N)	3,7,10,12	A
(T,C,N,A)	3,7,9,12	N
(T,N,A,C)	3,5,8,12	A
(T,N,C,A)	3,5,9,12	C
(N,A,C,T)	2,5,9,12	C
(N,A,T,C)	2,5,8,12	T
(N,C,A,T)	2,6,9,12	A
(N,C,T,A)	2,6,9,12	T
(N,T,A,C)	2,5,8,12	A
(N,T,C,A)	2,5,9,12	C

c) $$\frac{\text{The number of times A is pivotal in some permutation of voters}}{\text{The total number of permutations of voters}} = \frac{6}{24} = \frac{1}{4}$$

$$\frac{\text{The number of times F is pivotal in some permutation of voters}}{\text{The total number of permutations of voters}} = \frac{6}{24} = \frac{1}{4}$$

$$\frac{\text{The number of times T is pivotal in some permutation of voters}}{\text{The total number of permutations of voters}} = \frac{6}{24} = \frac{1}{4}$$

$$\frac{\text{The number of times N is pivotal in some permutation of voters}}{\text{The total number of permutations of voters}} = \frac{6}{24} = \frac{1}{4}$$

25. a) $4! = 24$

b) {F,R,W,C}, {F,R,C,W}, {F,W,R,C}, {F,W,C,R}, {F,C,R,W}, {F,C,W,R}, {R,F,C,W}, {R,F,W,C}, {R,W,C,F}, {R,W,F,C}, {R,C,W,F}, {R,C,F,W}, {W,F,R,C}, {W,F,C,R}, {W,R,F,C}, {W,R,C,F}, {W,C,F,R}, {W,C,R,F}, {C,F,W,R}, {C,F,R,W}, {C,R,W,F}, {C,R,F,W}, {C,W,R,F}, {C,W,F,R}

27.

Coalition	Sum of votes	Pivotal Voter
{FL, IL, NJ}	27 + 18 + 12	IL
{FL, NJ, IL}	27 + 12 + 18	NJ
{IL, FL, NJ}	18 + 27 + 12	FL
{IL, NJ, FL}	18 + 12 + 27	NJ
{NJ, FL, IL}	12 + 27 + 18	FL
{NJ, IL, FL}	12 + 18 + 27	IL

$$\frac{\text{The number of times FL is pivotal in some permutation of voters}}{\text{The total number of permutations of voters}} = \frac{2}{6} = \frac{1}{3}$$

$$\frac{\text{The number of times IL is pivotal in some permutation of voters}}{\text{The total number of permutations of voters}} = \frac{2}{6} = \frac{1}{3}$$

$$\frac{\text{The number of times NJ is pivotal in some permutation of voters}}{\text{The total number of permutations of voters}} = \frac{2}{6} = \frac{1}{3}$$

29. The Banzhaf index for voter A is the number of times that A is a critical member of a coalition divided by the total number of times all voters are critical members of some coalition. The Shapley-Shubik index is the number of times A is pivotal in some coalition divided by the total number of permutations of all voters.

31. Before B can be pivotal, the managing editor and one other member of the board must already have been chosen.

33. $20! \approx 2.4329 \times 10^{18}$;

$$\frac{2.4329 \times 10^{18}}{1,000,000} = \frac{2.43 \times 10^{18}}{1 \times 10^{6}} = 2.43 \times 10^{12}\,sec$$

$$2.4329 \times 10^{12}\,sec \times \frac{1\ min}{60\ sec} \times \frac{1\ hr}{60\ min} \times \frac{1\ day}{24\ hr} \times \frac{1\ year}{365\ day} \approx 77,147\ years$$

Chapter Review Exercises

1. a) No. The total number of votes is 2,156 + 1,462 + 986 + 428 = 5,032. $2156 \div 5032 \approx 0.428$, which is less than 50%.

   b) Myers

2. A with a total of 93 points

Preference	1st place votes × 4 (points)	2nd place votes × 3 (points)	3rd place votes × 2 (points)	4th place votes × 1 (points)	Total Points
A	15×4	6×3	3×2	9×1	93
B	7×4	13×3	6×2	7×1	86
C	6×4	10×3	17×2	0×1	88
D	5×4	4×3	7×2	17×1	63

3. Since education in the future (E) didn't have any first-place votes, it is removed. The new table would be

Preference	1,531	1,102	906	442	375
1st	G	R	S	S	G
2nd	R	S	G	R	S
3rd	S	G	R	G	R

Since role of government in a free society ® has the fewest first-place votes, it is then eliminated

Preference	1,906	1,102	906	442
1st	G	S	S	S
2nd	S	G	G	G

By combining identical columns, we have

Preference	1,906	2,450
1st	G	S
2nd	S	G

Since social justice (S) now has the most first-place votes, it wins the election.

4. We must compare a) A with B, b) A with C, c) A with D, d) B with C, e) B with D, and f) C with D.

a) In comparing A with B, we ignore C and D. Thus, 8 prefer A over B, and $4 + 5 + 6 = 15$ prefer B over A. Thus, we award 1 point to option B.

b) In comparing A with C, we ignore B and D. Thus, 8 prefer A over C, and $4 + 5 + 6 = 15$ prefer C over A. Thus, we award 1 point to option C.

c) In comparing A with D, we ignore B and C. Thus, $8 + 6 = 14$ prefer A over D, and $4 + 5 = 9$ prefer D over A. Thus, we award 1 point to option A.

d) In comparing B with C, we ignore A and D. Thus, $8 + 4 + 5 = 17$ prefer B over C, and 6 prefer C over B. Thus, we award 1 point to option B.

e) In comparing B with D, we ignore A and C. Thus, $8 + 5 + 6 = 19$ prefer B over D, and 4 prefer D over B. Thus, we award 1 point to option B.

f) In comparing C with D, we ignore A and C. Thus, $8 + 6 = 14$ prefer C over D, and $4 + 5 = 9$ prefer D over C. Thus, we award 1 point to option C.

Since B has 3 points, C has 2, and A has 1, option B is declared the winner.

5.

Preference	1st place votes × 4 (points)	2nd place votes × 3 (points)	3rd place votes × 2 (points)	4th place votes × 1 (points)	Total Points
R	2×4	0×3	0×2	1×1	9
D	1×4	2×3	0×2	0×1	10
P	0×4	1×3	1×2	1×1	6
Q	0×4	0×3	2×2	1×1	5

Option D is the winner by the Borda method. No, this election does not satisfy the majority criterion. Option R had the majority of the 1st place votes.

6.

Preference	1st place votes × 4 (points)	2nd place votes × 3 (points)	3rd place votes × 2 (points)	4th place votes × 1 (points)	Total Points
E	18×4	8×3	0×2	9×1	105
A	1×4	14×3	8×2	12×1	74
N	8×4	12×3	15×2	0×1	98
S	8×4	1×3	12×2	14×1	73

Option E is the winner by the Borda count. The Condorcet's criterion is satisfied in this example, because option E can beat all of the other options in a head-to-head competition and is the winner by Borda count. 26 out of 35 prefer E to A, 18 out of 35 prefer E to N, and 26 out of 35 prefer E to S.

7.

Preference	1st place votes × 4 (points)	2nd place votes × 3 (points)	3rd place votes × 2 (points)	4th place votes × 1 (points)	Total Points
A	3×4	10×3	23×2	0×1	88
B	13×4	14×3	3×2	6×1	106
C	17×4	3×3	4×2	12×1	97
D	3×4	9×3	6×2	18×1	69

Option B is the winner by the Borda count. The independence-of-irrelevant-alternatives criterion is not satisfied in this example because if option A were taken out, then the preference table would be

Preference	11	3	6	9	4	3
1st	C	D	C	B	B	C
2nd	B	B	D	D	C	B
3rd	D	C	B	C	D	D

According to the Borda count, C would now be declared the winner.

Preference	1st place votes × 3 (points)	2nd place votes × 2 (points)	3rd place votes × 1 (points)	Total Points
C	20×3	4×2	12×1	80
B	13×3	17×2	6×1	79
D	3×3	15×2	18×1	57

8. Remove option D, since it had the least number of first-place votes. The new table would be

Preference	10	7	2	4
1st	D	B	D	B
2nd	B	D	B	D

By combining identical columns, we have

Preference	12	11
1st	D	B
2nd	B	D

Since option D now has the most first-place votes, it wins the election.

Now if we remove option B, the new table would be

Preference	10	7	2	4
1st	D	A	D	D
2nd	C	D	A	A

By combining identical columns we have

Preference	17	6
1st	D	C
2nd	C	D

Since option D now has the most first-place votes, it still wins the election. The independence-of-irrelevant-alternatives criterion is satisfied.

9. The quota is 17; There are 4 voters. Voter A has 1 vote, voter B has 5, voter C has 7, and voter D has 8; There are no dictators; Voters B, C, and D each have veto power.

10. Let voter A be the voter with 2 votes, voter B with 3, voter C with 5, and voter D with 7.

Winning Coalition	Weight
{C,D}	12
{A,B,D}	12
{A,C,D}	14
{B,C,D}	15
{A,B,C,D}	17

11. Let voter A be the voter with 2 votes, voter B with 3, voter C with 5, and voter D with 7.

Coalition	Weight	Critical Voters
{C,D}	12	C,D
{A,B,D}	12	A,B,D
{A,C,D}	14	C,D
{B,C,D}	15	C,D
{A,B,C,D}	17	D

$$\frac{\text{The number of times A is critical in winning coalitions}}{\text{The total number of times voters are critical in winning coalitions}} = \frac{1}{10}$$

$$\frac{\text{The number of times B is critical in winning coalitions}}{\text{The total number of times voters are critical in winning coalitions}} = \frac{1}{10}$$

$$\frac{\text{The number of times C is critical in winning coalitions}}{\text{The total number of times voters are critical in winning coalitions}} = \frac{3}{10}$$

$$\frac{\text{The number of times D is critical in winning coalitions}}{\text{The total number of times voters are critical in winning coalitions}} = \frac{5}{10} = \frac{1}{2}$$

12. a) Let voter A be the voter with 1 vote, voter B with 2, voter C with 3, and voter D with 4.

Coalition	Weight	Critical Voters
{A,B,C,D}	10	A,B,C,D

Each voter would have the same Banzhaf power index as voter A

$$\frac{\text{The number of times A is critical in winning coalitions}}{\text{The total number of times voters are critical in winning coalitions}} = \frac{1}{4}$$

This is expected, because the sum of the voters' weights is the same as the quota. Only when they vote together will a winning coalition be formed. Each voter is a critical voter in this winning coalition.

12. (continued)

b) Let voter A be the voter with 11 votes, voter B with 1, voters C and D with 3, and voter E with 2.

Coalition	Weight	Critical Voters
{A}	11	A
{A,B}	12	A
{A,C}	14	A
{A,D}	14	A
{A,E}	13	A
{A,B,C}	15	A
{A,B,D}	15	A
{A,B,E}	14	A
{A,C,D}	17	A
{A,C,E}	16	A
{A,D,E}	16	A
{A,B,C,D}	18	A
{A,B,C,E}	17	A
{A,B,D,E}	17	A
{A,C,D,E}	19	A
{A,B,C,D,E}	20	A

Voter A has a Banzhaf power index of 1

$$\frac{\text{The number of times A is critical in winning coalitions}}{\text{The total number of times voters are critical in winning coalitions}} = \frac{16}{16} = 1$$

while the other four voter have a Banzhaf power index of 0

$$\frac{\text{The number of times B,C,D,or E is critical in winning coalitions}}{\text{The total number of times voters are critical in winning coalitions}} = \frac{0}{16} = 0$$

This result is expected, because the voter A is a dictator.

13.

Coalition	Weight	Critical Voters
{A,B}	12	A,B
{A,C}	14	A,C
{A,D}	14	A,D
{A,B,C}	15	A
{A,B,D}	15	A
{A,C,D}	17	A
{A,B,C,D}	18	A

Voter A has a Banzhaf power index of $\frac{7}{10}$

$$\frac{\text{The number of times A is critical in winning coalitions}}{\text{The total number of times voters are critical in winning coalitions}} = \frac{7}{10}$$

while the other four voter have a Banzhaf power index of $\frac{1}{10}$

$$\frac{\text{The number of times B,C,or D is critical in winning coalitions}}{\text{The total number of times voters are critical in winning coalitions}} = \frac{1}{10}$$

14. $7! = 5,040$

15.

Sum of Weights of Coalition Members until Quota of 6 is Reached	Pivotal voter
(A, B, C) 4 + 3	B
(A, C, B) 4 + 2	C
(B, A, C) 3 + 4	A
(B, C, A) 3 + 2 + 4	A
(C, A, B) 2 + 4	A
(C, B, A) 2 + 3 + 4	A

16. From the table in Exercise 15:

$$\frac{\text{The number of times A is pivotal in some permutation of voters}}{\text{The total number of permutations of voters}} = \frac{4}{6} = \frac{2}{3}$$

$$\frac{\text{The number of times B is pivotal in some permutation of voters}}{\text{The total number of permutations of voters}} = \frac{1}{6}$$

$$\frac{\text{The number of times C is pivotal in some permutation of voters}}{\text{The total number of permutations of voters}} = \frac{1}{6}$$

17. The Shapley–Shubik index for each person will be $\frac{1}{6}$.

18.

Sum of Weights of Coalition Members until Quota of 9 is Reached	Pivotal voter
(MN, WI, DE) 7 + 8	WI
(MN, DE, WI) 8 + 1 + 7	DE
(WI, MN, DE) 8 + 7	MN
(WI, DE, MN) 7 + 1 + 8	MN
(DE, MN, WI) 1 + 8	MN
(DE, WI, MN) 1 + 7 + 8	MN

$$\frac{\text{The number of times MN is pivotal in some permutation of voters}}{\text{The total number of permutations of voters}} = \frac{4}{6} = \frac{2}{3}$$

$$\frac{\text{The number of times WI is pivotal in some permutation of voters}}{\text{The total number of permutations of voters}} = \frac{1}{6}$$

$$\frac{\text{The number of times DE is pivotal in some permutation of voters}}{\text{The total number of permutations of voters}} = \frac{1}{6}$$

Chapter Test

1. a) No. The total number of votes is 2,543 + 1,532 + 892 + 473 = 5,440. $2{,}543 \div 5{,}440 \approx 0.467$, which is less than 50%.

   b) Molina

2. Since quality of life (Q) didn't have any first-place votes, it is removed. The new table would be

Preference	327	130	149	85	234
1st	E	R	A	E	R
2nd	R	E	E	R	E
3rd	A	A	R	A	A

By combining identical columns, we have

Preference	412	365	149
1st	E	R	A
2nd	R	E	E
3rd	A	A	R

Since attracting more women to science (A) has the fewest first-place votes, it is then eliminated

Preference	412	364	149
1st	E	R	E
2nd	R	E	R

By combining identical columns, we have

Preference	561	364
1st	E	R
2nd	R	E

Since equality in the workplace (E) now has the most first-place votes, it wins the election.

3.

Preference	1st place votes × 4 (points)	2nd place votes × 3 (points)	3rd place votes × 2 (points)	4th place votes × 1 (points)	Total Points
A	$1{,}612 \times 4$	$1{,}754 \times 3$	849×2	0×1	13,409
B	$1{,}754 \times 4$	$1{,}612 \times 3$	0×2	849×1	12,701
C	849×4	0×3	$2{,}457 \times 2$	909×1	9,219
D	0×4	849×3	909×2	$2{,}457 \times 1$	6,822

Option A is the winner by the Borda count. The Condorcet's criterion is satisfied in this example, because option A can beat all of the other options in a head-to-head competition and is the winner by Borda count. 2,461 out of 4,215 prefer A to B, 3,366 out of 4,215 prefer A to C, and 3,366 out of 4,215 prefer A to D.

4. The quota is 15; Voter A has 5 vote, voter B has 3, voter C has 1, voter D has 3, voter E has 4, and voter F has 2; There are no dictators; Voters A and E each have veto power.

5. $6! = 720$

6. B with a total of 77 points

Preference	1st place votes × 4 (points)	2nd place votes × 3 (points)	3rd place votes × 2 (points)	4th place votes × 1 (points)	Total Points
A	5×4	11×3	8×2	5×1	74
B	11×4	5×3	5×2	8×1	77
C	8×4	5×3	5×2	11×1	68
D	5×4	8×3	11×2	5×1	71

7.

Preference	1st place votes × 4 (points)	2nd place votes × 3 (points)	3rd place votes × 2 (points)	4th place votes × 1 (points)	Total Points
A	2×4	0×3	0×2	1×1	9
B	1×4	2×3	0×2	0×1	10
C	0×4	0×3	3×2	0×1	6
D	0×4	1×3	0×2	2×1	5

Option B is the winner by the Borda method. No, this election does not satisfy the majority criterion. Option A had the majority of the 1st place votes.

8. The Shapley–Shubik index for each person will be $\frac{1}{8}$.

9. Remove option D, since it had the least number of first-place votes. The new table would be

Preference	35	71	36	14
1st	A	B	C	A
2nd	B	A	A	C
3rd	C	C	B	B

Then remove option C, since it had the least number of first-place votes. The new table would be

Preference	35	71	36	14
1st	A	B	A	A
2nd	B	A	B	B

By combining identical columns, we have

Preference	85	71
1st	A	B
2nd	B	A

Since option A now has the most first-place votes, it wins the election.

Now if we remove option C, the new table would be

Preference	35	71	36	14
1st	A	B	D	D
2nd	B	A	A	A
3rd	D	D	B	B

By combining identical columns we have

Preference	35	71	50
1st	A	B	D
2nd	B	A	A
3rd	D	D	B

9. (continued)

Remove option A, since it had the least number of first-place votes. The new table would be

Preference	35	71	50
1st	B	B	D
2nd	D	D	B

By combining identical columns we have

Preference	104	50
1st	B	D
2nd	D	B

Since option B now has the most first-place votes, it wins the election instead of A. The independence-of-irrelevant-alternatives criterion is not satisfied.

10. a) Let voter A be the voter with 2 votes, voter B with 8, voter C with 3, and voter D with 2.

Coalition	Weight	Critical Voters
{A,B,C,D}	15	A,B,C,D

Each voter would have the same Banzhaf power index as voter A

$$\frac{\text{The number of times A is critical in winning coalitions}}{\text{The total number of times voters are critical in winning coalitions}} = \frac{1}{4}$$

This is expected, because the sum of the voters' weights is the same as the quota. Only when they vote together will a winning coalition be formed. Each voter is a critical voter in this winning coalition.

b) Let voter A be the voter with 15 votes, voter B with 2, voter C with 4, voter D with 1, and voter E with 3.

Coalition	Weight	Critical Voters
{A}	15	A
{A,B}	17	A
{A,C}	19	A
{A,D}	16	A
{A,E}	18	A
{A,B,C}	21	A
{A,B,D}	18	A
{A,B,E}	20	A
{A,C,D}	20	A
{A,C,E}	22	A
{A,D,E}	19	A
{A,B,C,D}	22	A
{A,B,C,E}	24	A
{A,B,D,E}	21	A
{A,C,D,E}	23	A
{A,B,C,D,E}	25	A

Voter A has a Banzhaf power index of 1

$$\frac{\text{The number of times A is critical in winning coalitions}}{\text{The total number of times voters are critical in winning coalitions}} = \frac{16}{16} = 1$$

while the other four voter have a Banzhaf power index of 0

$$\frac{\text{The number of times B,C,D,or E is critical in winning coalitions}}{\text{The total number of times voters are critical in winning coalitions}} = \frac{0}{16} = 0$$

This result is expected, because the voter A is a dictator.

11.

Sum of Weights of Coalition Members	Pivotal voter
(A, B, C) 5 + 3	B
(A, C, B) 5 + 3	C
(B, A, C) 3 + 5	A
(B, C, A) 3 + 3 + 5	A
(C, A, B) 3 + 5	A
(C, B, A) 3 + 3 + 5	A

12. We must compare a) A with B, b) A with C, c) A with D, d) B with C, e) B with D, and f) C with D.

a) In comparing A with B, we ignore C and D. Thus, 23 + 83 = 106 prefer A over B, and 47 + 21 = 68 prefer B over A. Thus, we award 1 point to option A.

b) In comparing A with C, we ignore B and D. Thus, 23 + 47 = 70 prefer A over C, and 83 + 21 = 104 prefer C over A. Thus, we award 1 point to option C.

c) In comparing A with D, we ignore B and C. Thus, 23 + 47 + 21 = 91 prefer A over D, and 83 prefer D over A. Thus, we award 1 point to option A.

d) In comparing B with C, we ignore A and D. Thus, 23 + 47 = 70 prefer B over C, and 83 + 21 = 104 prefer C over B. Thus, we award 1 point to option C.

e) In comparing B with D, we ignore A and C. Thus, 23 + 47 + 21 = 91 prefer B over D, and 83 prefer D over B. Thus, we award 1 point to option B.

f) In comparing C with D, we ignore A and C. Thus, 23 + 47 + 21 = 91 prefer C over D, and 83 prefer D over C. Thus, we award 1 point to option C.

Since C has 3 points, A has 2, and B has 1, option C is declared the winner.

13.

Preference	1st place votes × 4 (points)	2nd place votes × 3 (points)	3rd place votes × 2 (points)	4th place votes × 1 (points)	Total Points
A	19×4	21×3	6×2	8×1	159
B	21×4	7×3	20×2	6×1	151
C	6×4	8×3	23×2	17×1	111
D	8×4	18×3	5×2	23×1	119

Option A is the winner by the Borda count. The independence-of-irrelevant-alternatives criterion is not satisfied in this example because if option D were taken out, then the preference table would be

Preference	7	5	6	12	16	8
1st	A	B	C	A	B	C
2nd	B	A	A	B	A	B
3rd	C	C	B	C	C	A

According to the Borda count, B would now be declared the winner.

Preference	1st place votes × 3 (points)	2nd place votes × 2 (points)	3rd place votes × 1 (points)	Total Points
A	19×3	27×2	8×1	119
B	21×3	27×2	6×1	123
C	14×3	0×2	40×1	96

14. Let voter A be the voter with 3 votes, voter B with 4, voter C with 6, and voter D with 8.

Winning Coalition	Weight
{A,B,D}	15
{A,C,D}	17
{B,C,D}	18
{A,B,C,D}	21

15.

Sum of Weights of Coalition Members	Pivotal voter
(A, B, C) 4 + 4	B
(A, C, B) 4 + 2 + 4	B
(B, A, C) 4 + 4	A
(B, C, A) 4 + 2 + 4	A
(C, A, B) 2 + 4 + 4	B
(C, B, A) 2 + 4 + 4	A

$$\frac{\text{The number of times A is pivotal in some permutation of voters}}{\text{The total number of permutations of voters}} = \frac{3}{6} = \frac{1}{2}$$

$$\frac{\text{The number of times B is pivotal in some permutation of voters}}{\text{The total number of permutations of voters}} = \frac{3}{6} = \frac{1}{2}$$

$$\frac{\text{The number of times C is pivotal in some permutation of voters}}{\text{The total number of permutations of voters}} = \frac{0}{6} = 0$$

16.

Winning Coalition	Critical Voters
{A,B,C,D}	C
{A,B,D,C}	D
{A,C,B,D}	B
{A,C,D,B}	B
{A,D,B,C}	B
{A,D,C,B}	B
{B,A,C,D}	C
{B,A,D,C}	D
{B,C,A,D}	A
{B,C,D,A}	A
{B,D,A,C}	A
{B,D,C,A}	A

Winning Coalition	Critical Voters
{C,A,B,D}	B
{C,A,D,B}	B
{C,B,A,D}	A
{C,B,D,A}	A
{C,D,A,B}	B
{C,D,B,A}	A
{D,A,B,C}	B
{D,A,C,B}	B
{D,B,A,C}	A
{D,B,C,A}	A
{D,C,A,B}	B
{D,C,B,A}	A

16. (continued)

$$\frac{\text{The number of times A is critical in winning coalitions}}{\text{The total number of times voters are critical in winning coalitions}} = \frac{10}{24} = \frac{5}{12}$$

$$\frac{\text{The number of times B is critical in winning coalitions}}{\text{The total number of times voters are critical in winning coalitions}} = \frac{10}{24} = \frac{5}{12}$$

$$\frac{\text{The number of times C is critical in winning coalitions}}{\text{The total number of times voters are critical in winning coalitions}} = \frac{2}{24} = \frac{1}{12}$$

$$\frac{\text{The number of times D is critical in winning coalitions}}{\text{The total number of times voters are critical in winning coalitions}} = \frac{2}{24} = \frac{1}{12}$$

Chapter 12
Counting: Just How Many Are There?

Section 12.1 Introduction to Counting Methods

1. UC, UD, UB, UA, CD, CB, CA, DB, DA, BA

3. UC, CU, UD, DU, UB, BU, UA, AU, CD, DC, CB, BC, CA, AC, DB, BD, DA, AD, BA, AB

Use the diagram on the next page for the solutions to Exercises 5 – 7.

5. 4, the ways are HTTT, THTT, TTHT, and TTTH

7. 6, the ways are HHTT, HTHT, HTTH, THHT, THTH, and TTHH

9. Without drawing the tree, you can imagine that the tree begins and there are six branches on the first number choice. Each of those branches split into five branches for the second number choice for a total of thirty branches. There would therefore be **thirty** two-digit numbers that could be formed.

11. Without drawing the tree, you can imagine that the tree begins and there are six branches on the first number choice. Each of those branches split into five branches for the second number choice for a total of thirty branches. Each of those branches split into four branches for the third number choice for a total of 120 branches. There would therefore be **120** three-digit numbers that could be formed.

Use the diagram on the page 297 for the solutions to Exercises 13 – 17.

13. 4 ways; (1,4), (2,3), (3,2), (4,1)

15. 6 ways; (1,1), (2,2), (3,3), (4,4), (5,5), (6,6)

17. 10 ways; (1,1), (1,2), (1,3), (1,4), (2,1), (2,2), (2,3), (3,1), (3,2), (4,1)

19. Without drawing the tree, you can imagine that the tree begins and there are six branches on the first top choice. Each of those branches split into five branches for the second choice of pants for a total of thirty branches. Each of those branches split into four branches for the choice of a jacket for a total of 120 branches. There would therefore be **120** outfits that could be formed.

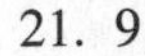

21. 9

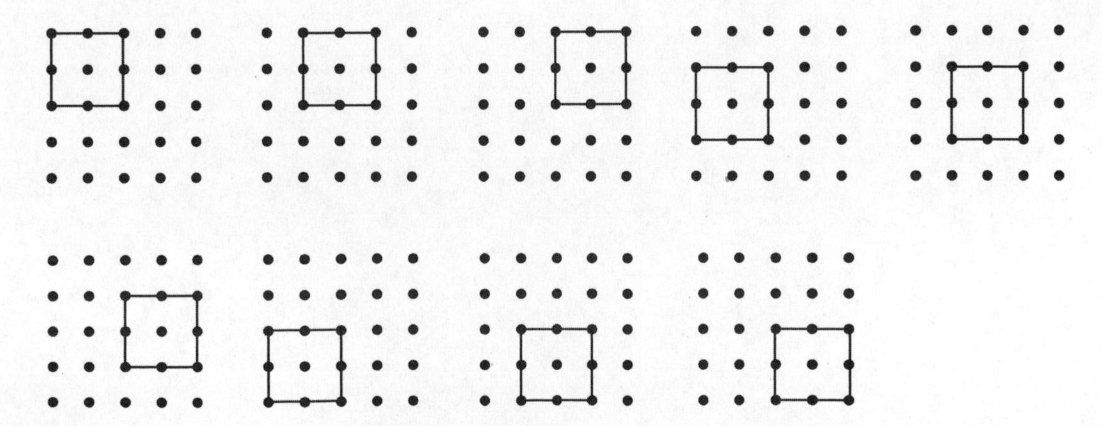

Use the following diagram for the solutions to Exercises 5 – 7.

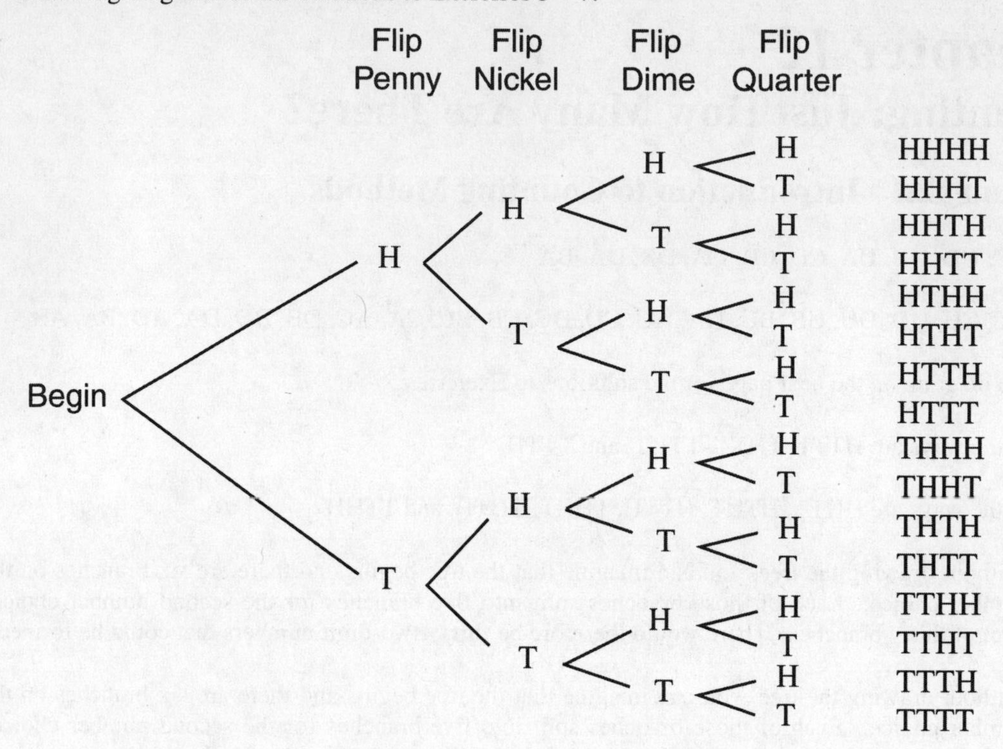

Use this diagram for the solutions to Exercises 13 – 17.

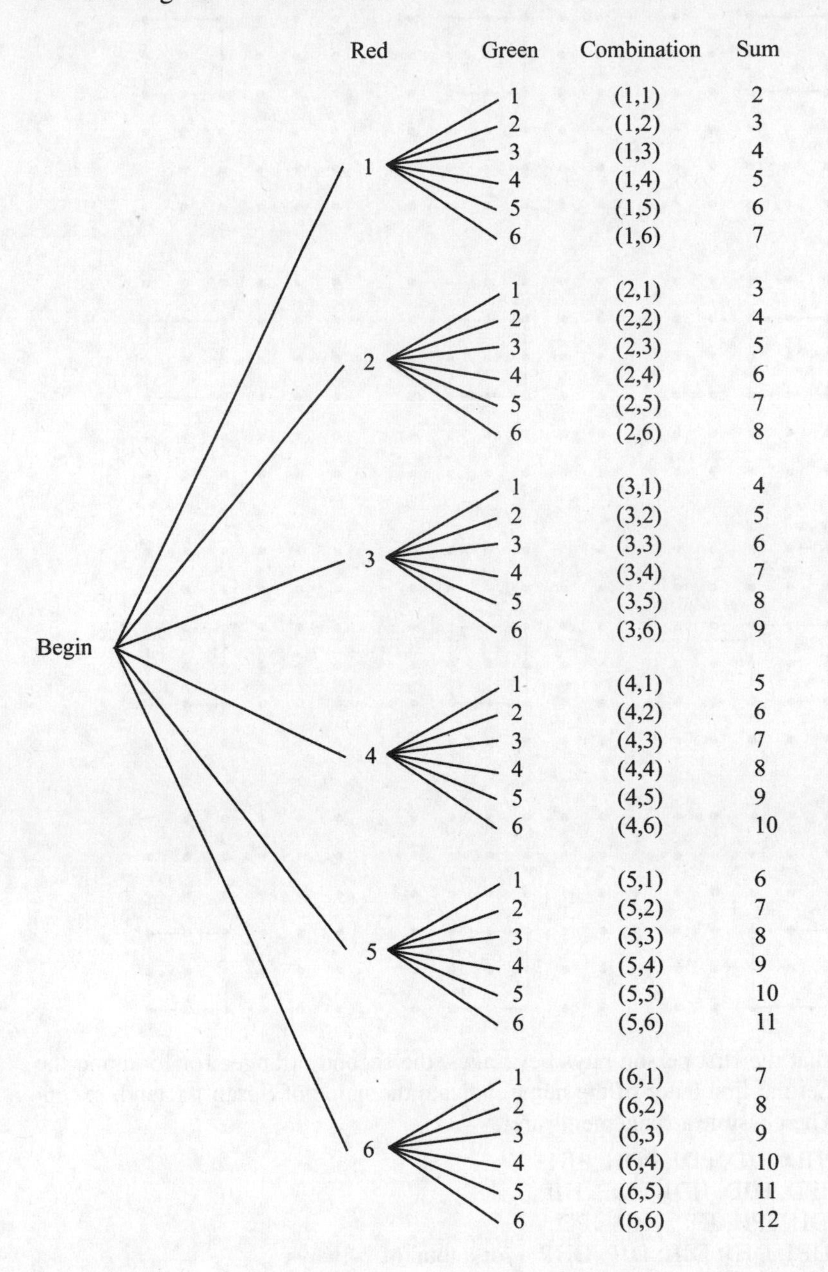

23. 16

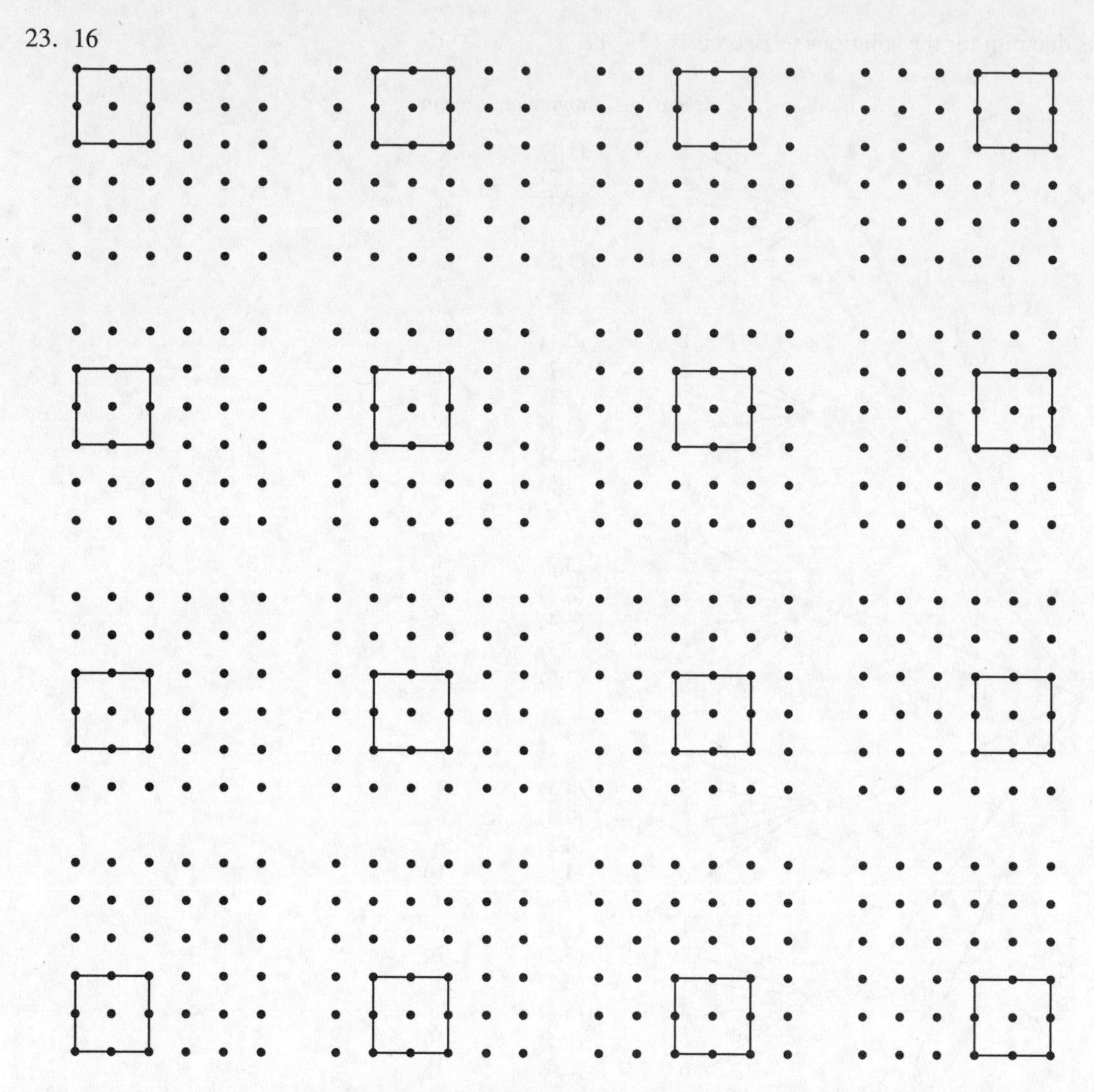

25. Assume that the ordering is that the first person buys beverages, the second arranges for food and the third sends out invitations. Let the first letter of the name indicate the name of Susan's friend, except for H representing Phyllis . The possible arrangements are:

    PHJ, PHD, PJD, PDJ, PDH, PJH
    HPJ, HPD, HJD, HDJ, HDP, HJP
    JDP, JDH, JPH, JHP, JHD, JPD
    DPH, DPJ, DHJ, DJH, DJP, DHP for a total of 24 ways.

27. Without drawing the tree, you can imagine that the tree begins and there are two branches on the first flip. Each of those branches split into two branches on the second flip for a total of four branches. Each of those branches split into two branches on the third flip for a total of eight branches. Each of those branches split into two branches on the fourth flip for a total of sixteen branches. Finally, each of those branches split into two branches on the fifth flip for a total of thirty-two branches. There would therefore be **thirty-two** ways five coins could be flipped.

29. Without drawing the tree, you can imagine that the tree begins and there are four branches on the first roll. Each of those branches split into four branches on the second roll for a total of sixteen branches. There would therefore be **sixteen** ways two tetrahedral dice could be rolled.

31. Without drawing the tree, you can imagine that the tree begins and there are three branches for the first choice of letters. Each of those branches split into two branches for the second choice of the letter not chosen on the first choice. Each of those branches extends to the third and last letter choice. Each of these six branches split into three branches for the first number choice. Each of these 18 branches splits into two branches for the second number choice and each of these 36 branches extends for the third and final number choice. There would therefore be **thirty-six** license plates to be investigated.

33. a) $2^5 = 32$

b) 1; 5; 10

c) 16 out of 32

35. At each stage, there are 8 times the number of shaded squares in the previous stage, so there will be $8^3 = 512$ shaded squares.

37. 24

Seat 1	Seat 2	Seat 3	Seat 4
William	David	Elton	Kate
William	David	Kate	Elton
William	Elton	David	Kate
William	Elton	Kate	David
William	Kate	David	Elton
William	Kate	Elton	David
David	William	Elton	Kate
David	William	Kate	Elton
David	Elton	William	Kate
David	Elton	Kate	William
David	Kate	William	Elton
David	Kate	Elton	William

Seat 1	Seat 2	Seat 3	Seat 4
Elton	William	David	Kate
Elton	William	Kate	David
Elton	David	William	Kate
Elton	David	Kate	William
Elton	Kate	William	David
Elton	Kate	David	William
Kate	William	Elton	David
Kate	William	David	Elton
Kate	David	William	Elton
Kate	David	Elton	William
Kate	Elton	William	David
Kate	Elton	David	William

39. 4

Seat 1	Seat 2	Seat 3	Seat 4
Elton	William	Kate	David
David	William	Kate	Elton
Elton	Kate	William	David
David	Kate	William	Elton

41. $1+3+6+10+15+21+28+36+45+55+66+78=364$

Use the diagram on the next page for the solutions to Exercises 43 – 45.

43. 27

45. 18

Vanilla – Vanilla – Strawberry, Vanilla – Vanilla – Chocolate, Vanilla – Strawberry – Vanilla, Vanilla – Strawberry – Strawberry, Vanilla – Chocolate – Vanilla, Vanilla-Chocolate – Chocolate, Strawberry – Vanilla – Vanilla, Strawberry – Vanilla – Strawberry, Strawberry – Strawberry – Vanilla, Strawberry – Strawberry – Chocolate, Strawberry – Chocolate – Strawberry, Strawberry – Chocolate – Chocolate, Chocolate – Vanilla – Vanilla, Chocolate – Vanilla – Chocolate, Chocolate – Strawberry – Strawberry, Chocolate – Strawberry – Chocolate, Chocolate – Chocolate – Vanilla, and Chocolate – Chocolate – Strawberry

47. In the bottom row, there are $5\times5=25$ oranges. In the next row there will be $4\times4=16$ oranges. This is followed by oranges, then $2\times2=4$ oranges, and finally one orange on top. There is a total of $25+16+9+4+1=55$ oranges.

Use the following diagram for the solutions to Exercises 43 – 45.

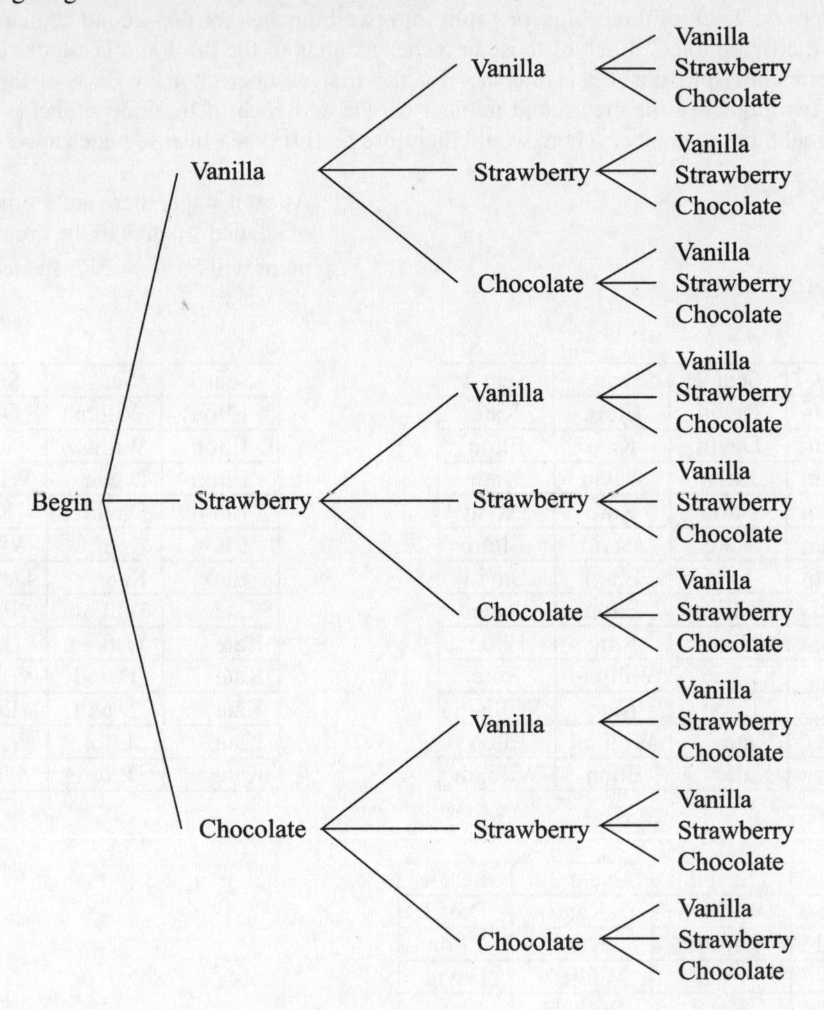

49. 14

51. drawing pictures, being systematic, looking for patterns

53. Follow the branches in the tree below to make the schedules.

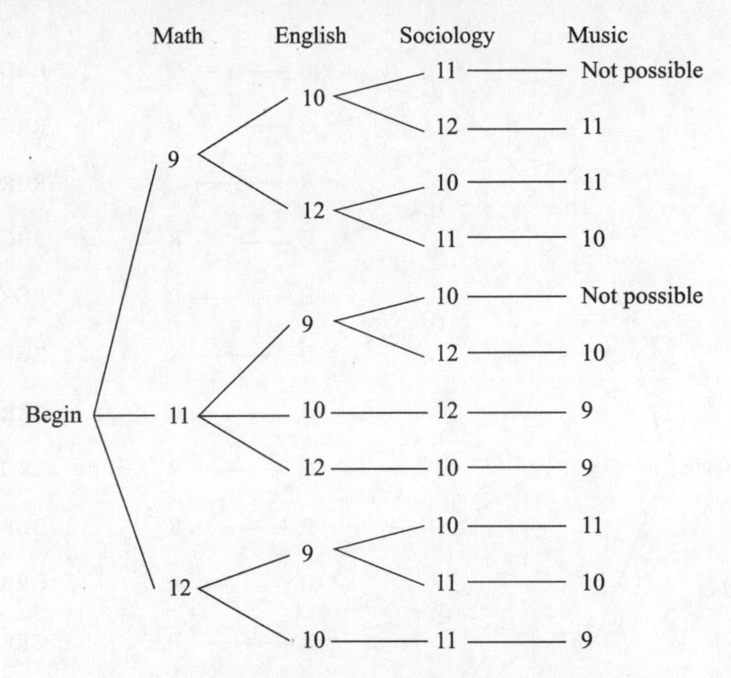

His nine possible schedules are:

9:00	10:00	11:00	12:00
Math	English	Music	Sociology
Math	Sociology	Music	English
Math	Music	Sociology	English
English	Music	Math	Sociology
Music	English	Math	Sociology
Music	Sociology	Math	English
English	Sociology	Music	Math
English	Music	Sociology	Math
Music	English	Sociology	Math

Use the diagram on the next page for the solutions to Exercises 55 – 59.

55. RRBG, RBRG, BRRG

57. RBGR, RGBR, BRGR, BGRR, GRBR, GBRR

59. 120

Use the following diagram for the solutions to Exercises 55 – 59.

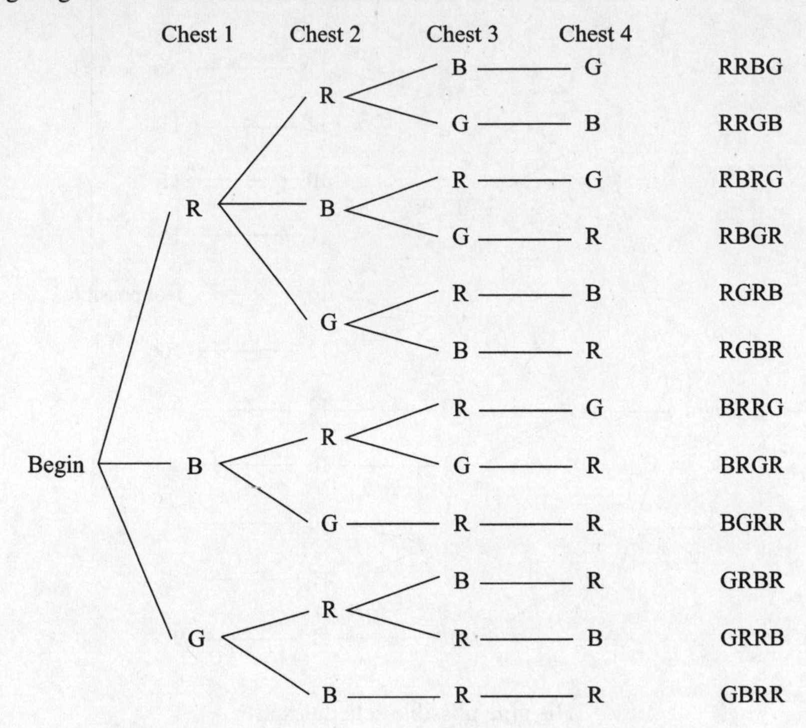

Section 12.2 The Fundamental Counting Principle

1. $7\times6=42$

3. $8\times7\times6=336$

5. $4\times8\times3\times5=480$

7. $5\times6\times13\times4=1{,}560$

9. $8\times8=64$

11. $8\times12=96$

13. $9\times10\times10\times10=9{,}000$

15. If the number must be greater than 5,000 and odd, the first digit must be 5 or greater and the last digit must be odd, so $5\times10\times10\times5=2{,}500$.

17. $2\times2\times2\times2\times2\times2\times2\times2=256$

19. $3\times4\times6=72$

21. $1\times26\times26\times26=17{,}576$

23. $12\times11\times10\times9\times8\times7\times6\times5\times4\times3\times2\times1=479{,}001{,}600$

25. $3\times3\times3\times3\times3=243$

27. To choose direct routes, you must always be traveling down and/or to the right. The numbers in the diagram below indicate how many ways there are to get to the adjacent intersection, starting from each stop.

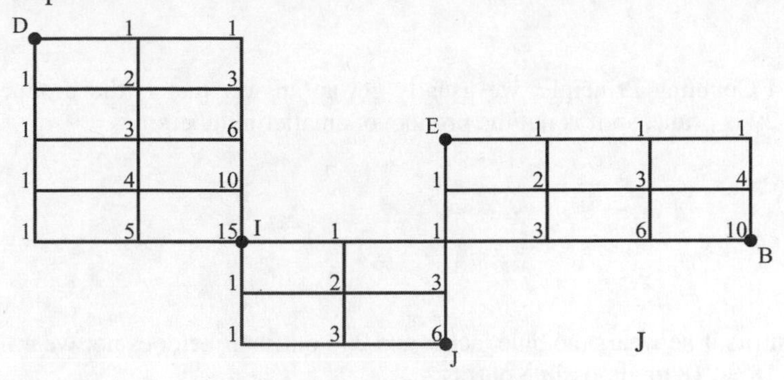

a) There are 15 ways to go directly from D to I.

b) There are 6 ways to go directly from I to J.

c) There is only one way to go directly from J to E, directly up.

d) There are 10 ways to go directly from E to B.

e) There are $15\times6\times1\times10=900$ direct routes from D to B, passing through J and E.

29. a) $2\times2\times2\times2\times2\times2\times2\times2=256$

b) $256\cdot2=512$ minutes which is $\dfrac{512}{60}=8$ hours and 32 minutes or over $8\frac{1}{2}$ hours.

31. 6; Explanations may vary.
Let n be the number of combinations of toppings.
Consider that there are two main choices, crust and toppings. There are three ways to choose the crust and n ways to choose the toppings. By the fundamental counting principle we have

$$3\times n<200$$
$$n<66\tfrac{2}{3}$$

From Chapter 1, we have that there will be 2^t combinations, where t is the number of toppings. Since $2^6=64=n$, there must be 6 toppings for $3\cdot64=192$ combinations.

33. a) Louise and her tutor can sit in the front row in 7 different ways.

Seat 1	Seat 2	Seat 3	Seat 4	Seat 5	Seat 6	Seat 7	Seat 8

b) Louise can sit to the right or left of her tutor, so there are two ways they can sit in the two seats.

c) Since there are 10 other students, the other seats can be filled in $10\times9\times8\times7\times6\times5=151,200$ ways.

d) There are a total of $7\times2\times151,200=$ 2,116,800 ways to seat students in the front row.

35. $6 \times 5 \times 4 \times 3 \times 2 \times 1 = 720$

37. $6 \times 2 \times 1 \times 3 \times 2 \times 1 = 72$

39. All examples use this technique.

41. Answers may vary.
When we use the Fundamental Counting Principle, we usually get an answer that is the product of smaller numbers. The number 29 is prime, so it is not the product of smaller numbers.

43. No solution provided.

45. a) $3 \times 3 \times 2 = 18$
   b) $2 \times 5 \times 4 \times 5 = 200$
   c) Since Julio can choose 18 outfits if he wears the blue jacket and 200 outfits if he does not wear the blue jacket, there are 200 + 18 = 218 total possible outfits.

47. Answers will vary.

Section 12.3 Permutations and Combinations

1. $4! = 4 \cdot 3 \cdot 2 \cdot 1 = 24$

3. $(8-5)! = 3! = 3 \cdot 2 \cdot 1 = 6$

5. $$\frac{10!}{7!} = \frac{10 \cdot 9 \cdot 8 \cdot 7 \cdot 6 \cdot 5 \cdot 4 \cdot 3 \cdot 2 \cdot 1}{7 \cdot 6 \cdot 5 \cdot 4 \cdot 3 \cdot 2 \cdot 1} = 10 \cdot 9 \cdot 8 = 720$$

7. $$\frac{10!}{7!3!} = \frac{10 \cdot 9 \cdot 8 \cdot 7 \cdot 6 \cdot 5 \cdot 4 \cdot 3 \cdot 2 \cdot 1}{7 \cdot 6 \cdot 5 \cdot 4 \cdot 3 \cdot 2 \cdot 1 \cdot 3 \cdot 2 \cdot 1} = \frac{10 \cdot 9 \cdot 8}{3 \cdot 2 \cdot 1} = 10 \cdot 3 \cdot 4 = 120$$

9. $$P(6,2) = \frac{6!}{(6-2)!} = \frac{6!}{4!} = \frac{6 \cdot 5 \cdot 4 \cdot 3 \cdot 2 \cdot 1}{4 \cdot 3 \cdot 2 \cdot 1} = 6 \cdot 5 = 30$$

11. $$C(10,3) = \frac{10!}{3!(10-3)!} = \frac{10!}{3!7!} = \frac{10 \cdot 9 \cdot 8 \cdot 7 \cdot 6 \cdot 5 \cdot 4 \cdot 3 \cdot 2 \cdot 1}{3 \cdot 2 \cdot 1 \cdot 7 \cdot 6 \cdot 5 \cdot 4 \cdot 3 \cdot 2 \cdot 1} = \frac{10 \cdot 9 \cdot 8}{3 \cdot 2 \cdot 1} = 10 \cdot 3 \cdot 4 = 120$$

13. The number of permutations of three selections that can be made from 10 objects.

15. $$P(8,3) = \frac{8!}{(8-3)!} = \frac{8!}{5!} = \frac{8 \cdot 7 \cdot 6 \cdot 5 \cdot 4 \cdot 3 \cdot 2 \cdot 1}{5 \cdot 4 \cdot 3 \cdot 2 \cdot 1} = 8 \cdot 7 \cdot 6 = 336$$

17. $$P(10,8) = \frac{10!}{(10-8)!} = \frac{10!}{2!} = \frac{10 \cdot 9 \cdot 8 \cdot 7 \cdot 6 \cdot 5 \cdot 4 \cdot 3 \cdot 2 \cdot 1}{2 \cdot 1} = 10 \cdot 9 \cdot 8 \cdot 7 \cdot 6 \cdot 5 \cdot 4 \cdot 3 = 1{,}814{,}400$$

19. $$C(8,3) = \frac{8!}{3!(8-3)!} = \frac{8!}{3!5!} = \frac{8 \cdot 7 \cdot 6 \cdot 5 \cdot 4 \cdot 3 \cdot 2 \cdot 1}{3 \cdot 2 \cdot 1 \cdot 5 \cdot 4 \cdot 3 \cdot 2 \cdot 1} = \frac{8 \cdot 7 \cdot 6}{3 \cdot 2 \cdot 1} = 8 \cdot 7 = 56$$

21. $$C(10,8) = \frac{10!}{8!(10-8)!} = \frac{10!}{8!2!} = \frac{10 \cdot 9 \cdot 8 \cdot 7 \cdot 6 \cdot 5 \cdot 4 \cdot 3 \cdot 2 \cdot 1}{8 \cdot 7 \cdot 6 \cdot 5 \cdot 4 \cdot 3 \cdot 2 \cdot 1 \cdot 2 \cdot 1} = \frac{10 \cdot 9}{2 \cdot 1} = 5 \cdot 9 = 45$$

23.

$$\begin{array}{c} 1 \\ 1\ \ 1 \\ 1\ \ 2\ \ 1 \\ 1\ \ 3\ \ 3\ \ 1 \\ 1\ \ 4\ \ 6\ \ 4\ \ 1 \\ 1\ \ 5\ \ 10\ \ 10\ \ 5\ \ 1 \\ 1\ \ 6\ \ 15\ \ 20\ \ 15\ \ 6\ \ 1 \\ 1\ \ 7\ \ 21\ \ 35\ \ 35\ \ 21\ \ 7\ \ 1 \\ \mathbf{1\ \ 8\ \ 28\ \ 56\ \ 70\ \ 56\ \ 28\ \ 8\ \ 1} \end{array}$$

25. Seventh Row: 1 7 **<u>21</u>** 35 35 21 7 1

27. the second entry of the 18th row

29. the sixth entry of the 20th row

31. $P(8,8)$

33. $C(17,3)$

35. $P(5,5)$

37. $P(4,4)$

39. $C(10,3)$

41. $C(9,5)$

43. $C(17,8)$

45. $C(25,6)=\dfrac{25!}{6!(25-6)!}=\dfrac{25!}{6!19!}$

$$=\frac{25\cdot 24\cdot 23\cdot 22\cdot 21\cdot 20\cdot 19\cdot 18\cdot 17\cdot 16\cdot 15\cdot 14\cdot 13\cdot 12\cdot 11\cdot 10\cdot 9\cdot 8\cdot 7\cdot 6\cdot 5\cdot 4\cdot 3\cdot 2\cdot 1}{6\cdot 5\cdot 4\cdot 3\cdot 2\cdot 1\cdot 19\cdot 18\cdot 17\cdot 16\cdot 15\cdot 14\cdot 13\cdot 12\cdot 11\cdot 10\cdot 9\cdot 8\cdot 7\cdot 6\cdot 5\cdot 4\cdot 3\cdot 2\cdot 1}$$

$$=\frac{25\cdot 24\cdot 23\cdot 22\cdot 21\cdot 20}{6\cdot 5\cdot 4\cdot 3\cdot 2\cdot 1}=5\cdot 23\cdot 11\cdot 7\cdot 20=177{,}100$$

47. $P(17,8)=\dfrac{17!}{(17-8)!}=\dfrac{17!}{9!}=\dfrac{17\cdot 16\cdot 15\cdot 14\cdot 13\cdot 12\cdot 11\cdot 10\cdot 9\cdot 8\cdot 7\cdot 6\cdot 5\cdot 4\cdot 3\cdot 2\cdot 1}{9\cdot 8\cdot 7\cdot 6\cdot 5\cdot 4\cdot 3\cdot 2\cdot 1}$

$=17\cdot 16\cdot 15\cdot 14\cdot 13\cdot 12\cdot 11\cdot 10=980{,}179{,}200$ minutes

There are $60\cdot 24\cdot 365=525{,}600$ minutes per year, so it would take $\dfrac{980{,}179{,}200}{525{,}600}\approx 1{,}864.9$ years or about 1,865 years.

49. Order is not important, but Anna used $P(8,3)$ instead of $C(8,3)$.

51. $P(15,5)=\dfrac{15!}{(15-5)!}=\dfrac{15!}{10!}=\dfrac{15\cdot 14\cdot 13\cdot 12\cdot 11\cdot 10\cdot 9\cdot 8\cdot 7\cdot 6\cdot 5\cdot 4\cdot 3\cdot 2\cdot 1}{10\cdot 9\cdot 8\cdot 7\cdot 6\cdot 5\cdot 4\cdot 3\cdot 2\cdot 1}=15\cdot 14\cdot 13\cdot 12\cdot 11=360{,}360$

53. $P(15,5)\times P(15,5)\times P(15,4)\times P(15,5)\times P(15,5)=552{,}446{,}474{,}061{,}128{,}648{,}601{,}600{,}000$ (a VERY large number)

55. $C(6,2)\times C(8,3)=\dfrac{6!}{2!(6-2)!}\cdot\dfrac{8!}{3!(8-3)!}=\dfrac{6!}{2!4!}\cdot\dfrac{8!}{3!5!}=\dfrac{6\cdot5\cdot4\cdot3\cdot2\cdot1}{2\cdot1\cdot4\cdot3\cdot2\cdot1}\cdot\dfrac{8\cdot7\cdot6\cdot5\cdot4\cdot3\cdot2\cdot1}{3\cdot2\cdot1\cdot5\cdot4\cdot3\cdot2\cdot1}$

$$=\frac{6\cdot5\cdot}{2\cdot1}\cdot\frac{8\cdot7\cdot6}{3\cdot2\cdot1}=(3\cdot5)\cdot(8\cdot7)=15\cdot56=840$$

57. $C(7,3)\times C(5,2)\times C(11,3)=\dfrac{7!}{3!(7-3)!}\cdot\dfrac{5!}{2!(5-2)!}\cdot\dfrac{11!}{3!(11-3)!}=\dfrac{7!}{3!4!}\cdot\dfrac{5!}{2!3!}\cdot\dfrac{11!}{3!8!}=$

$$\frac{7\cdot6\cdot5\cdot4\cdot3\cdot2\cdot1}{3\cdot2\cdot1\cdot4\cdot3\cdot2\cdot1}\cdot\frac{5\cdot4\cdot3\cdot2\cdot1}{2\cdot1\cdot3\cdot2\cdot1}\cdot\frac{11\cdot10\cdot9\cdot8\cdot7\cdot6\cdot5\cdot4\cdot3\cdot2\cdot1}{3\cdot2\cdot1\cdot8\cdot7\cdot6\cdot5\cdot4\cdot3\cdot2\cdot1}=\frac{7\cdot6\cdot5}{3\cdot2\cdot1}\cdot\frac{5\cdot4}{2\cdot1}\cdot\frac{11\cdot10\cdot9}{3\cdot2\cdot1}=$$

$$(7\cdot5)\cdot(5\cdot2)\cdot(11\cdot5\cdot3)=35\cdot10\cdot165=57{,}750$$

59. $C(5,2)\times C(4,2)=\dfrac{5!}{2!(5-2)!}\cdot\dfrac{4!}{2!(4-2)!}=\dfrac{5!}{2!3!}\cdot\dfrac{4!}{2!2!}=\dfrac{5\cdot4\cdot3\cdot2\cdot1}{2\cdot1\cdot3\cdot2\cdot1}\cdot\dfrac{4\cdot3\cdot2\cdot1}{2\cdot1\cdot2\cdot1}=\dfrac{5\cdot4}{2\cdot1}\cdot\dfrac{4\cdot3}{2\cdot1}$

$$=(5\cdot2)\cdot(2\cdot3)=10\cdot6=60$$

61. $C(24,8)\times C(16,8)=\dfrac{24!}{8!(24-8)!}\cdot\dfrac{16!}{8!(16-8)!}=\dfrac{24!}{8!(16)!}\cdot\dfrac{16!}{8!(12)!}=9{,}465{,}511{,}770$

63. $P(12,2)\times C(10,2)=\dfrac{12!}{(12-2)!}\cdot\dfrac{10!}{2!(10-2)!}=\dfrac{12!}{10!}\cdot\dfrac{10!}{2!8!}$

$$=\frac{12\cdot11\cdot10\cdot9\cdot8\cdot7\cdot6\cdot5\cdot4\cdot3\cdot2\cdot1}{10\cdot9\cdot8\cdot7\cdot6\cdot5\cdot4\cdot3\cdot2\cdot1}\cdot\frac{10\cdot9\cdot8\cdot7\cdot6\cdot5\cdot4\cdot3\cdot2\cdot1}{2\cdot1\cdot8\cdot7\cdot6\cdot5\cdot4\cdot3\cdot2\cdot1}=12\cdot11\cdot\frac{10\cdot9}{2\cdot1}$$

$$=132\cdot(5\cdot9)=132\cdot45=5{,}940$$

65. $P(5,2)\cdot C(11,3)=\dfrac{5!}{(5-2)!}\cdot\dfrac{11!}{3!(11-3)!}=\dfrac{5!}{3!}\cdot\dfrac{11!}{3!8!}$

$$=\frac{5\cdot4\cdot3\cdot2\cdot1}{3\cdot2\cdot1}\cdot\frac{11\cdot10\cdot9\cdot8\cdot7\cdot6\cdot5\cdot4\cdot3\cdot2\cdot1}{3\cdot2\cdot1\cdot8\cdot7\cdot6\cdot5\cdot4\cdot3\cdot2\cdot1}=5\cdot4\cdot\frac{11\cdot10\cdot9}{3\cdot2\cdot1}$$

$$=20\cdot(11\cdot5\cdot3)=20\cdot165=3{,}300$$

67. As you look at the intersections, you'll see that there is a pattern as to how many routes can be created as you leave the mall on the way to the bank. To choose direct routes, you must always be traveling down and/or to the right. The numbers indicate how many ways there are to get to the intersection below and to the right of the number. The total length of the route is 8 blocks. There are 70 possible routes.

Mall	1	1	1	1
1	2	3	4	5
1	3	6	10	15
1	4	10	20	35
1	5	15	35	70

Bank

Also, since there are 8 blocks to be traveled and you could turn right at any of these intersections four times. So there are $C(8,4)=\dfrac{8!}{4!(8-4)!}=\dfrac{8!}{4!4!}=\dfrac{8\cdot7\cdot6\cdot5\cdot4\cdot3\cdot2\cdot1}{4\cdot3\cdot2\cdot1\cdot4\cdot3\cdot2\cdot1}=\dfrac{8\cdot7\cdot6\cdot5}{4\cdot3\cdot2\cdot1}=7\cdot2\cdot5=70$ paths.

Also, since there are 8 blocks to be traveled and you could go down at any of these intersections four times. So there are $C(8,4)=\dfrac{8!}{4!(8-4)!}=\dfrac{8!}{4!4!}=\dfrac{8\cdot7\cdot6\cdot5\cdot4\cdot3\cdot2\cdot1}{4\cdot3\cdot2\cdot1\cdot4\cdot3\cdot2\cdot1}=\dfrac{8\cdot7\cdot6\cdot5}{4\cdot3\cdot2\cdot1}=7\cdot2\cdot5=70$ paths.

69. $C(18,6)\times C(12,6)=\dfrac{18!}{6!(18-6)!}\cdot\dfrac{12!}{6!(12-6)!}=\dfrac{18!}{6!(12)!}\cdot\dfrac{12!}{6!(6)!}=17{,}153{,}136$

71. In permutation problems, the order of the arrangement matters; in combination problems, the order of arrangement does not matter.

73. a) There is only one way to choose no objects from a set of five objects.
    b) There are eight ways to omit one object from a set of eight objects.

75. It would cost \$1,000 dollars to buy the tickets and you would be guaranteed to win \$500. Since \$500 – \$1000 = –\$500, you would lose \$500 dollars.

77. Because of the special conditions on the numbers, we cannot use all digits at each position in the number.

79. $C(13,2)\cdot C(13,3)=\dfrac{13!}{2!(13-2)!}\cdot\dfrac{13!}{3!(13-3)!}=\dfrac{13!}{2!11!}\cdot\dfrac{13!}{3!10!}=$

$\dfrac{13\cdot12\cdot11\cdot10\cdot9\cdot8\cdot7\cdot6\cdot5\cdot4\cdot3\cdot2\cdot1}{2\cdot1\cdot11\cdot10\cdot9\cdot8\cdot7\cdot6\cdot5\cdot4\cdot3\cdot2\cdot1}\cdot\dfrac{13\cdot12\cdot11\cdot10\cdot9\cdot8\cdot7\cdot6\cdot5\cdot4\cdot3\cdot2\cdot1}{3\cdot2\cdot1\cdot10\cdot9\cdot8\cdot7\cdot6\cdot5\cdot4\cdot3\cdot2\cdot1}=\dfrac{13\cdot12}{2\cdot1}\cdot\dfrac{13\cdot12\cdot11}{3\cdot2\cdot1}=$

$(13\cdot6)\cdot(13\cdot2\cdot11)=78\cdot286=22{,}308$

81.

83. 6; for six lines

85. $C(n-1,r-1)$

87. $C(n-1,r-1)+C(n-1,r)=C(n,r)$

89. The number of ways that we can choose an r-element set from n elements is the same as the number of ways that we can choose $n-r$ elements from n elements; for example, the number of ways that we can choose two elements from five elements is the same as the number of ways that we can choose three elements from five elements.

Section 12.4 Looking Deeper: Counting and Gambling

1. $2\times5\times12=120$

3. $5\times2\times2=20$

5. $1\times5\times2=10$

7. 4, since there are four suits

9. a) 4

   b) $C(13,5)=\dfrac{13!}{5!(13-5)!}=\dfrac{13!}{5!8!}=\dfrac{13\cdot12\cdot11\cdot10\cdot9\cdot8\cdot7\cdot6\cdot5\cdot4\cdot3\cdot2\cdot1}{5\cdot4\cdot3\cdot2\cdot1\cdot8\cdot7\cdot6\cdot5\cdot4\cdot3\cdot2\cdot1}$

   $=\dfrac{13\cdot12\cdot11\cdot10\cdot9}{5\cdot4\cdot3\cdot2\cdot1}=13\cdot11\cdot9=1,287$

   c) Although there are $4\cdot1,287=5,148$ ways to choose five cards from the same suit, 40 of these ways are either straight flushes or royal flushes. So, there are $5,148-40=5,108$ ways to construct a flush.

11. We can construct a two-pair hand in stages:

    Stage 1: Pick the rank of the card of which we will have a pair. This decision can be made in 13 ways.

    Stage 2: Now choose these two cards from the four cards of a rank. This can be done in $C(4,2)=6$ ways

    Stage 3: Now that you have chosen two cards, we select the next two for a pair. This decision can be made in 12 ways, because we don't want four-of-a-kind.

    Stage 4: Now choose these two cards from the four cards of a rank. This can be done in $C(4,2)=6$ ways

    Stage 5: Now the fifth card can be chosen in 44 ways since eight of the cards must be excluded.

    Since there are two ways to choose the two ranks, we must divide by two.

    So there are $\dfrac{13\times6\times12\times6}{2}\times44=\dfrac{5,616}{2}\times44=2,808\times44=123,552$ ways to choose a two-pair hand.

13. We are not simply choosing three cards from a possible 52.

15. The number of ways we can obtain the payoff for three cherries is $2\times5\times8=80$.
    The number of ways we can obtain the payoff for three oranges is $5\times2\times2=20$.
    Since there are fewer ways to obtain three oranges, we would expect the payoff to be higher.

17. a) There are 4 ways to make a Royal Flush (Exercise 7). There are 36 ways to make a Straight Flush (Exercise 8). There are 624 ways to make a Four of a Kind (Example 4). There are 3,744 ways to make a Full House (Example 6). There are 5,108 ways to make a Flush (Exercise 9). There are 10,200 ways to make a Straight (Exercise 10). There are 54,912 ways to make a Three of a Kind (Example 7). There are 123,552 ways to make Two Pair (Exercise 11). There are 1,098,240 ways to make One Pair (Exercise 12). This totals to be 1,296,420 winning hands.

    b) There are 2,598,960 ways to make a poker hand (Example 3). 2,598,960 – 1,296,420 = 1,302,540.

Chapter Review Exercises

1. PQ, PR, PS, QP, QR, QS, RP, RQ, RS, SP, SQ, SR

2. Without drawing the tree, you can imagine that the tree begins and there are four branches on the first roll. Each of those branches split into four branches on the second roll for a total sixteen branches. Each of those branches split into four branches on the third roll for a total of sixty-four branches. There would therefore be **sixty-four** ways three four-sided dice could be rolled.

3. Without drawing the tree, you can imagine that the tree begins and there are five branches on the first shirt choice. Each of those branches split into four branches for the second choice of pants for a total of twenty branches. Each of those branches split into six branches for the choice of a tie for a total of 120 branches. Each of those branches split into two branches for the choice of a jacket for a total of 240 branches. There would therefore be **240** outfits that could be formed.

4. $14\times13\times12=2{,}184$

5. $4\times12\times6=288$

6. $3\times6\times3\times4=216$

7.

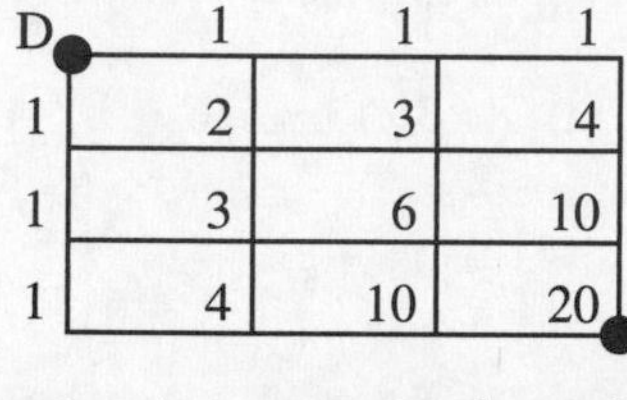

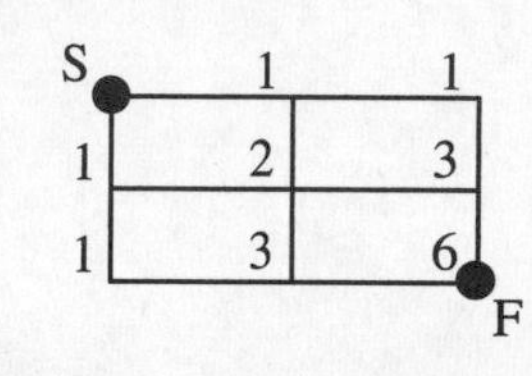

To choose direct routes, you must always be traveling down and/or to the right. The numbers indicate how many ways there are to get to the intersection below and to the right of the number. We see there are 20 ways to go directly from D to S and 6 ways to go directly from S to F. Therefore, there are $20\times6=120$ ways to go directly from D to F.

8.

Seat 1	Seat 2	Seat 3	Seat 4	Seat 5	Seat 6
Alex	Bonnie	X	X	X	X
X	Alex	Bonnie	X	X	X
X	X	Alex	Bonnie	X	X
X	X	X	Alex	Bonnie	X
X	X	X	X	Alex	Bonnie

Alex and Bonnie can occupy five different groupings of two seats. There are two different arrangements for each of these five possibilities. The number of ways that the students could be arranged in these six seats is $5\times2\times4\times3\times2\times1=240$.

9. In a permutation, order is important; in a combination, it is not.

10. $P(8,8)=\dfrac{8!}{(8-0)!}=\dfrac{8!}{0!}=\dfrac{8\cdot7\cdot6\cdot5\cdot4\cdot3\cdot2\cdot1}{1}=40{,}320$

11. $C(17,3)=\dfrac{17!}{3!(17-3)!}=\dfrac{17!}{3!14!}=\dfrac{17\cdot16\cdot15\cdot14\cdot13\cdot12\cdot11\cdot10\cdot9\cdot8\cdot7\cdot6\cdot5\cdot4\cdot3\cdot2\cdot1}{3\cdot2\cdot1\cdot14\cdot13\cdot12\cdot11\cdot10\cdot9\cdot8\cdot7\cdot6\cdot5\cdot4\cdot3\cdot2\cdot1}$

$=\dfrac{17\cdot16\cdot15}{3\cdot2\cdot1}=17\cdot8\cdot5=680$

12. $P(26,3)\cdot P(10,2)=\dfrac{26!}{(26-3)!}\cdot\dfrac{10!}{(10-2)!}=\dfrac{26!}{23!}\cdot\dfrac{10!}{8!}=$

$\dfrac{26\cdot25\cdot24\cdot23\cdot22\cdot21\cdot20\cdot19\cdot18\cdot17\cdot16\cdot15\cdot14\cdot13\cdot12\cdot11\cdot10\cdot9\cdot8\cdot7\cdot6\cdot5\cdot4\cdot3\cdot2\cdot1}{23\cdot22\cdot21\cdot20\cdot19\cdot18\cdot17\cdot16\cdot15\cdot14\cdot13\cdot12\cdot11\cdot10\cdot9\cdot8\cdot7\cdot6\cdot5\cdot4\cdot3\cdot2\cdot1}\cdot$

$\dfrac{10\cdot9\cdot8\cdot7\cdot6\cdot5\cdot4\cdot3\cdot2\cdot1}{8\cdot7\cdot6\cdot5\cdot4\cdot3\cdot2\cdot1}=(26\cdot25\cdot24)\cdot(10\cdot9)=15{,}600\cdot90=1{,}404{,}000$

13. $C(n,r)=\dfrac{P(n,r)}{r!}$

14. 21

```
            1
           1 1
          1 2 1
         1 3 3 1
        1 4 6 4 1
      1 5 10 10 5 1
    1 6 15 20 15 6 1
  1 7 (21) 35 35 21 7 1
```

15. Beginning with the zeroth entry, $C(18,2)$ represents the 2nd entry of the 18th row.

16. a) $4\times3\times13=156$ b) $4\times3\times7=84$

17. First, we need to pick five cards from the same suit:

$C(13,5)=\dfrac{13!}{5!(13-5)!}=\dfrac{13!}{5!8!}=\dfrac{13\cdot12\cdot11\cdot10\cdot9\cdot8\cdot7\cdot6\cdot5\cdot4\cdot3\cdot2\cdot1}{5\cdot4\cdot3\cdot2\cdot1\cdot8\cdot7\cdot6\cdot5\cdot4\cdot3\cdot2\cdot1}$

$=\dfrac{13\cdot12\cdot11\cdot10\cdot9}{5\cdot4\cdot3\cdot2\cdot1}=13\cdot11\cdot9=1{,}287$

Since there are four suites, there are $4\cdot1{,}287=5{,}148$ ways to choose five cards from the same suit. However, 40 of these ways are either straight flushes or royal flushes. So, there are $5{,}148-40=5{,}108$ ways to construct a flush.

Chapter Test

1. AB, AC, AD, BA, BC, BD, CA, CB, CD, DA, DB, DC

2. Without drawing the tree, you can imagine that the tree begins and there are eight branches on the first roll. Each of those branches split into eight branches on the second roll for a total of sixty-four branches. There would therefore be **64** ways two octahedral dice could be rolled.

3.

Seat 1	Seat 2	Seat 3	Seat 4	Seat 5
Devaun	Emily	X	X	X
X	Devaun	Emily	X	X
X	X	Devaun	Emily	X
X	X	X	Devaun	Emily

Devaun and Emily can occupy four different groupings of two seats. There are two different arrangements for each of these four possibilities. The number of ways that the students could be arranged in these six seats is $4\times2\times3\times2\times1=48$.

4. 20

1
1 1
1 2 1
1 3 3 1
1 4 6 4 1
1 5 10 10 5 1
1 6 15 (20) 15 6 1

5. $1\times5\times2=10$

6. $C(12,3)=\frac{12!}{3!(12-3)!}=\frac{12!}{3!(9!)}=\frac{12\cdot11\cdot10\cdot9\cdot8\cdot7\cdot6\cdot5\cdot4\cdot3\cdot2\cdot1}{3\cdot2\cdot1\cdot9\cdot8\cdot7\cdot6\cdot5\cdot4\cdot3\cdot2\cdot1}=\frac{12\cdot11\cdot10}{3\cdot2\cdot1}=4\cdot11\cdot5=220$

7. $P(15,4)=15\times14\times13\times12=32,760$

8. Since the ace can be considered either as a one or higher than a king, there are 10 ways to choose a sequence of five cards within a suit.

A	2	3	4	5	6	7	8	9	10	J	Q	K	A

Although there are $4\cdot10=40$ ways to choose a sequence of five cards, four of these ways are also royal flushes. So there are $40-4=36$ ways to construct a straight flush.

9. Without drawing the tree, you can imagine that the tree begins and there are three branches for the biology class choice. Each of those branches split into four branches for the diversity class choice for a total of twelve branches. Each of those branches split into six branches for the choice of a writing class for a total of seventy-two branches. There would therefore be **72** class schedules that could be formed.

10. In a permutation, order is important; in a combination, it is not.

11. $5 \times 4 \times 11 \times 6 = 1,320$

12. $P(6,6) = \dfrac{6!}{(6-0)!} = \dfrac{6!}{0!} = \dfrac{6 \cdot 5 \cdot 4 \cdot 3 \cdot 2 \cdot 1}{1} = 720$

13. $C(n,r) = \dfrac{P(n,r)}{r!}$

14. $C(12,5)$ represents the 5th entry of the 12th row.

15. $3 \times 9 \times 2 \times 2 = 108$

16. $P(26,3) \cdot P(10,4) = \dfrac{26!}{(26-3)!} \cdot \dfrac{10!}{(10-4)!} = \dfrac{26!}{23!} \cdot \dfrac{10!}{6!} =$

$$\frac{26 \cdot 25 \cdot 24 \cdot 23 \cdot 22 \cdot 21 \cdot 20 \cdot 19 \cdot 18 \cdot 17 \cdot 16 \cdot 15 \cdot 14 \cdot 13 \cdot 12 \cdot 11 \cdot 10 \cdot 9 \cdot 8 \cdot 7 \cdot 6 \cdot 5 \cdot 4 \cdot 3 \cdot 2 \cdot 1}{23 \cdot 22 \cdot 21 \cdot 20 \cdot 19 \cdot 18 \cdot 17 \cdot 16 \cdot 15 \cdot 14 \cdot 13 \cdot 12 \cdot 11 \cdot 10 \cdot 9 \cdot 8 \cdot 7 \cdot 6 \cdot 5 \cdot 4 \cdot 3 \cdot 2 \cdot 1} \cdot$$

$$\frac{10 \cdot 9 \cdot 8 \cdot 7 \cdot 6 \cdot 5 \cdot 4 \cdot 3 \cdot 2 \cdot 1}{6 \cdot 5 \cdot 4 \cdot 3 \cdot 2 \cdot 1} = (26 \cdot 25 \cdot 24) \cdot (10 \cdot 9 \cdot 8 \cdot 7) = 15,600 \cdot 5040 = 78,624,000$$

Chapter 13
Probability: What Are The Chances?

Section 13.1 The Basics of Probability Theory

1. {(1, 6), (2, 5), (3, 4), (4, 3), (5, 2), (6, 1)}

3. If we let "H" represent a heads and "T" represent tails, then the outcomes would be {HHH, HHT, HTH, THH}.

5. If we let "r" represent red, "b" represent blue, and "y" represent yellow, then the outcomes would be {(r, b), (r, y), (b, r), (y, r)}.

7. If we let "r" represent red, "b" represent blue, and "y" represent yellow, then the outcomes would be {(b, b, r), (b, r, b), (r, b, b), (b, b, y), (b, y, b), (y, b, b)}.

9. a) The sample space would have sixteen pairs. The sample space would be {(1, 1), (1, 2), (1, 3), (1, 4), (2, 1), (2, 2), (2, 3), (2, 4), (3, 1), (3, 2), (3, 3), (3, 4), (4, 1), (4, 2), (4, 3), (4, 4)}.
   b) The event would be {(1, 1), (1, 3), (2, 2), (2, 4), (3, 1), (3, 3), (4, 2), (4, 4)}.
   c) From (b), an even total can occur in eight ways. Since there are 16 pairs that can occur, the probability of rolling an even sum on two four-sided dice is $\frac{8}{16}=\frac{1}{2}$
   d) A total greater than six can occur in three ways. They are: {(3,4), (4,3), (4,4)}. Since there are 16 pairs that can occur, the probability of rolling a total greater than six on two four-sided dice is $\frac{3}{16}$.

11. a) $4\times3\times2\times1=24$
   b) {KREB, KRBE, RKEB, RKBE, EKRB, BKRE, ERKB, BRKE, EBKR, BEKR, EBRK, BERK}
   c) From (b), there are twelve ways where the women perform consecutively. Since there are 24 ways for the people to perform, the probability that the women perform consecutively is $\frac{12}{24}=\frac{1}{2}=0.5$.

13. a) 20; The sample space is {(s, .c), (s, w), (s, d), (s, h), (c, s), (c, w), (c, d), (c, h), (w, s), (w, c), (w, d), (w, h), (d, s), (d, c), (d, w), (d, h), (h, s), (h, c), (h, w), (h, d)}. Also, we could use $P(5,2)=20$.
   b) {(s, c), (s, w), (s, d), (s, h), (c, s), (w, s), (d, s), (h, s)}.
   c) Since exactly one star can occur in eight ways, the probability is $\frac{8}{20}=\frac{2}{5}$.
   d) No hearts can occur in twelve ways. They are: {(s,c), (s,w), (s,d), (c,s), (c,w), (c,d), (w,s), (w,c), (w,d), (d,s), (d,c), (d,w)}, thus, the probability is $\frac{12}{20}=\frac{3}{5}$.

15. Let E represent the event that a total of nine shows on the roll of two fair dice.

a) A total of nine can occur in four ways. They are: {(3,6), (4,5), (5,4), (6,3)}. Since there are 36 pairs that can occur, $P(E)=\frac{4}{36}=\frac{1}{9}\approx 0.11$

b) Since $P(E)=\frac{1}{9}$, $P(E')=1-\frac{1}{9}=\frac{9}{9}-\frac{1}{9}=\frac{8}{9}$. Thus, the odds against the event are $\frac{P(E')}{P(E)}=\frac{8/9}{1/9}=\frac{8}{1}$, or 8 to 1.

17. Let E represent the event of drawing a heart.

a) Since there are 13 hearts out of the 52 cards, $P(E)=\frac{13}{52}=\frac{1}{4}=0.25$.

b) Since $P(E)=\frac{1}{4}$, $P(E')=1-\frac{1}{4}=\frac{4}{4}-\frac{1}{4}=\frac{3}{4}$. Thus, the odds against the event are $\frac{P(E')}{P(E)}=\frac{3/4}{1/4}=\frac{3}{1}$, or 3 to 1.

19. $\frac{C(13,5)}{C(52,5)}=\frac{1,287}{2,598,960}\approx 0.000495$

21. $4\times\frac{C(13,5)}{C(52,5)}=4\times\frac{1,287}{2,598,960}\approx 0.002$

23. a) Since 5,040 people support the racetrack out of the total of 9,072 people, the probability of selecting a person that supports the racetrack is $\frac{5,040}{9,072}=\frac{5}{9}\approx 0.5556$

b) Since $P(E)=\frac{5}{9}$, $P(E')=1-\frac{5}{9}=\frac{9}{9}-\frac{5}{9}=\frac{4}{9}$. Thus, the odds for the event are $\frac{P(E)}{P(E')}=\frac{5/9}{4/9}=\frac{5}{4}$, or 5 to 4.

25. Since there are 1,951,379 females out of the 1,951,379 + 2,048,861 = 4,000,240 children, the probability of selecting a female is $\frac{1,951,379}{4,000,240}\approx 0.49$

27. Since there are 15 Thin Mints out of $16+28+12+24=80$ cookies, the probability that a Thin Mint is selected is $\frac{16}{80}=\frac{1}{5}=0.2$.

29. Since there are $28+12=40$ out of 80 cookies that are either a Caramel DeLite or Shortbread, the probability that a Caramel DeLite or Shortbread is selected is $\frac{40}{80}=\frac{1}{2}=0.5$.

31. Since there are 277 short plants out of the 787 + 277 = 1,064 plants, the probability that the plant will be short is $\frac{277}{1,064}\approx 0.26$. This is consistent with the theoretical results that there should be a 3 to 1 ratio in the second generation of dominant trait to recessive trait; that is, the theoretical probability of that a plant selected from the second generation will be short is 0.25.

33. The probability that the child is a carrier is $\frac{2}{4}=\frac{1}{2}$.

		Second Parent	
		s	*n*
First	*s*	*ss*	*sn*
Parent	*s*	*ss*	*sn*

35. a)

		First Gen.	
		r	*r*
First	*w*	*wr*	*wr*
Gen.	*w*	*wr*	*wr*

b) All flowers will be pink. *P*(pink) = 1; all other probabilities are 0.

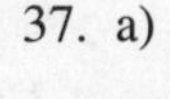

37. a)

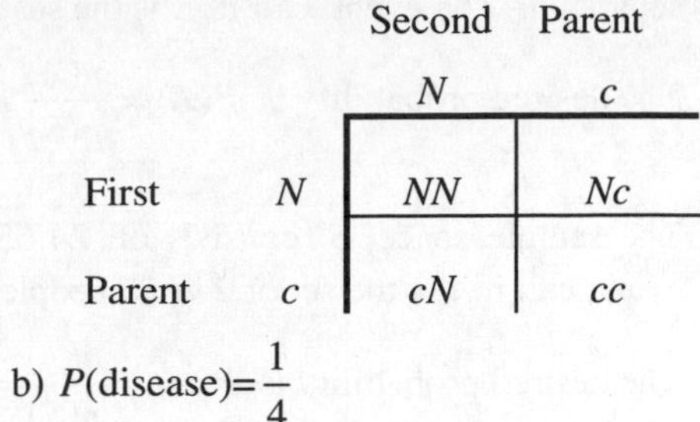

		Second Parent	
		N	*c*
First	*N*	*NN*	*Nc*
Parent	*c*	*cN*	*cc*

b) $P(\text{disease})=\frac{1}{4}$

39. The total area of the diagram is $8\times4=32 \text{ in}^2$ and the yellow area is $4\times2-2\times1=6 \text{ in}^2$. $P(\text{Yellow})=\frac{6}{32}=\frac{3}{16}=0.1875$.

41. The total area of the diagram is $8\times4=32 \text{ in}^2$, the green area is $3\times2+4\times1+2\times1=12 \text{ in}^2$, and the blue area is $8\times4-6\times3=14 \text{ in}^2$. $P(\text{Yellow or Blue})=\frac{6+14}{32}=\frac{20}{32}=\frac{5}{8}=0.625$.

43. There are 109 + 29 = 138 students out of the 320 surveyed that has a GPA of at least 2.5. The probability of selecting a student with a GPA of at least 2.5 is $\frac{138}{320}=\frac{69}{160}\approx0.43$.

45. Since you are rolling two dice, there are 36 possible combinations. St. James Place, Tennessee Avenue, and New York Avenue are within 12 spaces from the Electric Company. St. James Place is 4 spaces from the Electric Company. You can land on St. James Place if you roll (1,3), (2,2), or (3,1). Tennessee Avenue is 6 spaces from the Electric Company. You can land on Tennessee Avenue if you roll (1,5), (2,4), (3,3), (4,2), or (5,1). New York Avenue is 7 spaces from the Electric Company. You can land on New York Avenue if you roll (1,6), (2,5), (3,4), (4,3), (5,2), or (6,1). Since there are 14 ways to go bankrupt, there are $36-14=22$ ways to avoid going bankrupt. Thus, the probability of avoiding bankruptcy is $\frac{22}{36}=\frac{11}{18}\approx0.61$.

47. Since you are rolling two dice, there are 36 possible combinations. There are two railroads within 12 spaces from Virginia Avenue. The closest one is one space away. You cannot land on it, since the minimum roll would be a total of 2. The second railroad is 11 spaces from Virginia Avenue. You can land on it if you roll (5,6) or (6,5). Thus the probability of landing on a railroad is $\frac{2}{36}=\frac{1}{18}\approx0.06$.

49. The sample space, S, consists of $1,621+2,328=3,949$ thousand people that are divorced or separated. The event, call it A, is the set of 2,328 women that are divorced or separated.

The desired probability is $P(A)=\frac{n(A)}{n(S)}=\frac{2,328}{3,949}\approx 0.59$.

51. The sample space, S, consists of $24,655+20,449=45,104$ people that were never married. The event, call it A is the set of 24,655 people that are men.

The desired probability is $P(A)=\frac{n(A)}{n(S)}=\frac{24,655}{45,104}\approx 0.55$.

53. You should choose spinner A.
The sample space is {(8, 5), (8, 4), (8, 1), (3, 5), (3, 4), (3, 1), (2, 5), (2, 4), (2, 1)}, where the first number represents the value of Spinner A and the second number represents the value of Spinner B.
The events are equally likely, so $P(A \text{ wins})=\frac{5}{9}$ and $P(B \text{ wins})=\frac{4}{9}$.

55. Since order is important, the number of ways of choosing the horses would be a permutation. There are $P(8,3)=336$ ways to choose the horses. Since only one way is a winner, the probability of winning is $\frac{1}{336}\approx 0.002976$.

57. The odds against E are $\frac{5}{2}=\frac{5/7}{2/7}=\frac{P(E')}{P(E)}$. Thus, the probability of E is $\frac{2}{7}$.

59. The odds against W are $\frac{7}{5}=\frac{7/12}{5/12}=\frac{P(W')}{P(W)}$. Thus, the probability of winning is $\frac{5}{12}$.

61. a) The odds in favor of E (winning) are $\frac{P(E)}{P(E')}=\frac{0.30}{1-0.30}=\frac{0.30}{0.70}=\frac{3/10}{7/10}=\frac{3}{7}$, or 3 to 7.

b) The odds against E (winning) are $\frac{P(E')}{P(E)}=\frac{1-0.30}{0.30}=\frac{0.70}{0.30}=\frac{7/10}{3/10}=\frac{7}{3}$, or 7 to 3.

63. There are $C(59,6)=45,057,474$ ways to choose six numbers from the numbers 1 to 59. Since there is only one way to pick the correct numbers, the odds against winning are 45,057,474 to 1.

65. P(a randomly selected member of the senate has a law degree) = $\frac{1}{1.75}\approx 0.571$

67. No, the likelihood of a frame being defective or non-defective is probably not equal. The outcomes are not equally likely, so the formula does not apply.

69. Answers may vary.
An outcome is an element in a sample space; an event is a subset of a sample space.

71. Since the hacker can enter a password every 10 seconds, he will have 6 opportunities each minute. Since the hacker is allowed three minutes, he has a total of 18 opportunities. Since the hacker is randomly trying passwords, it is conceivable that the hacker might try the same password more than once. Since there are $26\times26\times10\times10\times10=676,000$ possible passwords, the probability that the hacker would be successful in discovering a valid password is $18\times\frac{1}{676,000}\approx0.000027$.

73. Since the hacker can enter a password every 10 seconds, he will have 6 opportunities each minute. Since the hacker is allowed three minutes, he has a total of 18 opportunities. Since the hacker is randomly trying passwords, it is conceivable that the hacker might try the same password more than once. There are $26\times26\times10\times10\times10=676,000$ possible passwords that are two letters followed by 3 digits. Since these can appear in any order, there are 10 possible ordering of two letters and 3 digits: (LLDDD, LDLDD, LDDLD, LDDDL, DLLDD, DLDLD, DLDDL, DDLLD, DDLDL, DDDLL). The probability that the hacker would be successful in discovering a valid password is $18\times\frac{1}{676,000\times10}\approx0.0000027$.

75. – 77. No solution provided.

Section 13.2 Complements and Unions Of Events

1. $1-0.015=0.985$

3. $1-\frac{1}{1,000}=\frac{1,000}{1,000}-\frac{1}{1,000}=\frac{999}{1,000}$

5. Let E be the event that either die shows a five on it. This can happen in 11 ways: {(1,5), (2,5), (3,5), (4,5), (5,5), (6,5), (5,1), (5,2), (5,3), (5,4), (5,6)}. Since there are 36 ways to roll two dice, the probability that neither die shows a five on it is $P(E')=1-P(E)=1-\frac{11}{36}=\frac{25}{36}$.

7. Let E be the event that no coin shows a head. This can happen in only one way: {TTTTT}. Since there are $2\times2\times2\times2\times2=32$ ways to flip five coins, the probability of obtaining at least one head is $P(E')=1-P(E)=1-\frac{1}{32}=\frac{31}{32}$.

9. Let F be the event that we choose a five, and let R be the event that we draw a red card. There are 4 fives in a deck, 26 red cards, and 2 red fives.

$$P(F\cup R)=P(F)+P(R)-P(F\cap R)=\frac{4}{52}+\frac{26}{52}-\frac{2}{52}=\frac{28}{52}=\frac{7}{13}$$

11. Let R be the event of rain and F be the event of for.

$$P(R\cup F)=P(R)+P(F)-P(R\cap F)=0.6+0.4-0.15=0.85$$

13.
$$\begin{aligned}P(A\cup B)&=P(A)+P(B)-P(A\cap B)\\0.85&=0.55+0.40-P(A\cap B)\\0.85&=0.95-P(A\cap B)\\-0.10&=-P(A\cap B)\\P(A\cap B)&=0.10\end{aligned}$$

15.
$$\begin{aligned}P(A\cup B)&=P(A)+P(B)-P(A\cap B)\\0.70&=0.40+P(B)-0.25\\0.70&=0.15+P(B)\\P(B)&=0.55\end{aligned}$$

17. There are 1,483 (in thousands) surveyed. The total (in thousands) of those that are 55 or older is $61+53=114$. The probability of selecting a person less than 55 years old is $P(E)=1-P(E')=1-\frac{114}{1,483}=\frac{1,369}{1,483}\approx 0.923$.

19. Let T be the event that we select a consumer that spends 0-2 hours per month shopping on the Internet, and let A be the event that we select a consumer that has an annual income below \$40,000.

$$P(T\cup A)=P(T)+P(A)-P(T\cap A)=\frac{544}{1,600}+\frac{592}{1,600}-\frac{272}{1,600}=\frac{864}{1,600}=\frac{27}{50}=0.54$$

21. Let T be the event that we select a consumer that spends more than 2 hours per month shopping on the Internet, and let A be the event that we select a consumer that has an annual income of \$60,000 or less.

$$P((T\cup A)')=1-P(T\cup A)=1-[P(T)+P(A)-P(T\cap A)]$$

$$=1-\left[\frac{480+576}{1,600}+\frac{512+592}{1,600}-\frac{160+128+208+192}{1,600}\right]$$

$$=1-\left[\frac{1,056}{1,600}+\frac{1,104}{1,600}-\frac{688}{1,600}\right]=1-\frac{1,472}{1,600}=1-\frac{23}{25}=\frac{2}{25}=0.08$$

23. Let E be the event that Joanna will earn less than \$1,000 in commissions.

$$P(E')=1-P(E)=1-0.08=0.92$$

25. Let E be the event that Joanna will earn more than \$1,999 in commissions.

$$P(E')=1-P(E)=1-(0.05+0.08+0.03)=1-0.16=0.84$$

27. Let H be the event that we select a heart and Q be the event that we select a queen.

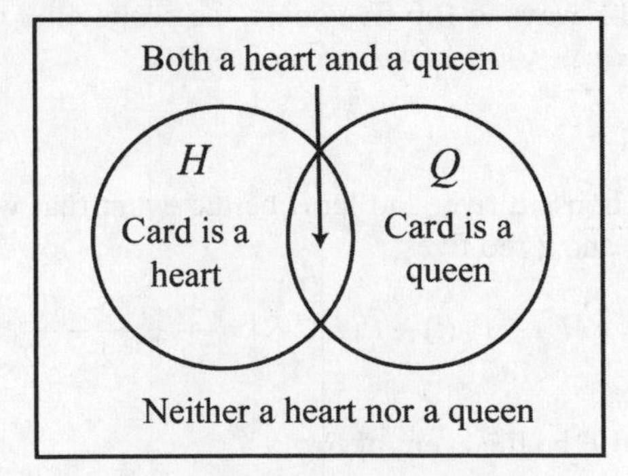

$$P\left((H\cup Q)'\right)=1-P(H\cup Q)=1-[P(H)+P(Q)-P(H\cap Q)]$$

$$=1-\left[\frac{13}{52}+\frac{4}{52}-\frac{1}{52}\right]=1-\frac{16}{52}=\frac{36}{52}=\frac{9}{13}$$

29. Let M be the event that you need a new motor and S be the event that you need a new switch.

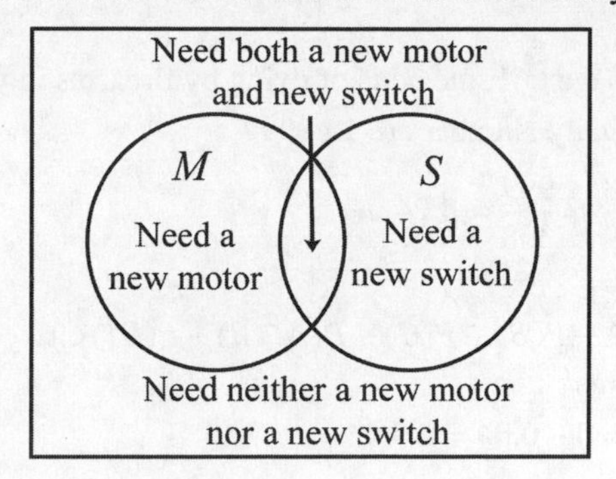

$$P\left((M \cup S)'\right) = 1 - P(M \cup S) = 1 - [P(M) + P(S) - P(M \cap S)]$$

$$= 1 - [0.55 + 0.4 - 0.15] = 1 - 0.8 = 0.2$$

Use the following table for Exercises 31 – 33. The bolded information needed to be determined.

	Probation	Not on Probation	Totals
Satisfied	**38**	**110**	**148**
Not Satisfied	32	20	**52**
Totals	70	**130**	200

31. Let E be the event that the student chosen is not on academic probation.

$$P(E) = \frac{130}{200} = 0.65$$

33. Let E be the event that the student chosen is on academic probation and is satisfied with advisement.

$$P(E) = \frac{38}{200} = 0.19$$

35. $P(\text{At least 1 defective camera sold}) = 1 - P(\text{No defective cameras sold})$

$$= 1 - \frac{C(34,18)}{C(40,18)} = 1 - \frac{187}{9620} = \frac{9433}{9620} \approx 0.981$$

37. $P(\text{At least 1 customer receives spoiled food}) = 1 - P(\text{No customer receives spoiled food})$

$$= 1 - \frac{C(16,12)}{C(18,12)} = 1 - \frac{5}{51} = \frac{46}{51} \approx 0.902$$

39. False, this would be true only if the intersection of events A and B was empty.

41. If E and F are disjoint sets then $P(E \cap F) = 0$.

43. The chance of dying from assault by firearms is approximately 3.24 times greater than winning The Daily Number.

According to the table on page 661, the odds of dying by firearms in your lifetime are 1 to 308.

The odds of winning The Daily Number are 1 to 999

$$\frac{O(Firearms)}{O(Daily\ Number)} = \frac{1/308}{1/999} = \frac{999}{308} \approx 3.24$$

45. Explanations will vary; $P(A)+P(B)+P(C)-P(A\cap B)-P(B\cap C)$.

47. $P(yellow')=1-P(yellow)=1-0.04=0.96$

49. $P\left((odd \cup blue)'\right)=1-P\left((odd \cup blue)\right)=1-\left(P(odd)+P(blue)-P(odd \cap blue)\right)$

$=1-(0.50+0.48-0.12)=1-0.86=0.14$

Section 13.3 Conditional Probability and Intersections of Events

For Exercises 1 – 3, refer to the tree diagram on the next page.

1. $P(F)=\frac{6}{36}=\frac{1}{6}$

There are 18 times out of the 36 outcomes that yield an odd sum. Of these 18 occurrences, 6 of them yield a sum of 7.

$P(F\mid E)=\frac{6}{18}=\frac{1}{3}$

3. $P(F)=\frac{6}{36}=\frac{1}{6}$

There are 11 times out of the 36 outcomes that a three shows on at least one of the dice. Of these 11 occurrences, 2 of them yield a sum of less than 5.

$P(F\mid E)=\frac{2}{11}$

5. $\text{P(heart|red)} = \frac{\text{P(red} \cap \text{heart)}}{\text{P(red)}} = \frac{13/52}{26/52} = \frac{13}{26} = \frac{1}{2}$

7. $\text{P(seven|nonface card)} = \frac{\text{P(nonface card} \cap \text{seven)}}{\text{P(nonface card)}} = \frac{4/52}{40/52} = \frac{4}{40} = \frac{1}{10}$

9. $\text{P(heart | yellow)} = \frac{\text{P(heart} \cap \text{yellow)}}{\text{P(yellow)}} = \frac{2/10}{4/10} = \frac{2}{4} = \frac{1}{2}$

11. $\text{P(yellow | heart)} = \frac{\text{P(yellow} \cap \text{heart)}}{\text{P(heart)}} = \frac{2/10}{3/10} = \frac{2}{3}$

Use the following tree diagram for Exercises 1 – 3.

	First	Second	Outcome	Sum
Begin	1	1	(1,1)	2
		2	(1,2)	3
		3	(1,3)	4
		4	(1,4)	5
		5	(1,5)	6
		6	(1,6)	7
	2	1	(2,1)	3
		2	(2,2)	4
		3	(2,3)	5
		4	(2,4)	6
		5	(2,5)	7
		6	(2,6)	8
	3	1	(3,1)	4
		2	(3,2)	5
		3	(3,3)	6
		4	(3,4)	7
		5	(3,5)	8
		6	(3,6)	9
	4	1	(4,1)	5
		2	(4,2)	6
		3	(4,3)	7
		4	(4,4)	8
		5	(4,5)	9
		6	(4,6)	10
	5	1	(5,1)	6
		2	(5,2)	7
		3	(5,3)	8
		4	(5,4)	9
		5	(5,5)	10
		6	(5,6)	11
	6	1	(6,1)	7
		2	(6,2)	8
		3	(6,3)	9
		4	(6,4)	10
		5	(6,5)	11
		6	(6,6)	12

13. $P(\text{heart} \mid \text{pink}) = \dfrac{P(\text{heart} \cap \text{pink})}{P(\text{pink})} = \dfrac{0/10}{3/10} = 0$

15. a) $P(J \cap J) = P(J) \cdot P(J \mid J) = \dfrac{4}{52} \cdot \dfrac{3}{51} = \dfrac{1}{13} \cdot \dfrac{1}{17} = \dfrac{1}{221}$

    b) $P(J \cap J) = P(J) \cdot P(J) = \dfrac{4}{52} \cdot \dfrac{4}{52} = \dfrac{1}{13} \cdot \dfrac{1}{13} = \dfrac{1}{169}$

17. a) $P(F \cap N) = P(F) \cdot P(N \mid F) = \dfrac{12}{52} \cdot \dfrac{40}{51} = \dfrac{3}{13} \cdot \dfrac{40}{51} = \dfrac{120}{663} = \dfrac{40}{221}$

    b) $P(F \cap N) = P(F) \cdot P(N) = \dfrac{12}{52} \cdot \dfrac{40}{52} = \dfrac{3}{13} \cdot \dfrac{10}{13} = \dfrac{30}{169}$

19. a) $P(J \cap K) = P(J) \cdot P(K \mid J) = \dfrac{4}{52} \cdot \dfrac{4}{51} = \dfrac{1}{13} \cdot \dfrac{4}{51} = \dfrac{4}{663}$

    $P(K \cap J) = P(K) \cdot P(J \mid K) = \dfrac{4}{52} \cdot \dfrac{4}{51} = \dfrac{1}{13} \cdot \dfrac{4}{51} = \dfrac{4}{663}$

    $\dfrac{4}{663} + \dfrac{4}{663} = \dfrac{8}{663}$

    b) $P(J \cap K) = P(J) \cdot P(K) = \dfrac{4}{52} \cdot \dfrac{4}{52} = \dfrac{1}{13} \cdot \dfrac{1}{13} = \dfrac{1}{169}$

    $P(K \cap J) = P(K) \cdot P(J) = \dfrac{4}{52} \cdot \dfrac{4}{52} = \dfrac{1}{13} \cdot \dfrac{1}{13} = \dfrac{1}{169}$

    $\dfrac{1}{169} + \dfrac{1}{169} = \dfrac{2}{169}$

21. Let E be the event that we draw at least one face card. $\overline{E}$ is therefore the event that we draw no face cards. If N denotes a non-face card, then

$$P(E) = 1 - P(\overline{E}) = 1 - P(N \cap N) = 1 - P(N) \cdot P(N \mid N) = 1 - \frac{40}{52} \cdot \frac{39}{51} = 1 - \frac{1{,}560}{2{,}652} = 1 - \frac{10}{17} = \frac{7}{17}$$

23. $P(R \text{ then } R) = P(R) \cdot P(R \mid R) = \dfrac{4}{20} \cdot \dfrac{3}{19} = \dfrac{1}{5} \cdot \dfrac{3}{19} = \dfrac{3}{95} \approx 0.032$

25. $P(\text{Red and Blue}) = P(R \text{ then } B \text{ or } B \text{ then } R) = P(R) \cdot P(B \mid R) + P(B) \cdot P(R \mid B)$

$$= \frac{4}{20} \cdot \frac{6}{19} + \frac{6}{20} \cdot \frac{4}{19} = \frac{6}{95} + \frac{6}{95} = \frac{12}{95} \approx 0.126$$

27. $P(H \cap H \cap N) + P(H \cap N \cap H) + P(N \cap H \cap H) =$

$$\frac{13}{52} \cdot \frac{12}{51} \cdot \frac{39}{50} + \frac{13}{52} \cdot \frac{39}{51} \cdot \frac{12}{50} + \frac{39}{52} \cdot \frac{13}{51} \cdot \frac{12}{50} = \frac{6{,}084}{132{,}600} + \frac{6{,}084}{132{,}600} + \frac{6{,}084}{132{,}600} = \frac{18{,}252}{132{,}600} = \frac{117}{850}$$

29. Out of the 36 possible ways to throw two dice, half of the rolls will have an even total, so the probability of throwing an even total is $\frac{1}{2}$. Therefore, the probability of an even total on one of the three rolls is

$$P(O \cap O \cap E) + P(O \cap E \cap O) + P(E \cap O \cap O) =$$

$$\frac{1}{2} \cdot \frac{1}{2} \cdot \frac{1}{2} + \frac{1}{2} \cdot \frac{1}{2} \cdot \frac{1}{2} + \frac{1}{2} \cdot \frac{1}{2} \cdot \frac{1}{2} = \frac{1}{8} + \frac{1}{8} + \frac{1}{8} = \frac{3}{8} = 0.375$$

31. $P(A \mid J) = 0.25$

33. $P(E \cap A) = 0.40 \times 0.35 = 0.14$

35. Independent; Getting heads on the penny has no effect on the probability of getting tails on the nickel.

37. independent
Let R be the event "the card is red" and F be the event "the card is a face card."

$$P(R \mid F) = \frac{P(R \cap F)}{P(F)} = \frac{3/52}{12/52} = \frac{3}{12} = \frac{1}{4} = P(\text{Red}) = \frac{13}{52} = \frac{1}{4}$$

39. dependent
Let E be the event "the total is greater than nine" and F be the event "the total is even."

$$P(F \mid E) = \frac{P(F \cap E)}{P(E)} = \frac{4/36}{6/36} = \frac{4}{6} = \frac{2}{3} \neq P(F) = \frac{18}{36} = \frac{1}{2}$$

For Exercises 41 – 43, let M = the student has Mono and T = the test was positive.

41. $P(M \mid T) = \frac{P(M \cap T)}{P(\text{T})} = \frac{72/140}{76/140} = \frac{18}{19} \approx 0.947$

43. $P(T \mid M) = \frac{P(T \cap M)}{P(M)} = \frac{72/140}{80/140} = \frac{9}{10} \approx 0.90$

For Exercises 45 – 47, let O = the student preferred Obama, M = the student preferred Romney, R = the student is a Republican, D = the student is a Democrat, and I = the student is an Independent.

45. $P(D \mid M) = \frac{P(D \cap M)}{P(M)} = \frac{15/240}{98/240} = \frac{15}{98} \approx 0.153$

47. $P(M \mid R) = \frac{P(M \cap R)}{P(R)} = \frac{68/240}{80/240} = \frac{68}{80} = 0.85$

49. $\frac{138}{138 + 481} = \frac{138}{619} \approx 0.223$

51. $\frac{196}{138 + 293 + 196} = \frac{196}{627} \approx 0.313$

53. $(0.30)(0.70)(0.70)(0.70) = 0.1029$

55. $6(0.30)(0.30)(0.70)(0.70) = 0.2646$

57. $\dfrac{0.06}{0.06+0.42}=\dfrac{0.06}{0.48}=0.125$

59.

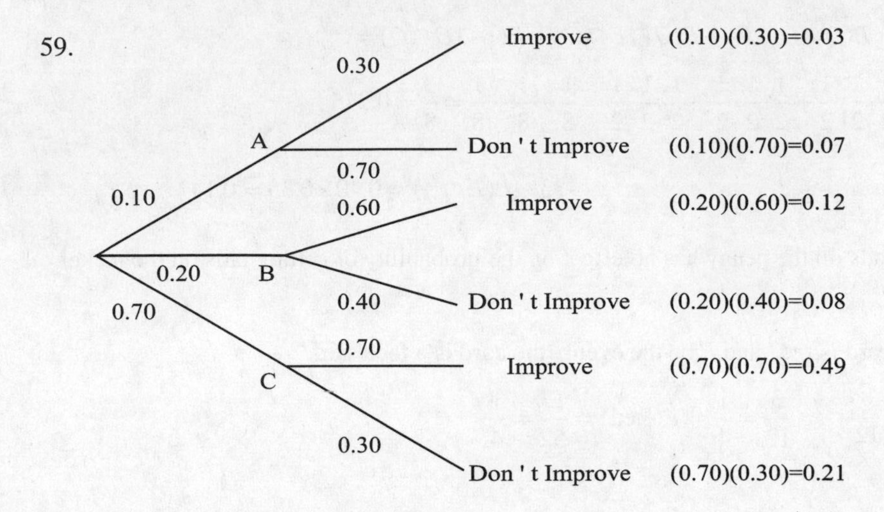

61. $0.03+0.12+0.49=0.64$

63. $\dfrac{0.18}{0.18+0.30}=\dfrac{0.18}{0.48}=0.375$

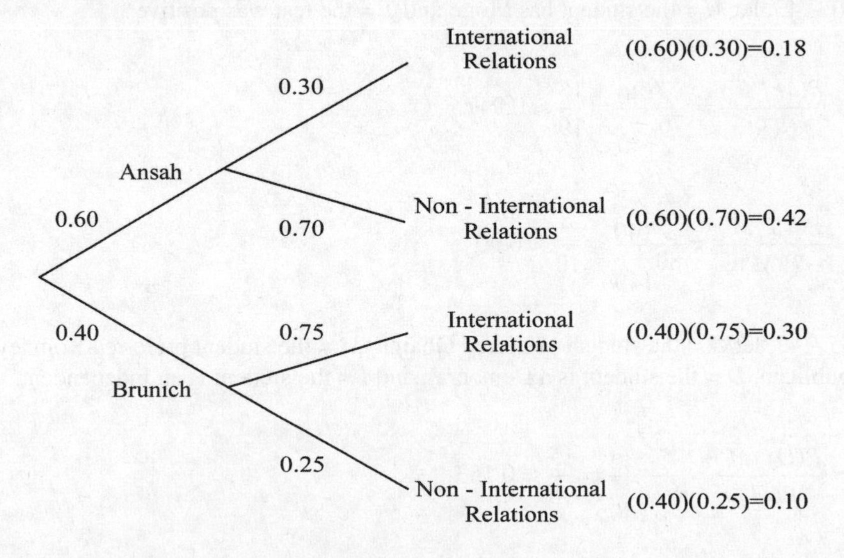

Use the tree diagram on the next page for Exercises 65 – 67.

65. 0.336

67. $\dfrac{0.039}{0.014+0.039}=\dfrac{0.039}{0.053}\approx 0.736$

Use the tree diagram on the next page for Exercises 69.

69. $\dfrac{0.0288}{0.0288+0.0392}=\dfrac{0.0288}{0.068}\approx 0.424$

Use the following tree diagram for Exercises 65 – 67.

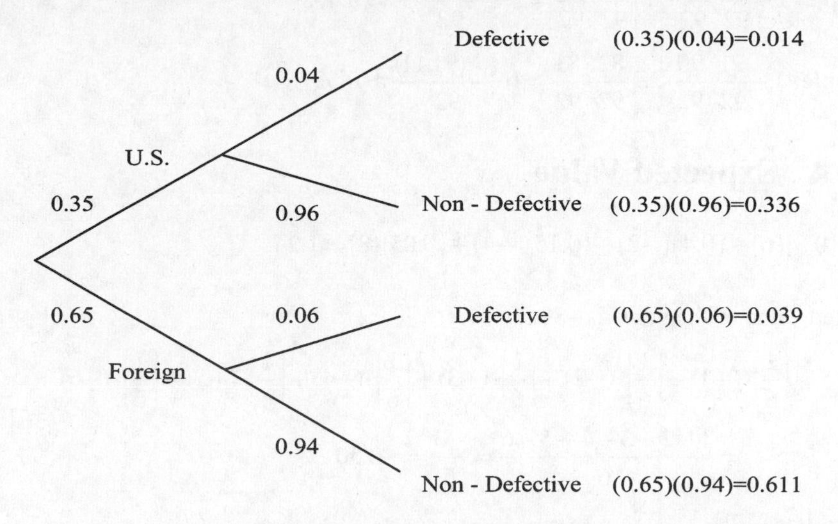

Use the following tree diagram for Exercise 69.

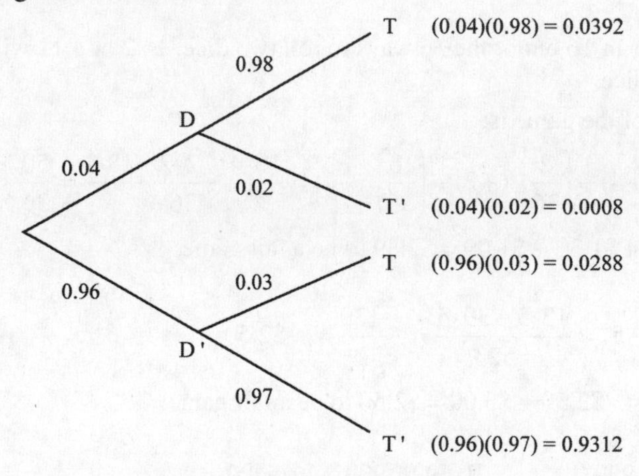

71. Multiply both sides of the equation by $P(E)$ to get $P(E \cap F) = P(F \mid E) \cdot P(E)$.

73. Answers will vary.

75. $P(\text{duplication of birthdays}) = 1 - P(\text{no duplication of birthdays}) =$

$$1 - \frac{364}{365} \cdot \frac{363}{365} \cdot \frac{362}{365} \cdot \frac{361}{365} \cdot \frac{360}{365} \cdot \frac{359}{365} \cdot \frac{358}{365} \cdot \frac{357}{365} \cdot \frac{356}{365} = 1 - \frac{P(364,9)}{365^9} \approx 0.117$$

77. For 21 people: $1 - \frac{364}{365} \cdot \frac{363}{365} \cdot \ldots \cdot \frac{346}{365} \cdot \frac{345}{365} = 1 - \frac{P(364,20)}{365^{20}} \approx 0.444$

For 22 people: $1 - \frac{364}{365} \cdot \frac{363}{365} \cdot \ldots \cdot \frac{345}{365} \cdot \frac{344}{365} = 1 - \frac{P(364,21)}{365^{21}} \approx 0.476$

For **23 people**: $1 - \frac{364}{365} \cdot \frac{363}{365} \cdot \ldots \cdot \frac{344}{365} \cdot \frac{343}{365} = 1 - \frac{P(364,22)}{365^{22}} \approx 0.507$

79. For 11 people: $1-\frac{91}{92}\cdot\frac{90}{92}\cdot\ldots\cdot\frac{83}{92}\cdot\frac{82}{92}=1-\frac{P(91,10)}{92^{10}}\approx 0.463$

For **12 people**: $1-\frac{91}{92}\cdot\frac{90}{92}\cdot\ldots\cdot\frac{82}{92}\cdot\frac{81}{92}=1-\frac{P(91,11)}{92^{11}}\approx 0.527$

Section 13.4 Expected Value

1. $(0.1)(4)+(0.3)(6)+(0.4)(-2)+(0.15)(-4)+(0.05)(8)=1.2$

3. The expected value of the game is –$0.50:

$$\left(\frac{1}{6}\right)\cdot(+1)+\left(\frac{1}{6}\right)\cdot(-2)+\left(\frac{1}{6}\right)\cdot(+3)+\left(\frac{1}{6}\right)\cdot(-4)+\left(\frac{1}{6}\right)\cdot(+5)+\left(\frac{1}{6}\right)\cdot(-6)$$

$$=\frac{1-2+3-4+5-6}{6}=\frac{-3}{6}=-0.50$$

This is not a fair game.

For Exercise 5, refer to the tree diagram for Exercises 1 – 3 in Section 13.3.

5. A 6, 7 or 8 will occur in 16 out of the 36 ways to roll two dice. A 2 or a 12 will occur in 2 out of the 36 ways to roll two dice.
The expected value of the game is:

$$\left(\frac{16}{36}\right)\cdot(+4)+\left(\frac{2}{36}\right)\cdot(+2)+\left(\frac{18}{36}\right)\cdot(-1)=\frac{16\cdot 4+2\cdot 2-18\cdot 1}{36}=\frac{50}{36}\approx \$1.39$$

The game should cost $1.39 + $1.00 = $2.39 to be a fair game.

7. $\left(\frac{13}{52}\right)\cdot(+5)+\left(\frac{39}{52}\right)\cdot(-5)=\frac{13\cdot 5-39\cdot 5}{52}=\frac{-130}{52}=-\$2.50$

The game should cost –$2.50 + $5.00 = $2.50 to be a fair game.

9. $\left(\frac{1}{1{,}000}\right)\cdot(+599)+\left(\frac{999}{1{,}000}\right)\cdot(-1)=\frac{1\cdot 599-999\cdot 1}{1{,}000}=\frac{-400}{1{,}000}=-\$0.40$

This is not a fair game. To make it a fair game, one should charge –$0.40 + $1.00 = $0.60

11. $\left(\frac{1}{500}\right)\cdot(+495)+\left(\frac{2}{500}\right)\cdot(+245)+\left(\frac{5}{500}\right)\cdot(+95)+\left(\frac{492}{500}\right)\cdot(-5)=$

$$\frac{1\cdot 495+2\cdot 245+5\cdot 95-492\cdot 5}{500}=\frac{-1{,}000}{500}=-\$2.00$$

This is not a fair game. To make it a fair game, one should charge –$2.00 + $5.00 = $3.00

13. The total number who earned a profit is $4+8+13+21+3+1=50$.

$$\left(\frac{4}{50}\right)\cdot(100)+\left(\frac{8}{50}\right)\cdot(200)+\left(\frac{13}{50}\right)\cdot(300)+\left(\frac{21}{50}\right)\cdot(400)+\left(\frac{3}{50}\right)\cdot(500)+\left(\frac{1}{50}\right)\cdot(600)$$

$$=\frac{400+1{,}600+3{,}900+8{,}400+1{,}500+600}{50}=\frac{16{,}400}{50}=\$328$$

15. $\left(\frac{5}{38}\right)\cdot(+6)+\left(\frac{33}{38}\right)\cdot(-1)=\frac{5\cdot 6-33\cdot 1}{38}=\frac{-3}{38}\approx -\$0.08$

17. $\left(\frac{3}{38}\right)\cdot(+8)+\left(\frac{35}{38}\right)\cdot(-1)=\frac{3\cdot 8-35\cdot 1}{38}=\frac{-11}{38}\approx -\$0.29$

19. The expected value is $\left(\frac{1}{4}\right)\cdot(+1)+\left(\frac{3}{4}\right)\cdot\left(-\frac{1}{4}\right)=\frac{4}{16}+\frac{-3}{16}=\frac{1}{16}$. Since this is positive, the student should guess.

21. The expected value when ruling out no option is: $\left(\frac{1}{5}\right)\cdot(+1)+\left(\frac{4}{5}\right)\cdot\left(-\frac{1}{2}\right)=\frac{2}{10}+\frac{-4}{10}=-\frac{2}{10}=-\frac{1}{5}$.

    The expected value when ruling out one option is: $\left(\frac{1}{4}\right)\cdot(+1)+\left(\frac{3}{4}\right)\cdot\left(-\frac{1}{2}\right)=\frac{2}{8}+\frac{-3}{8}=-\frac{1}{8}$.

    The expected value when ruling out **two options** is: $\left(\frac{1}{3}\right)\cdot(+1)+\left(\frac{2}{3}\right)\cdot\left(-\frac{1}{2}\right)=\frac{2}{6}+\frac{-2}{6}=0$.

23. $(600)(0.5)+(0)(0.3)+(-200)(0.2)=\$260.00$

25. $(1000)(0.45)+(0)(0.15)+(-600)(0.4)=\$210.0$

27. $(0.02)\cdot(+972.50)+(0.98)\cdot(-27.50)=-\$7.50$

29. \$44.00

$$
\begin{aligned}
(0.02)\cdot(+2{,}200-x)+(0.98)\cdot(-x)&=0\\
44-.02x-0.98x&=0\\
44-x&=0\\
44&=x
\end{aligned}
$$

31. $(0.60)\cdot(+45{,}000)+(0.40)\cdot(-5{,}000)=\$25{,}000$

33. Expected value predicts what we can expect to happen long term when an experiment is repeated many times. Expected value does not predict what will happen for any one repetition of an experiment.

35. Answers will vary.

37. a) $\left(\frac{3}{20}\right)\cdot(+0.25)\cdot 130+\left(\frac{6}{20}\right)(+0.25)\cdot 130+\left(\frac{5}{20}\right)(+0.25)\cdot 130+\left(\frac{6}{20}\right)(+0.25\cdot 120+(-0.65)\cdot 10)=$ \$29.80

    b) $\left(\frac{3}{20}\right)\cdot(+0.25)\cdot 140+\left(\frac{6}{20}\right)(+0.25)\cdot 140+\left(\frac{5}{20}\right)(+0.25\cdot 130+(-0.65)\cdot 10)+$ $\left(\frac{6}{20}\right)(+0.25\cdot 120+(-0.65)\cdot 20)=\$27.35$

39. a)

Sum of Dice

Green Die \ Red Die	1	2	3	4	5	6
1	2	3	4	5	6	7
2	3	4	5	6	7	8
3	4	5	6	7	8	9
4	5	6	7	8	9	10
5	6	7	8	9	10	11
6	7	8	9	10	11	12

$$\left(\frac{1}{36}\right)\cdot(2)+\left(\frac{2}{36}\right)\cdot(3)+\left(\frac{3}{36}\right)\cdot(4)+\left(\frac{4}{36}\right)\cdot(5)+$$
$$\left(\frac{5}{36}\right)\cdot(6)+\left(\frac{6}{36}\right)\cdot(7)+\left(\frac{5}{36}\right)\cdot(8)+\left(\frac{4}{36}\right)\cdot(9)+$$
$$\left(\frac{3}{36}\right)\cdot(10)+\left(\frac{2}{36}\right)\cdot(11)+\left(\frac{1}{36}\right)\cdot(12)=7$$

b)

Sum of Dice

Green Die \ Red Die	1	2	2	3	3	4
1	2	3	3	4	4	5
3	4	5	5	6	6	7
4	5	6	6	7	7	8
5	6	7	7	8	8	9
6	7	8	8	9	9	10
8	9	10	10	11	11	12

$$\left(\frac{1}{36}\right)\cdot(2)+\left(\frac{2}{36}\right)\cdot(3)+\left(\frac{3}{36}\right)\cdot(4)+\left(\frac{4}{36}\right)\cdot(5)+$$
$$\left(\frac{5}{36}\right)\cdot(6)+\left(\frac{6}{36}\right)\cdot(7)+\left(\frac{5}{36}\right)\cdot(8)+\left(\frac{4}{36}\right)\cdot(9)+$$
$$\left(\frac{3}{36}\right)\cdot(10)+\left(\frac{2}{36}\right)\cdot(11)+\left(\frac{1}{36}\right)\cdot(12)=7$$

41. No solution provided.

Section 13.5 Looking Deeper: Binomial Experiments

1. yes

3. No; the number of trials is not fixed.

5. yes

7. A total of five can be made in 4 ways, (1,4), (2,3), (3,2), (4,1). The probability of obtaining a total of five on a single roll is $\frac{4}{36}=\frac{1}{9}$. The probability of rolling a total of five exactly once is

$$B\left(3,1,\frac{1}{9}\right)=C(3,1)\cdot\left(\frac{1}{9}\right)^1\cdot\left(1-\frac{1}{9}\right)^{3-1}=\frac{3!}{1!2!}\cdot\frac{1}{9}\cdot\left(\frac{8}{9}\right)^2=3\cdot\frac{1}{9}\cdot\frac{64}{81}=\frac{1}{3}\cdot\frac{64}{81}=\frac{64}{243}\approx 0.2634.$$

9. An experiment is performed 5 times. $B\left(5,3,\frac{1}{4}\right)$ represents the probability of obtaining a certain outcome 3 times, where the probability that an individual trial results in that certain outcome is $\frac{1}{4}$.

$$B\left(5,3,\frac{1}{4}\right)=C(5,3)\cdot\left(\frac{1}{4}\right)^3\cdot\left(1-\frac{1}{4}\right)^{5-3}=\frac{5!}{3!2!}\cdot\frac{1}{64}\cdot\left(\frac{3}{4}\right)^2=\frac{5\cdot 4}{2}\cdot\frac{1}{64}\cdot\frac{9}{16}=\frac{180}{2048}=\frac{45}{512}\approx 0.0879$$

11. k is larger than n

13. $$B\left(12,9,\frac{1}{2}\right)=C(12,9)\cdot\left(\frac{1}{2}\right)^9\cdot\left(1-\frac{1}{2}\right)^{12-9}=\frac{12!}{9!3!}\cdot\frac{1}{512}\cdot\left(\frac{1}{2}\right)^3=\frac{12\cdot 11\cdot 10}{3\cdot 2}\cdot\frac{1}{512}\cdot\frac{1}{8}=\frac{1{,}320}{24{,}576}$$
$$=\frac{55}{1{,}024}\approx 0.0537$$

15. $B\left(12,9,\frac{1}{2}\right)=C(12,9)\cdot\left(\frac{1}{2}\right)^9\cdot\left(1-\frac{1}{2}\right)^{12-9}=\frac{12!}{9!3!}\cdot\frac{1}{512}\cdot\left(\frac{1}{2}\right)^3=\frac{12\cdot11\cdot10}{3\cdot2}\cdot\frac{1}{512}\cdot\frac{1}{8}=\frac{1,320}{24,576}$

$=\frac{55}{1,024}=0.0537109375$

$B\left(12,10,\frac{1}{2}\right)=C(12,10)\cdot\left(\frac{1}{2}\right)^{10}\cdot\left(1-\frac{1}{2}\right)^{12-10}=\frac{12!}{10!2!}\cdot\frac{1}{1,024}\cdot\left(\frac{1}{2}\right)^2=\frac{12\cdot11}{2}\cdot\frac{1}{1,024}\cdot\frac{1}{4}$

$=\frac{132}{8,192}\approx0.01611328125$

$B\left(12,11,\frac{1}{2}\right)=C(12,11)\cdot\left(\frac{1}{2}\right)^{11}\cdot\left(1-\frac{1}{2}\right)^{12-11}=\frac{12!}{11!1!}\cdot\frac{1}{2,048}\cdot\left(\frac{1}{2}\right)^1=12\cdot\frac{1}{2,048}\cdot\frac{1}{2}=\frac{12}{4,096}$

$=\frac{3}{1,024}=0.0029296875$

$B\left(12,12,\frac{1}{2}\right)=C(12,12)\cdot\left(\frac{1}{2}\right)^{12}\cdot\left(1-\frac{1}{2}\right)^{12-12}=1\cdot\frac{1}{4,096}\cdot\left(\frac{1}{2}\right)^0=\frac{1}{4,096}\approx0.0002441406$

$0.0537109375+0.01611328125+0.0029296875+0.0002441406=0.07299804685\approx0.0730$

17. $B(10,8,0.80)=C(10,8)\cdot(0.80)^8\cdot(1-0.80)^{10-8}=\frac{10!}{8!2!}\cdot0.16777216\cdot(0.20)^2$

$=\frac{10\cdot9}{2}\cdot0.16777216\cdot0.04=0.301989888$

$B(10,9,0.80)=C(10,9)\cdot(0.80)^9\cdot(1-0.80)^{10-9}=\frac{10!}{9!1!}\cdot0.1342217728\cdot(0.20)^1$

$=10\cdot0.134217728\cdot0.20=0.268435456$

$B(10,10,0.80)=C(10,10)\cdot(0.80)^{10}\cdot(1-0.80)^{10-10}=\frac{10!}{10!0!}\cdot0.1073741824\cdot(0.20)^0$

$=1\cdot0.1073741824\cdot1=0.1073741824$

$0.301989888+0.268435456+0.1073741824=0.6777995264\approx0.6778$

19. $B(5,2,0.250)=C(5,2)\cdot(0.250)^2\cdot(1-0.250)^{5-2}=\frac{5!}{2!3!}\cdot0.0625\cdot(0.750)^3$

$=\frac{5\cdot4}{2}\cdot0.0625\cdot0.421875=0.263671875\approx0.2637$

21. We need to determine the probability that the hospital does not meet the need so we need to determine the probability that the hospital has 0, 1, or 2 donors with type A+ blood waiting.

$$B(12,0,0.30) = C(12,0)\cdot(0.30)^0\cdot(1-0.30)^{12-0} = \frac{12!}{0!12!}\cdot 1\cdot(0.70)^{12}$$
$$\approx 1\cdot 1\cdot 0.013841287201 = 0.013841287201$$
$$B(12,1,0.30) = C(12,1)\cdot(0.30)^1\cdot(1-0.30)^{12-1} = \frac{12!}{1!11!}\cdot 0.30\cdot(0.70)^{11}$$
$$= 12\cdot 0.30\cdot 0.01977326743 = 0.071183762748$$
$$B(12,2,0.30) = C(12,2)\cdot(0.30)^2\cdot(1-0.30)^{12-2} = \frac{12!}{2!10!}\cdot 0.09\cdot(0.70)^{10}$$
$$\approx \frac{12\cdot 11}{2}\cdot 0.09\cdot 0.0282475249 \approx 0.167790297906$$

0.013841287201 + 0.071183762748 + 0.167790297906 = 0.252815347855

1 – 0.252815347855=0.747184652145 $\approx$ 0.7472

23. $\frac{3}{4}$

25. The total number of purchases to obtain a complete set T is: the number of purchases to obtain the first new figure + the number of purchases to obtain the second new figure + the number of purchases to obtain the third new figure + the number of purchases to obtain the fourth new figure + the number of purchases to obtain the fifth new figure + the number of purchases to obtain the sixth new figure =

$$1+\frac{1}{5/6}+\frac{1}{4/6}+\frac{1}{3/6}+\frac{1}{2/6}+\frac{1}{1/6}=1+\frac{6}{5}+\frac{6}{4}+\frac{6}{3}+\frac{6}{2}+\frac{6}{1}=\frac{147}{10}=14.7$$

The child should expect to purchase 15 boxes.

27. The experiment has a fixed number of trials. The experiment has two outcomes: "success" and "failure". The probability of success is the same from trial to trial. The trials are independent of each other.

29. We need to determine the probability that 4 or fewer people will come down with the cold without the vaccine.

$$B(10,4,0.50) = C(10,4)\cdot(0.50)^4\cdot(1-0.50)^{10-4} = \frac{10!}{4!6!}\cdot 0.0625\cdot(0.50)^6 = \frac{10\cdot 9\cdot 8\cdot 7}{4\cdot 3\cdot 2}\cdot 0.0625\cdot 0.015625$$
$$= 0.205078125$$
$$B(10,3,0.50) = C(10,3)\cdot(0.50)^3\cdot(1-0.50)^{10-3} = \frac{10!}{3!7!}\cdot 0.125\cdot(0.50)^7 = \frac{10\cdot 9\cdot 8}{3\cdot 2}\cdot 0.125\cdot 0.0078125$$
$$= 0.1171875$$
$$B(10,2,0.50) = C(10,2)\cdot(0.50)^2\cdot(1-0.50)^{10-2} = \frac{10!}{2!8!}\cdot 0.25\cdot(0.50)^8 = \frac{10\cdot 9}{2}\cdot 0.25\cdot 0.00390625$$
$$= 0.0439453125$$
$$B(10,1,0.50) = C(10,1)\cdot(0.50)^1\cdot(1-0.50)^{10-1} = \frac{10!}{1!9!}\cdot 0.50\cdot(0.50)^9 = 10\cdot 0.50\cdot 0.001953125$$
$$= 0.009765625$$

29. (continued)

$$B(10,0,0.50) = C(10,0)\cdot(0.50)^0\cdot(1-0.50)^{10-0} = \frac{10!}{0!10!}\cdot 1\cdot(0.50)^{10} = 1\cdot 1\cdot 0.0009765625$$
$$= 0.0009765625$$

$$0.205078125 + 0.1171875 + 0.0439453125 + 0.009765625 + 0.0009765625 = 0.376953125 \approx 0.3770$$

Chapter Review Exercises

1. a) If we let "H" represent a heads and "T" represent tails then we would have three outcomes. The set of outcomes would be {HHT, HTH, THH}.
   b) We would have five outcomes. The set of outcomes would be {(2, 6), (3, 5), (4, 4), (5, 3), (6, 2)}.

2. Since there are 6 red face cards in a deck, the probability of choosing one is $\frac{6}{52} = \frac{3}{26}$

3. To calculate the empirical of an event E, we divided the number of the times the event occurs by the number of the times the experiment is performed. To calculate theoretical probability, we use known mathematical techniques such as counting formulas.

4. The Punnett square is

		First Gen.	
		t	s
First	t	tt	ts
Gen.	s	st	ss

$P(\text{tall}) = \frac{3}{4}$

5. a) Let D be the event that the Dodgers win the World Series.

   The odds against D are $\frac{17}{2} = \frac{17/19}{2/19} = \frac{P(D')}{P(D)}$. Thus, the probability of D is $\frac{2}{19}$.

   b) Let R be the event that it rains.

   The odds against R are $\frac{P(E')}{P(E)} = \frac{1-0.55}{0.55} = \frac{0.45}{0.55} = \frac{0.45\times 100}{0.55\times 100} = \frac{45}{55} = \frac{9}{11}$, or 9 to 11.

6. a) $P(E') = 1 - P(E)$
   b) See Figure 13.6.
   c) We use the complement formula $P(E') = 1 - P(E)$ when it is easier to compute $P(E')$ than it is to compute $P(E)$ directly.

7. Let F be the event that we select a face card and R be the event that we select a red card.

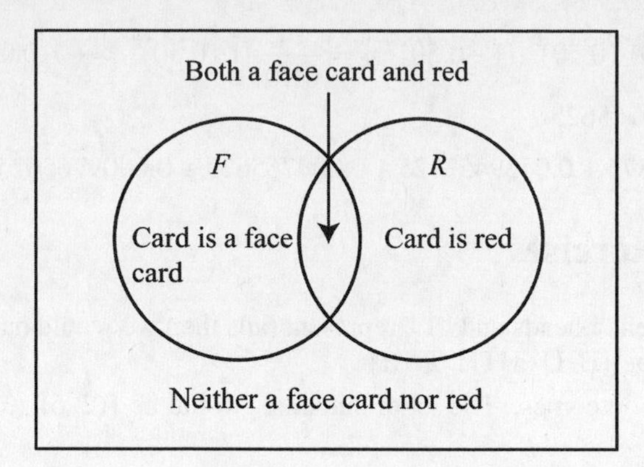

$$P(F \cup R) = P(F) + P(R) - P(F \cap R) = \frac{12}{52} + \frac{26}{52} - \frac{6}{52} = \frac{32}{52} = \frac{8}{13}$$

8. a) Let R be the event that the spinner stops on red.

   $P(R) = 0.09 + 0.09 = 0.18$, so $P(R') = 1 - P(R) = 1 - 0.18 = 0.82$

   b) Let E be the event spinner stops on an even number and B be the event the spinner stops on blue.

   $P(E \cup B) = P(E) + P(B) - P(E \cap B) = 0.50 + 0.48 - 0.36 = 0.62$

9. Answers will vary.

10. There are 26 ways to roll a total less than nine; they are {(1,1), (1,2), (1,3), **(1,4)**, (1,5), (1,6), (2,1), (2,2), **(2,3)**, (2,4), (2,5), (2,6), (3,1), **(3,2)**, (3,3), (3,4), (3,5), **(4,1)**, (4,2), (4,3), (4,4), (5,1), (5,2), (5,3), (6,1), (6,2)}. Of these, there are 4 ways to roll a total of 5; thus, the probability of rolling a total of five given that the total is less than nine is $\frac{4}{26} = \frac{2}{13}$.

11. a) $P(H \cap H) = P(H) \cdot P(H \mid H) = \frac{13}{52} \cdot \frac{12}{51} = \frac{1}{4} \cdot \frac{12}{51} = \frac{12}{204} = \frac{1}{17}$

   b) $P(Q \cap A) = P(Q) \cdot P(A \mid Q) = \frac{4}{52} \cdot \frac{4}{51} = \frac{1}{13} \cdot \frac{4}{51} = \frac{4}{663}$

12. There are 18 out of the 36 ways to roll an odd total. There are 10 ways to roll a total less than six, and they are {(1,1), (1,2), (1,3), (1,4), (2,1), (2,2), (2,3), (3,1), (3,2), (4,1)}.

$$P(F \mid E) = \frac{P(F \cap E)}{P(E)} = \frac{6/36}{18/36} = \frac{6}{18} = \frac{1}{3} \neq P(F) = \frac{10}{36} = \frac{5}{18}$$

These two events are therefore dependent.

13. $\dfrac{0.036}{0.036+0.0576}=\dfrac{0.036}{0.0936}\approx 0.385$

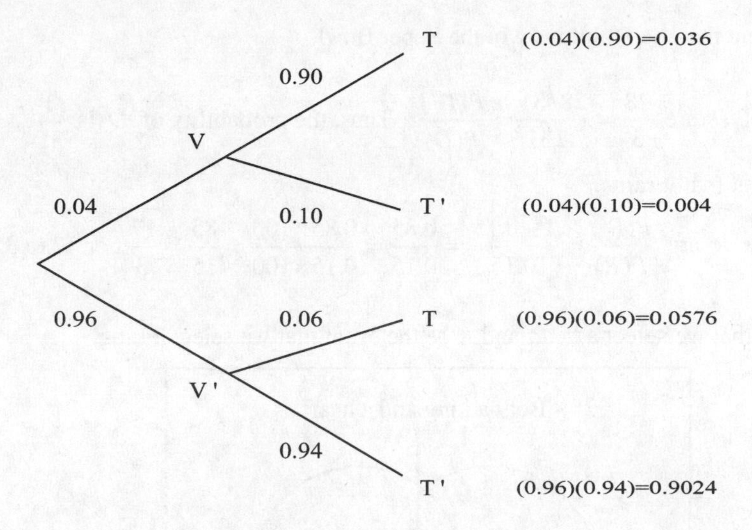

14. $(0.99)\cdot(+25.25)+(0.01)\cdot(-1{,}000)=24.9975-10=14.9975\approx \$15.00$

15. $\left(\dfrac{12}{52}\right)\cdot(+15)+\left(\dfrac{40}{52}\right)\cdot(-4)=\dfrac{12\cdot 15-40\cdot 4}{52}=\dfrac{20}{52}\approx \$0.38$

16. There are 16 ways to flip 4 coins. Of these 16 ways, two of them will yield all heads or all tails. The expected value of this game is $\left(\dfrac{2}{16}\right)\cdot(+5)+\left(\dfrac{14}{16}\right)\cdot(0)=\dfrac{2\cdot 5+14\cdot 0}{16}=\dfrac{10}{16}\approx \$0.63$

The price to play this game should be \$0.63 to make this game fair.

17. $B\left(8,3,\dfrac{1}{2}\right)=C(8,3)\cdot\left(\dfrac{1}{2}\right)^3\cdot\left(1-\dfrac{1}{2}\right)^{8-3}=\dfrac{8!}{3!5!}\cdot\dfrac{1}{8}\cdot\left(\dfrac{1}{2}\right)^5=\dfrac{8\cdot 7\cdot 6}{3\cdot 2}\cdot\dfrac{1}{8}\cdot\dfrac{1}{32}=\dfrac{336}{1{,}536}$

$=\dfrac{7}{32}\approx 0.219$

18. $B\left(10,8,\dfrac{1}{4}\right)=C(10,8)\cdot\left(\dfrac{1}{2}\right)^8\cdot\left(1-\dfrac{1}{2}\right)^{10-8}=\dfrac{10!}{8!2!}\cdot\dfrac{1}{256}\cdot\left(\dfrac{1}{2}\right)^2=\dfrac{10\cdot 9}{2}\cdot\dfrac{1}{256}\cdot\dfrac{1}{4}=\dfrac{90}{2{,}048}$

$=\dfrac{45}{1{,}024}\approx 0.044$

Chapter Test

1. a) There would be six outcomes. The set of outcomes would be {(4, 6), (6, 4), (5, 5), (5, 6), (6, 5), (6, 6)}.

   b) If "H" represents heads and "T" represent tails, there would be four outcomes. The set of outcomes would be {HHHT, HHTH, HTHH, THHH}.

2. Since there are 2 black kings in a deck, the probability of choosing one is $\frac{2}{52}=\frac{1}{26}$.

3. a) Let D be the event that the Dolphins win the Super Bowl.

The odds against D are $\frac{28}{3}=\frac{28/31}{3/31}=\frac{P(D')}{P(D)}$. Thus, the probability of D is $\frac{3}{31}$.

b) Let R be the event that it rains.

The odds against R are $\frac{P(R')}{P(R)}=\frac{1-0.15}{0.15}=\frac{0.85}{0.15}=\frac{0.85\times 100}{0.15\times 100}=\frac{85}{15}=\frac{17}{3}$, or 17 to 3.

4. Let H be the event that we select a heart and K be the event that we select a king.

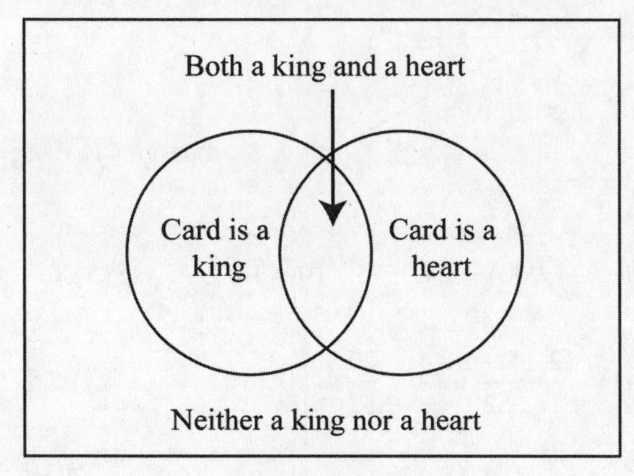

$$P(H\cup K)=P(H)+P(K)-P(H\cap K)=\frac{13}{52}+\frac{4}{52}-\frac{1}{52}=\frac{16}{52}=\frac{4}{13}.$$

5. $P(B\,|\,A)$ is the probability of B assuming that A has occurred. $P(A\,|\,B)$ is the probability of A assuming that B has occurred.

6. There are 6 out of the 36 ways to roll the same number. There are 8 ways to roll a total greater than 8, and they are {(4,5), (5,4), (5,5), (4,6), (6,4), (5, 6), (6, 5), (6,6)}.

$$P(F\,|\,E)=\frac{P(F\cap E)}{P(E)}=\frac{2/36}{8/36}=\frac{2}{8}=\frac{1}{4}\neq P(F)=\frac{8}{36}=\frac{4}{9}$$

These two events are therefore dependent.

7. Since it costs \$1 to play, the expected value of this game is $\left(\frac{6}{36}\right)\cdot(+4)+\left(\frac{30}{36}\right)\cdot(-1)=\frac{6\cdot 4-30\cdot 1}{36}$

$=-\frac{30}{36}=-\frac{1}{6}\approx -\$0.17.$

8. $B\left(10,2,\frac{1}{4}\right)=C(10,2)\cdot\left(\frac{1}{4}\right)^2\cdot\left(1-\frac{1}{4}\right)^{10-2}=\frac{10!}{2!8!}\cdot\frac{1}{16}\cdot\left(\frac{3}{4}\right)^8=\frac{10\cdot 9}{2}\cdot\frac{1}{16}\cdot\frac{6,561}{65,536}=\frac{295,245}{1,048,576}\approx 0.2816$

9. $B\left(5,3,\frac{1}{4}\right)=C(5,3)\cdot\left(\frac{1}{4}\right)^3\cdot\left(1-\frac{1}{4}\right)^{5-3}=\frac{5!}{3!2!}\cdot\frac{1}{64}\cdot\left(\frac{3}{4}\right)^2=\frac{5\cdot 4}{2}\cdot\frac{1}{64}\cdot\frac{9}{16}=\frac{180}{2,048}$

$=\frac{45}{512}\approx 0.0879$

10. a) $P(E)+P(E')=1$ b) See Figure 13.6.

11. The Punnett square is

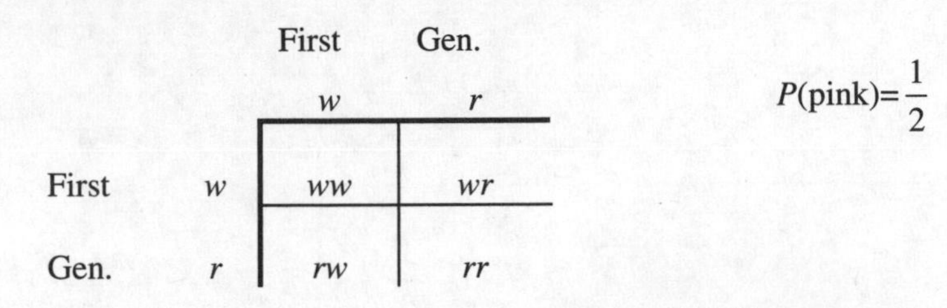

$P(\text{pink})=\frac{1}{2}$

12. $\frac{0.019}{0.019+0.049}=\frac{0.019}{0.068}\approx 0.279$

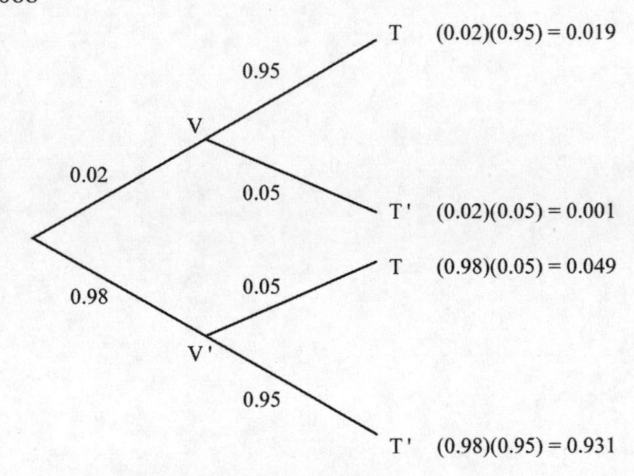

13. Since is costs $2 to play, the expected value of the raffle is

$\left(\frac{1}{500}\right)\cdot(+248)+\left(\frac{3}{500}\right)\cdot(98)+\left(\frac{5}{500}\right)\cdot(48)+\left(\frac{491}{500}\right)\cdot(-2)=\frac{1\cdot 248+3\cdot 98+5\cdot 48-2\cdot 491}{500}=$

$-\frac{200}{500}=-\$0.40.$

14. There are 6 ways to roll a total less than five; they are {(1,1), (1,2), (1,3), (3,1), (2,1), (2,2)}. Of these, there are 4 ways to roll an even total; thus, the probability of rolling an even total given that the total is less than five is $\frac{4}{6}=\frac{2}{3}$.

15. a) $P(F\cap F)=P(F)\cdot P(F\mid F)=\frac{12}{52}\cdot\frac{11}{51}=\frac{3}{13}\cdot\frac{11}{51}=\frac{1}{13}\cdot\frac{11}{17}=\frac{11}{221}$.

b) $P(K\,\&\,A)=P(K\cap A)+P(A\cap K)=P(K)\cdot P(A\mid K)+P(A)\cdot P(K\mid A)=$

$\frac{4}{52}\cdot\frac{4}{51}+\frac{4}{52}\cdot\frac{4}{51}=\frac{1}{13}\cdot\frac{4}{51}+\frac{1}{13}\cdot\frac{4}{51}=\frac{8}{663}$

Chapter 14
Descriptive Statistics: What a Data Set Tells Us

Section 14.1 Organizing and Visualizing Data

1.

x	Frequency
2	1
3	0
4	0
5	3
6	2
7	4
8	4
9	3
10	3

x	Relative Frequency
2	0.05
3	0.00
4	0.00
5	0.15
6	0.10
7	0.20
8	0.20
9	0.15
10	0.15

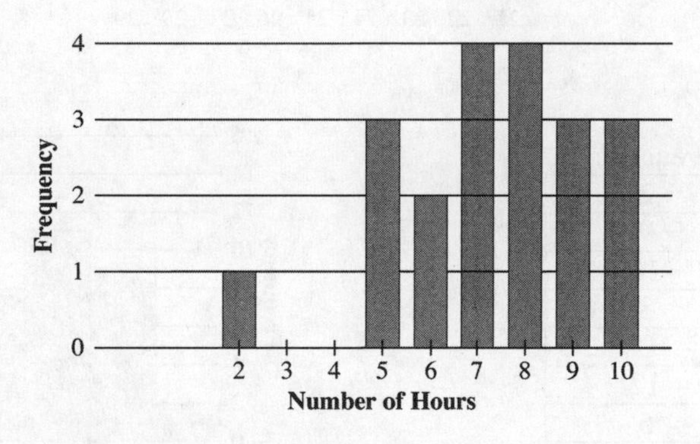

3.

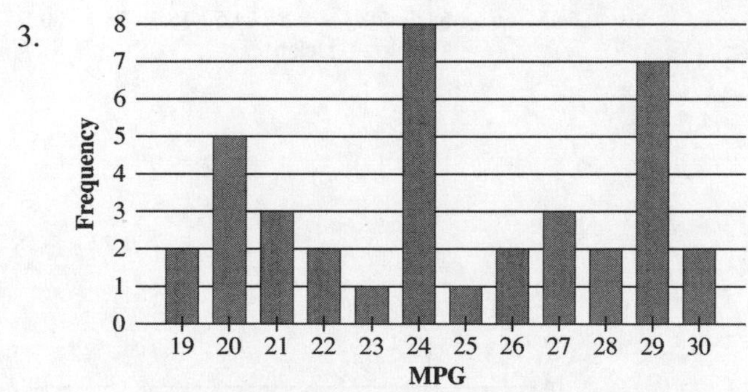

5.

x	Frequency
21	4
22	6
23	10
24	9
25	5
26	8
27	10
28	5
29	3

x	Relative Frequency
21	4/60≈0.067=6.7%
22	6/60=0.10=10%
23	10/60≈0.167=16.7%
24	9/60=0.15=15%
25	5/60≈0.083=8.3%
26	8/60≈0.133=13.3%
27	10/60≈0.167=16.7%
28	5/60≈0.083=8.3%
29	3/60=0.05=5%

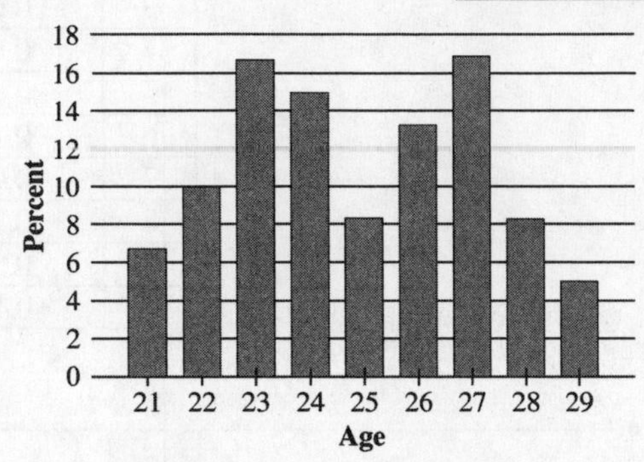

7.

x	Frequency
64.5+ to 66.5	2
66.5+ to 68.5	9
68.5+ to 70.5	11
70.5+ to 72.5	8
72.5+ to 74.5	15
74.5+ to 76.5	12
76.5+ to 78.5	5
78.5+ to 80.5	2

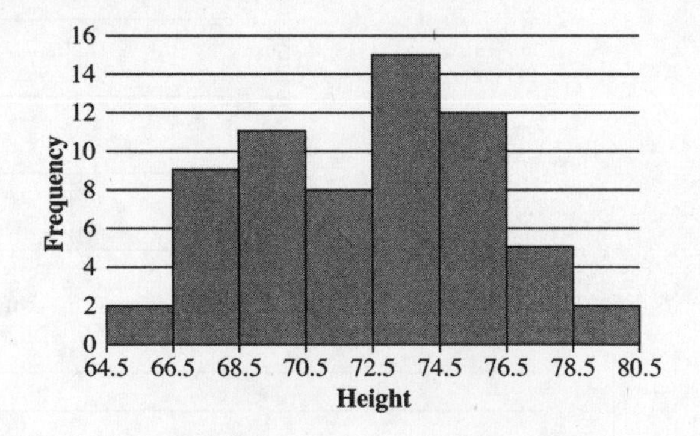

9.

A		B
8 8	1	3 6 8
9 6 2 2 1	2	0 1 1 2 2 2 3 4 6 8 9
9 9 8 7 4 3 2	3	2 3 3 8 9
7 3 3 3 2 2	4	7

11.

x	Frequency
1.75+ to 2.25	4
2.25+ to 2.75	13
2.75+ to 3.25	9
3.25+ to 3.75	8
3.75+ to 4.25	2

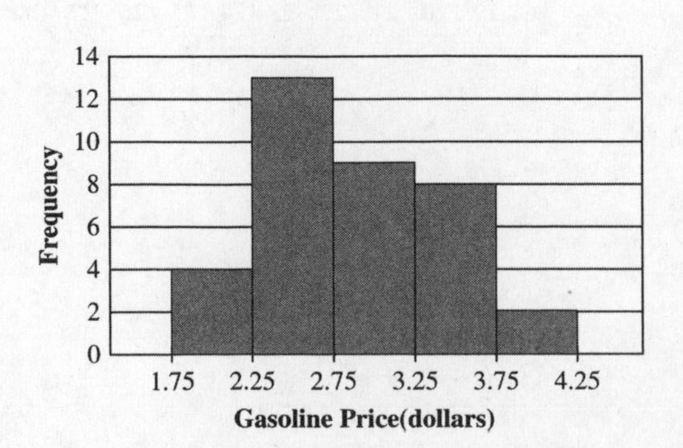

13. a) 12 occurred two times
    b) 6
    c) 13 + 11 + 5 + 8 + 12 + 9 + 6 + 5 + 2 = 71
    d) $\frac{9+6+5+2}{71}=\frac{22}{71}$

15. 2008

17. 2003 to 2005

19. Answers may vary.
    This exercise was based on real–world data. An approximation of 10% would be reasonable. The exact value is 10.2%.

21. 25 and older, earning at least $10.00 per hour

23. 16 to 19, earning less than $7.51 per hour

25. – 27. Answers will vary.

29. The distribution is shifted to higher numbers, so it appears that the orientation program is succeeding.

No Training		With Training
9	1	
9 8 8 7 6 6 3 2	2	1 8 8 9 9
4 2	3	2 2 3 6 6 7 9
3 3 2 1 0	4	1 3 4 5

31. The NFC may have an advantage.

NFC		AFC
7	0	
9 7 7 7 0	1	3 4 6 7 7 7 7
9 7 4 3 3 1 1	2	0 1 1 1 4 5 6 7 9
5 1 1 0	3	1 2 4 4
9 8	4	
2	5	

33.

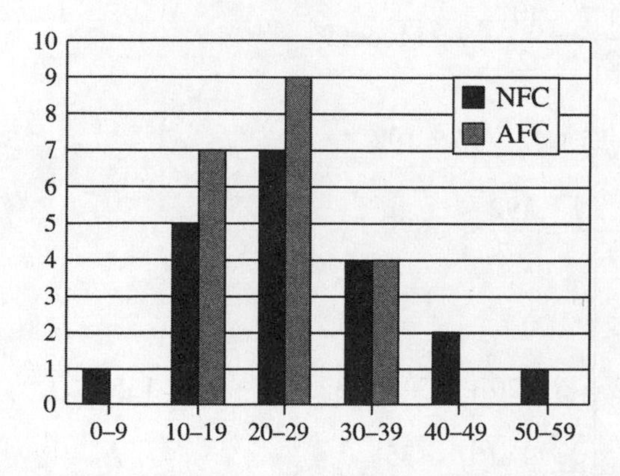

35.

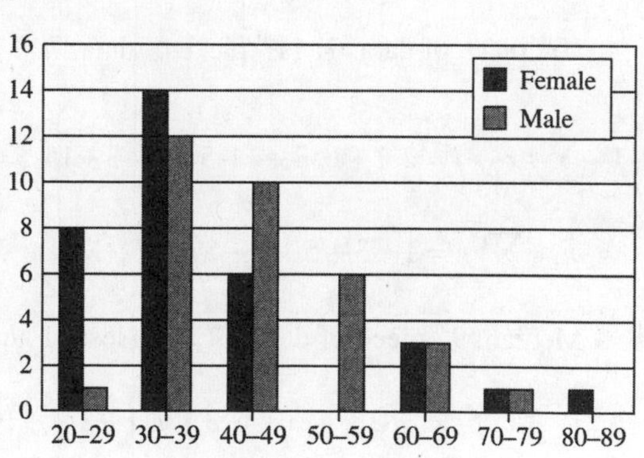

37. The population is the set of all objects being studied; a sample is a subset of the population; selection bias, leading question bias

39. A continuous variable can take on arbitrary values; a discrete variable cannot; a histogram

41. No solution provided.

43. Use a graph like (D) in Exercise 24, but with three bars instead of two.

45. No solution provided.

Section 14.2 Measures of Central Tendency

1. Place the data in order.
3, 4, 4, 5, 6, 7, 8, 9, 11
Mean: $\bar{x} = \frac{\sum x}{n} = \frac{57}{9} \approx 6.3$; Median: 6; Mode: 4

3. Place the data in order.
3, 4, 4, 6, 6, 7, 9, 9, 10, 11
Mean: $\bar{x} = \frac{\sum x}{n} = \frac{69}{10} = 6.9$; Median: $\frac{6+7}{2} = \frac{13}{2} = 6.5$; Mode: no mode

5. Place the data in order.
1, 2, 3, 4, 5, 5, 6, 7, 8, 9
Mean: $\bar{x} = \frac{\sum x}{n} = \frac{50}{10} = 5$; Median: $\frac{5+5}{2} = \frac{10}{2} = 5$; Mode: 5

7. Place the data in order.
2, 5, 5, 5, 6, 6, 7, 7, 7, 7, 8, 8, 8, 8, 9, 9, 9, 10, 10, 10
Mean: $\bar{x} = \frac{\sum x}{n} = \frac{146}{20} = 7.3$; Median: $\frac{7+8}{2} = \frac{15}{2} = 7.5$; Modes: 7, 8

9. $\sum(x \cdot f) = 5\cdot4 + 6\cdot3 + 7\cdot5 + 8\cdot2 + 9\cdot6 = 20 + 18 + 35 + 16 + 54 = 143$

$n = \sum f = 4+3+5+2+6 = 20$; Mean: $\bar{x} = \frac{\sum(x \cdot f)}{\sum f} = \frac{143}{20} = 7.15$

10th piece of data: 7; 11th piece of data: 7; Median: $\frac{7+7}{2} = \frac{14}{2} = 7$; Mode: 9

11. $\sum(x \cdot f) = 2\cdot1 + 6\cdot3 + 11\cdot5 + 14\cdot3 + 15\cdot5 = 2 + 18 + 55 + 42 + 75 = 192$

$n = \sum f = 1+3+5+3+5 = 17$; Mean: $\bar{x} = \frac{\sum(x \cdot f)}{\sum f} = \frac{192}{17} \approx 11.29$

Median: 9th piece of data: 11; Modes: 11 and 15

13. $\sum(x \cdot f) = 3\cdot2 + 4\cdot5 + 6\cdot3 + 7\cdot1 + 8\cdot4 + 10\cdot3 + 11\cdot2 = 6 + 20 + 18 + 7 + 32 + 30 + 22 = 135$

$n = \sum f = 2+5+3+1+4+3+2 = 20$; Mean: $\bar{x} = \frac{\sum(x \cdot f)}{\sum f} = \frac{135}{20} = 6.75$

10th piece of data: 6; 11th piece of data: 7; Median: $\frac{6+7}{2} = \frac{13}{2} = 6.5$; Mode: 4

15. $\sum(x\cdot f) = 5\cdot1+8\cdot3+11\cdot6+12\cdot2+14\cdot5+15\cdot4+18\cdot3 = 5+24+66+24+70+60+54 = 303$

$n=\sum f = 1+3+6+2+5+4+3 = 24$; Mean: $\bar{x} = \frac{\sum(x\cdot f)}{\sum f} = \frac{303}{24} \approx 12.6$

12th piece of data: 12; 13th piece of data: 14; Median: $\frac{12+14}{2} = \frac{26}{2} = 13$; Mode: 11

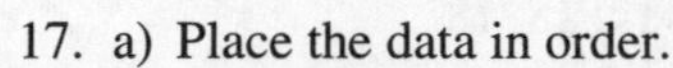

17. a) Place the data in order.

11, 17, 18, 23, 25, 26, 26, 31, 31, 33, 41, 44, 45, 48, 53

lower half: 11, 17, 18, 23, 25, 26, 26

upper half: 31, 33, 41, 44, 45, 48, 53

minimum: 11

first quartile is the median of the lower half: $Q_1 = 23$

median: 31

third quartile is the median of the upper half: $Q_3 = 44$

maximum: 53

b)

12 24 36 48

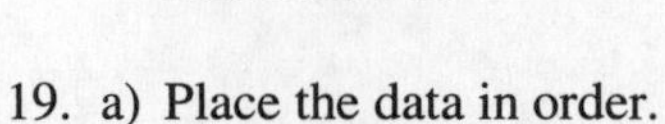

19. a) Place the data in order.

25, 28, 29, 30, 31, 32, 33, 33, 34, 37, 38, 41, 41, 45, 49

lower half: 25, 28, 29, 30, 31, 32, 33

upper half: 34, 37, 38, 41, 41, 45, 49

minimum: 25

first quartile is the median of the lower half: $Q_1 = 30$

median: 33

third quartile is the median of the upper half: $Q_3 = 41$

maximum: 49

b)

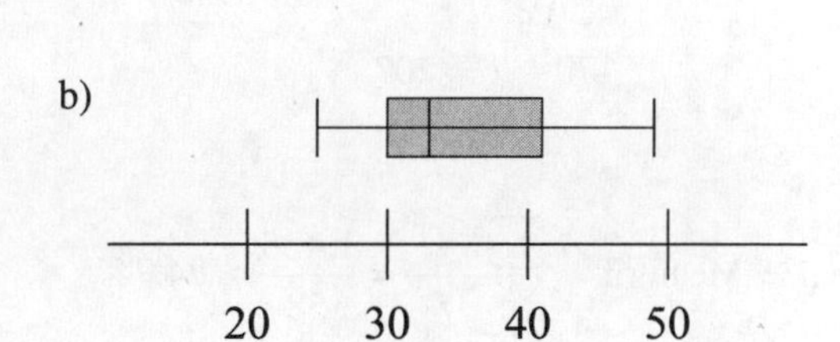

21. Place the data in order.

4, 5, 6, 6, 9, 9, 13, 13, 13, 14, 17, 18, 18, 20, 20, 20, 22, 23, 25, 25

Mean: $\bar{x} = \frac{\sum x}{n} = \frac{300}{20} = 15$; Median: $\frac{14+17}{2} = \frac{31}{2} = 15.5$; Modes: 13 and 20

No explanation provided as to which measure is typical

23. Place this data in order.

7, 7, 9, 10, 12, 12, 25, 27, 55, 170

Mean: $\bar{x} = \frac{\sum x}{n} = \frac{334}{10} = 33.4$; Median: $\frac{12+12}{2} = \frac{24}{2} = 12$; Modes: 7 and 12

No explanation provided as to which measure is typical

25. Place the data in order.

367, 369, 371, 373, 375, 377, 379, 381, 383, 385

Mean: $\overline{x}=\frac{\sum x}{n}=\frac{3760}{10}=376$; Median: $\frac{375+377}{2}=\frac{752}{2}=376$; Mode: no mode

No explanation provided as to which measure is typical

27. Place the data in order.

4, 4, 5, 5, 5, 6, 6, 7, 7, 7, 7, 8, 8, 8, 8, 9, 10, 11, 12, 13, 14, 15, 16, 17, 17, 17, 18, 18

Mean: $x=\frac{\sum x}{n}=\frac{282}{28}\approx 10.07$; Median: $\frac{8+8}{2}=\frac{16}{2}=8$; Modes: 7 and 8

No explanation provided as to which measure is typical

29. $\frac{3\times 4+2\times 3+3\times 4+3\times 0+4\times 3}{15}=\frac{42}{15}\approx 2.8$

31. Let x be the incorrect final exam score. Assuming that the three exams are equally weighted we have

$$\frac{84+86+x}{3}=69$$
$$\frac{170+x}{3}=69$$
$$170+x=207$$
$$x=37.$$

The correct grade was 73.

33. Mean: $\overline{x}=\frac{\sum x\cdot f}{\sum f}=\frac{1,435}{58}\approx 24.74$

Median: 25; You would need to add together the 29th and the 30th values and divide by two. Both of these numbers are 25.

Mode: 24, 25

No explanation provided as to which measure is typical

35. a) $\frac{78+82+56+72+x}{5}=70$

$$\frac{288+x}{5}=70$$
$$288+x=350$$
$$x=62$$

b) $\frac{78+82+56+72+x}{5}=80$

$$\frac{288+x}{5}=80$$
$$288+x=400$$
$$x=112$$

37. $\frac{9\cdot 50+15\cdot 125+7\cdot 245}{31}=\frac{4040}{31}\approx \$130.32$

39. 1975–1989: Place the data in order.

31, 36, 37, 37, 37, 38, 38, 39, 40, 40, 47, 48, 48, 49, 52

lower half: 31, 36, 37, 37, 37, 38, 38

upper half: 40, 40, 47, 48, 48, 49, 52

39. (continued)

minimum: 31

first quartile is the median of the lower half: $Q_1 = 37$

median: 39

third quartile is the median of the upper half: $Q_3 = 48$

maximum: 52

1990–2004: Place the data in order.

35, 38, 40, 40, 43, 46, 47, 47, 48, 49, 49, 50, 65, 70, 73

lower half: 35, 38, 40, 40, 43, 46, 47

upper half: 48, 49, 49, 50, 65, 70, 73

minimum: 35

first quartile is the median of the lower half: $Q_1 = 40$

median: 47

third quartile is the median of the upper half: $Q_3 = 50$

maximum: 73

The top box-and-whisker plot is for 1975–1989, the bottom is for 1990–2004

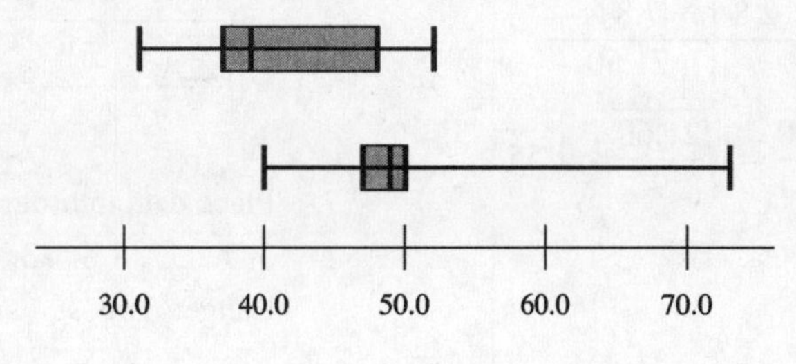

41. a) Place the data in order.

40, 41, 44, 48, 58, 59, 66, 98, 105, 120

lower half: 40, 41, 44, 48, 58

upper half: 59, 66, 98, 105, 120

minimum: 40

first quartile is the median of the lower half: $Q_1 = 44$

median: 58.5

third quartile is the median of the upper half: $Q_3 = 98$

maximum: 120

b)

30 40 50 60 70 80 90 100 110 120 130

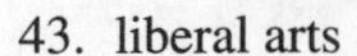

43. liberal arts

45. education; no; No discussion provided.

47. The maximum and minimum salaries are not too far from the middle 50% of the salaries.

49. With $\sum x$, we are adding the individual scores in the distribution; with $\sum x \cdot f$, we are multiplying each scores by its frequency before adding.

51. No solution provided.

53. – 65. No solution provided.

Section 14.3 Measures of Dispersion

1. Place data in order.
17, 18, 18, 18, 19, 19, 20, 20, 21, 22
Range: largest – smallest = 22 – 17 = 5

Mean: $\bar{x} = \frac{\sum x}{n} = \frac{192}{10} = 19.2$

Standard deviation:

Data Value	$x - \bar{x}$	$(x - \bar{x})^2$
17	17 – 19.2 = – 2.2	4.84
18	18 – 19.2 = – 1.2	1.44
18	18 – 19.2 = – 1.2	1.44
18	18 – 19.2 = – 1.2	1.44
19	19 – 19.2 = – 0.2	0.04
19	19 – 19.2 = – 0.2	0.04
20	20 – 19.2 = 0.8	0.64
20	20 – 19.2 = 0.8	0.64
21	21 – 19.2 = 1.8	3.24
22	22 – 19.2 = 2.8	7.84
Total	0	21.60

$$s = \sqrt{\frac{\sum (x - \bar{x})^2}{n-1}} = \sqrt{\frac{21.60}{10-1}} = \sqrt{\frac{21.60}{9}} \approx 1.55$$

3. Place data in order.
4, 5, 6, 7, 7, 8, 9, 10
Range: largest – smallest = 10 – 4 = 6

Mean: $\bar{x} = \frac{\sum x}{n} = \frac{56}{8} = 7$

Standard deviation:

Data Value	$x - \bar{x}$	$(x - \bar{x})^2$
4	4 – 7 = – 3	9
5	5 – 7 = – 2	4
6	6 – 7 = – 1	1
7	7 – 7 = 0	0
7	7 – 7 = 0	0
8	8 – 7 = 1	1
9	9 – 7 = 2	4
10	10 – 7 = 3	9
Total	0	28

$$s = \sqrt{\frac{\sum (x - \bar{x})^2}{n-1}} = \sqrt{\frac{28}{8-1}} = \sqrt{\frac{28}{7}} = 2$$

5. Place data in order.
2, 3, 5, 8, 9, 11, 12, 14
Range: largest – smallest = 14 – 2 = 12

Mean: $\bar{x} = \frac{\sum x}{n} = \frac{64}{8} = 8$

Standard deviation:

Data Value	$x - \bar{x}$	$(x - \bar{x})^2$
2	2 – 8 = – 6	36
3	3 – 8 = – 5	25
5	5 – 8 = – 3	9
8	8 – 8 = 0	0
9	9 – 8 = 1	1
11	11 – 8 = 3	9
12	12 – 8 = 4	16
14	14 – 8 = 6	36
Total	0	132

$$s = \sqrt{\frac{\sum (x - \bar{x})^2}{n-1}} = \sqrt{\frac{132}{8-1}} = \sqrt{\frac{132}{7}} \approx 4.34$$

7. Place data in order.
3, 7, 7, 7, 8, 9, 13, 18
Range: largest – smallest = 18 – 3 = 15

Mean: $\bar{x} = \frac{\sum x}{n} = \frac{72}{8} = 9$

Standard deviation:

Data Value	$x - \bar{x}$	$(x - \bar{x})^2$
3	3 – 9 = – 6	36
7	7 – 9 = – 2	4
7	7 – 9 = – 2	4
7	7 – 9 = – 2	4
8	8 – 9 = – 1	1
9	9 – 9 = 0	0
13	13 – 9 = 4	16
18	18 – 9 = 9	81
Total	0	146

$$s = \sqrt{\frac{\sum (x - \bar{x})^2}{n-1}} = \sqrt{\frac{146}{8-1}} = \sqrt{\frac{146}{7}} \approx 4.57$$

9. Place data in order.

3, 3, 3, 3, 3, 3, 3

Range: largest – smallest = 3 – 3 = 0

Mean: $\bar{x} = \frac{\sum x}{n} = \frac{21}{7} = 3$

Standard deviation:

Data Value	$x - \bar{x}$	$(x-\bar{x})^2$
3	3 – 3 = 0	0
3	3 – 3 = 0	0
3	3 – 3 = 0	0
3	3 – 3 = 0	0
3	3 – 3 = 0	0
3	3 – 3 = 0	0
3	3 – 3 = 0	0
Total	0	0

$$s = \sqrt{\frac{\sum (x-\bar{x})^2}{n-1}} = \sqrt{\frac{0}{7-1}} = \sqrt{\frac{0}{6}} = 0$$

11. $\bar{x} = \frac{\sum x \cdot f}{\sum f} = \frac{140}{20} = 7$ and $s = \sqrt{\frac{\sum (x-\bar{x})^2 \cdot f}{n-1}} = \sqrt{\frac{86}{20-1}} = \sqrt{\frac{86}{19}} \approx 2.13$

Number x	Frequency f	Product $x \cdot f$	Deviation $x-\bar{x}$	Deviation[2] $(x-\bar{x})^2$	Product $(x-\bar{x})^2 \cdot f$
2	1	2	2 – 7 = – 5	25	25·1 = 25
3	1	3	3 – 7 = – 4	16	16·1 = 16
4	0	0	4 – 7 = – 3	9	9·0 = 0
5	2	10	5 – 7 = – 2	4	4·2 = 8
6	3	18	6 – 7 = – 1	1	1·3 = 3
7	4	28	7 – 7 = 0	0	0·4 = 0
8	4	32	8 – 7 = 1	1	1·4 = 4
9	3	27	9 – 7 = 2	4	4·3 = 12
10	2	20	10 – 7 = 3	9	9·2 = 18
	$\sum f = 20$	$\sum x \cdot f = 140$			$\sum (x-\bar{x})^2 \cdot f = 86$

13. $\bar{x} = \frac{\sum x \cdot f}{\sum f} = \frac{48}{12} = 4$ and $s = \sqrt{\frac{\sum (x-\bar{x})^2 \cdot f}{n-1}} = \sqrt{\frac{28}{12-1}} = \sqrt{\frac{28}{11}} \approx 1.60$

Number x	Frequency f	Product $x \cdot f$	Deviation $x-\bar{x}$	Deviation[2] $(x-\bar{x})^2$	Product $(x-\bar{x})^2 \cdot f$
2	2	4	2 – 4 = – 2	4	4·2 = 8
3	4	12	3 – 4 = – 1	1	1·4 = 4
4	2	8	4 – 4 = 0	0	0·2 = 0
5	0	0	5 – 4 = 1	1	1·0 = 0
6	4	24	6 – 4 = 2	4	4·4 = 16
	$\sum f = 12$	$\sum x \cdot f = 48$			$\sum (x-\bar{x})^2 \cdot f = 28$

15. $\overline{x}=\dfrac{\sum x\cdot f}{\sum f}=\dfrac{105}{15}=7$ and $s=\sqrt{\dfrac{\sum(x-\overline{x})^2\cdot f}{n-1}}=\sqrt{\dfrac{42}{15-1}}=\sqrt{\dfrac{42}{14}}\approx 1.73$

Number x	Frequency f	Product $x\cdot f$	Deviation $x-\overline{x}$	Deviation² $(x-\overline{x})^2$	Product $(x-\overline{x})^2\cdot f$
3	1	3	$3-7=-4$	16	$16\cdot 1=16$
4	1	4	$4-7=-3$	9	$9\cdot 1=9$
5	0	0	$5-7=-2$	4	$4\cdot 0=0$
6	2	12	$6-7=-1$	1	$1\cdot 2=2$
7	5	35	$7-7=\ 0$	0	$0\cdot 5=0$
8	3	24	$8-7=\ 1$	1	$1\cdot 3=3$
9	3	27	$9-7=\ 2$	4	$4\cdot 3=12$
	$\sum f=15$	$\sum x\cdot f=105$			$\sum(x-\overline{x})^2\cdot f=42$

17. $\overline{x}=\dfrac{\sum x\cdot f}{\sum f}=\dfrac{108}{18}=6$ and $s=\sqrt{\dfrac{\sum(x-\overline{x})^2\cdot f}{n-1}}=\sqrt{\dfrac{78}{18-1}}=\sqrt{\dfrac{78}{17}}\approx 2.14$

Number x	Frequency f	Product $x\cdot f$	Deviation $x-\overline{x}$	Deviation² $(x-\overline{x})^2$	Product $(x-\overline{x})^2\cdot f$
3	2	6	$3-6=-3$	9	$2\cdot 9=18$
4	4	16	$4-6=-2$	4	$4\cdot 4=16$
5	3	15	$5-6=-1$	1	$3\cdot 1=3$
6	1	6	$6-6=0$	0	$1\cdot 0=0$
7	2	14	$7-6=1$	1	$2\cdot 1=2$
8	3	24	$8-6=2$	4	$3\cdot 4=12$
9	3	27	$9-6=3$	9	$3\cdot 9=27$
	$\sum f=18$	$\sum x\cdot f=108$			$\sum(x-\overline{x})^2\cdot f=78$

19. $\overline{x}=\dfrac{\sum x\cdot f}{\sum f}=\dfrac{1{,}580}{20}=79$ and $s=\sqrt{\dfrac{\sum(x-\overline{x})^2\cdot f}{n-1}}=\sqrt{\dfrac{826}{20-1}}=\sqrt{\dfrac{826}{19}}\approx 6.59$

Number x	Frequency f	Product $x\cdot f$	Deviation $x-\overline{x}$	Deviation² $(x-\overline{x})^2$	Product $(x-\overline{x})^2\cdot f$
72	7	504	$72-79=-7$	49	$49\cdot 7=343$
73	1	73	$73-79=-6$	36	$36\cdot 1=36$
78	3	234	$78-79=-1$	1	$1\cdot 3=3$
84	6	504	$84-79=\ 5$	25	$25\cdot 6=150$
86	2	172	$86-79=\ 7$	49	$49\cdot 2=98$
93	1	93	$93-79=14$	196	$196\cdot 1=196$
	$\sum f=20$	$\sum x\cdot f=1{,}580$			$\sum(x-\overline{x})^2\cdot f=826$

21. a) $\bar{x}=\frac{\sum x}{n}=\frac{45.89}{8}\approx 5.74$ and $s=\sqrt{\frac{\sum(x-\bar{x})^2}{n-1}}=\sqrt{\frac{32.9445}{8-1}}=\sqrt{\frac{32.9445}{7}}\approx 2.17$

Number x	Deviation $x-\bar{x}$	Deviation^2 $(x-\bar{x})^2$
4.24	4.24 – 5.74 = – 1.5	2.25
9.55	9.55 – 5.74 = 3.81	14.5161
8.25	8.25 – 5.74 = 2.51	6.3001
6	6 – 5.74 = 0.26	0.0676
5.53	5.53 – 5.74 = – 0.21	0.0441
4.9	4.9 – 5.74 = – 0.84	0.7056
3.07	3.07 – 5.74 = – 2.67	7.1289
4.35	4.35 – 5.74 = – 1.39	1.9321
$\sum x = 45.89$		$\sum(x-\bar{x})^2 = 32.9445$

b) No; The states have different size populations. The mean only represents the average of the tax rates in effect for these states.

23. $\bar{x}=\frac{\sum x\cdot f}{\sum f}=\frac{313}{71}\approx 4.41$ and $s=\sqrt{\frac{\sum(x-\bar{x})^2\cdot f}{n-1}}=\sqrt{\frac{801.1551}{71-1}}=\sqrt{\frac{801.1551}{70}}\approx 3.38$

Number x	Frequency f	Product $x\cdot f$	Deviation $x-\bar{x}$	Deviation^2 $(x-\bar{x})^2$	Product $(x-\bar{x})^2\cdot f$
0	13	0	0 – 4.41 = – 4.41	19.4481	252.8253
1	11	11	1 – 4.41 = – 3.41	11.6281	127.9091
2	5	10	2 – 4.41 = – 2.41	5.8081	29.0405
3	0	0	3 – 4.41 = – 1.41	1.9881	0
4	0	0	4 – 4.41 = – 0.41	0.1681	0
5	8	40	5 – 4.41 = 0.59	0.3481	2.7848
6	12	72	6 – 4.41 = 1.59	2.5281	30.3372
7	9	63	7 – 4.41 = 2.59	6.7081	60.3729
8	6	48	8 – 4.41 = 3.59	12.8881	77.3286
9	5	45	9 – 4.41 = 4.59	21.0681	105.3405
10	0	0	10 – 4.41 = 5.59	31.2481	0
11	0	0	11 – 4.41 = 6.59	43.4281	0
12	2	24	12 – 4.41 = 7.59	57.6081	115.2162
	$\sum f = 71$	$\sum x\cdot f =$ 313			$\sum(x-\bar{x})^2\cdot f =$ 801.1551

25. $\bar{x} = \frac{\sum x}{n} = \frac{392}{8} = 49$ and $s = \sqrt{\frac{\sum(x-\bar{x})^2}{n-1}} = \sqrt{\frac{18}{8-1}} = \sqrt{\frac{18}{7}} \approx 1.60$

Number x	Deviation $x-\bar{x}$	Deviation² $(x-\bar{x})^2$
47	47 – 49 = – 2	4
48	48 – 49 = – 1	1
50	50 – 49 = 1	1
49	49 – 49 = 0	0
51	51 – 49 = 2	4
47	47 – 49 = – 2	4
49	49 – 49 = 0	0
51	51 – 49 = 2	4
$\sum x = 392$		$\sum(x-\bar{x})^2 = 18$

$\frac{51-49}{1.6} = 1.25$; Family H's income is 1.25 standard deviations above the mean.

27. $\bar{x} = \frac{\sum x}{n} = \frac{800}{10} = 80$ and $s = \sqrt{\frac{\sum(x-\bar{x})^2}{n-1}} = \sqrt{\frac{184}{10-1}} = \sqrt{\frac{184}{9}} \approx 4.52$

Number x	Deviation $x-\bar{x}$	Deviation² $(x-\bar{x})^2$
80	80 – 80 = 0	0
76	76 – 80 = – 4	16
81	81 – 80 = 1	1
84	84 – 80 = 4	16
79	79 – 80 = – 1	1
80	80 – 80 = 0	0
90	90 – 80 = 10	100
75	75 – 80 = – 5	25
75	75 – 80 = – 5	25
80	80 – 80 = 0	0
$\sum x = 800$		$\sum(x-\bar{x})^2 = 184$

$\frac{76-80}{4.52} \approx -0.88$; A grade of 76 is 0.88 standard deviations below the mean. This would be a grade of "D".

29. $\bar{x} = \frac{\sum x}{n} = \frac{714}{9} \approx 79.3$ and $s = \sqrt{\frac{\sum(x-\bar{x})^2}{n-1}} = \sqrt{\frac{77.5265}{9-1}} = \sqrt{\frac{77.5265}{8}} \approx 3.11$

Number x	Deviation $x-\bar{x}$	Deviation² $(x-\bar{x})^2$
80	80 – 79.75 = 0.25	0.0625
75	75 – 79.75 = – 4.75	22.5625
81	81 – 79.75 = 1.25	1.5625
83	83 – 79.75 = 3.25	10.5625
80	80 – 79.75 = 0.25	0.0625
78	78 – 79.75 = – 1.75	3.0625
84	84 – 79.75 = 4.25	18.0625
77	77 – 79.75 = – 2.75	7.5625
76	76 – 79.75 = – 3.75	14.0625
$\sum x = 714$		$\sum(x-\bar{x})^2 = 77.5265$

A grade of "A" would have to be greater than or equal to $79.8 + 1.5(3.11) \approx 84$. Only the score of 84 would earn an "A", so there would be only one "A".

31. Apple Computer is more volatile.

Apple Computer: $CV = \frac{s}{\overline{x}} \cdot 100\% = \frac{12.3}{123.76} \cdot 100\% \approx 9.9\%$

Dell: $CV = \frac{s}{\overline{x}} \cdot 100\% = \frac{7.2}{78.6} \cdot 100\% \approx 9.2\%$

33. WebMaster is more volatile.

DJIA: $CV = \frac{s}{\overline{x}} \cdot 100\% = \frac{72.17}{11261.12} \cdot 100\% \approx 0.6\%$;

WebMaster: $CV = \frac{s}{\overline{x}} \cdot 100\% = \frac{1.7}{37.6} \cdot 100\% \approx 4.5\%$

35. Coffee

$$\overline{x} = \frac{\sum x}{n} = \frac{62.3}{12} \approx 5.19 \text{ and } \sigma = \sqrt{\frac{\sum (x-\overline{x})^2}{n}} = \sqrt{\frac{2.8496}{12}} \approx 0.49$$

Number x	Deviation $x-\overline{x}$	Deviation² $(x-\overline{x})^2$
4.42	4.42 – 5.19 = – 0.77	0.5929
4.22	4.22 – 5.19 = – 0.97	0.9409
4.64	4.64 – 5.19 = – 0.55	0.3025
5.10	5.10 – 5.19 = – 0.09	0.0081
5.13	5.13 – 5.19 = – 0.06	0.0036
5.23	5.23 – 5.19 = 0.04	0.0016
5.55	5.55 – 5.19 = 0.36	0.1296
5.77	5.77 – 5.19 = 0.58	0.3364
5.65	5.65 – 5.19 = 0.46	0.2116
5.51	5.51 – 5.19 = 0.32	0.1024
5.64	5.64 – 5.19 = 0.45	0.2025
5.44	5.44 – 5.19 = 0.25	0.0625
$\sum x = 62.3$		$\sum (x-\overline{x})^2 = 2.8496$

35. (continued)

Gasoline

$$\overline{x}=\frac{\sum x}{n}=\frac{42.32}{12}\approx 3.53 \text{ and } \sigma=\sqrt{\frac{\sum(x-\overline{x})^2}{n}}=\sqrt{\frac{0.7056}{12}}\approx 0.24$$

Number x	Deviation $x-\overline{x}$	Deviation^2 $(x-\overline{x})^2$
3.09	3.09 – 3.53 = – 0.44	0.1936
3.17	3.17 – 3.53 = – 0.36	0.1296
3.55	3.55 – 3.53 = 0.02	0.0004
3.82	3.82 – 3.53 = 0.29	0.0841
3.93	3.93 – 3.53 = 0.40	0.1600
3.70	3.70 – 3.53 = 0.17	0.0289
3.65	3.65 – 3.53 = 0.12	0.0144
3.63	3.63 – 3.53 = 0.10	0.0100
3.61	3.61 – 3.53 = 0.08	0.0064
3.47	3.47 – 3.53 = – 0.06	0.0036
3.42	3.42 – 3.53 = – 0.11	0.0121
3.28	3.28 – 3.53 = – 0.25	0.0625
$\sum x = 42.32$		$\sum(x-\overline{x})^2 = 0.7056$

37. G went from the mean income to 0.935 (0.94 rounded) standard deviations above the mean.

Family	Year One $\frac{x-\overline{x}}{s}$	Year Two $\frac{x-\overline{x}}{s}$	Change in Standard Deviations
A	$\frac{47-49}{1.60}=-1.25$	$\frac{49-51}{1.07}\approx-1.87$	Down 0.62
B	$\frac{48-49}{1.60}=-0.625$	$\frac{51-51}{1.07}=0$	Up 0.625
C	$\frac{50-49}{1.60}=0.625$	$\frac{52-51}{1.07}\approx 0.935$	Up 0.31
D	$\frac{49-49}{1.60}=0$	$\frac{51-51}{1.07}=0$	No Change
E	$\frac{51-49}{1.60}=1.25$	$\frac{51-51}{1.07}=0$	Down 1.25
F	$\frac{47-49}{1.60}=-1.25$	$\frac{50-51}{1.07}\approx-0.935$	Up 0.315
G	$\frac{49-49}{1.60}=0$	$\frac{52-51}{1.07}\approx 0.935$	**Up 0.935**
H	$\frac{51-49}{1.60}=1.25$	$\frac{52-51}{1.07}\approx 0.935$	Down 0.315

39. $CV=\frac{s}{\overline{x}}\cdot 100\%=\frac{1.78}{7.8}\cdot 100\%\approx 22.82\%$

41. a) false; The mean is 0, but the standard deviation is not. Additional explanations may vary.

$$\overline{x}=\frac{\sum x}{n}=\frac{0}{8}=0 \text{ and } s=\sqrt{\frac{\sum(x-\overline{x})^2}{n-1}}=\sqrt{\frac{32}{8-1}}=\sqrt{\frac{32}{7}}\approx 2.14$$

Number x	Deviation $x-\overline{x}$	Deviation^2 $(x-\overline{x})^2$
–2	–2 – 0 = – 2	4
2	2 – 0 = 2	4
–2	–2 – 0 = – 2	4
2	2 – 0 = 2	4
–2	–2 – 0 = – 2	4
2	2 – 0 = 2	4
–2	–2 – 0 = – 2	4
2	2 – 0 = 2	4
$\sum x=0$		$\sum(x-\overline{x})^2=32$

b) true; Explanations/examples may vary.

43. false; Explanations/examples may vary.

45. a) Examples will vary.

b) The mean will be 20 more than the original mean; the standard deviation will stay the same.

c) The mean will be 5 less than the original mean; the standard deviation will stay the same.

d) The new mean will be more or less than the old mean by the amount added to each value in the original data set; the standard deviation will remain the same.

e) You could calculate the mean and standard deviation of –2, –3, –1, 0, 1, 2, and 3 instead and thus work with smaller numbers.

47. The distribution in Exercise 23 appears to be more spread out so we would expect it to have the larger standard deviation.

Section 14.4 The Normal Distribution

1. Since 12 is one standard deviation above the mean, we would expect $\frac{68\%}{2}=34\%$ of the values to lie between 10 and 12.

3. Since 14 is two standard deviations above the mean, we would expect $\frac{95\%}{2}=47.5\%$ of the values to lie between 10 and 14. Therefore, 50% – 47.5%=2.5% of the values should fall above 14.

5. Since 12 is one standard deviation above the mean, we would expect $\frac{68\%}{2}=34\%$ of the values to lie between 10 and 12. Therefore, 50% – 34%=16% of the values should fall above 12.

7. Since 9 is one standard deviation below the mean, we would expect $\frac{68\%}{2}=34\%$ of the values to lie between 9 and 8. Therefore, 50% – 34%=16% of the values should fall below 9.

9. Since 6 is two standard deviations below the mean, we would expect $\frac{95\%}{2} = 47.5\%$ of the values to lie between 6 and 12. Therefore, 50% + 47.5% = 97.5% of the values should fall above 6.

11. Since 18 is two standard deviations above the mean, we would expect $\frac{95\%}{2} = 47.5\%$ of the values to lie between 12 and 18. Similarly, since 15 is one standard deviation above the mean, we would expect $\frac{68\%}{2} = 34\%$ of the values to lie between 12 and 15. Therefore, we would expect 47.5% – 34% = 13.5% of the values to lie between 15 and 18.

13. a) 0.436
    b) 0.496
    c) 0.496 – .436 = 0.060
    d) 0.500 – 0.496 = 0.004

15. 39.1%

17. 27.3%

19. 47.4% – 39.4% = 8%

21. 27.6% –14.8% = 12.8%

23. 50% – 42.7% = 7.3%

25. 50% – 41.9% = 8.1%

27. 50% + 40.8% = 90.8%

29. We need to find a *z*–score that corresponds to 0.5 – 0.1 = 0.4. The closest *z*–score is 1.28.

31. We need to find a *z*–score that corresponds to 0.5 – 0.12 = 0.38. Since 0.38 is halfway between 0.379 and 0.381, the *z*–score would be 1.175. This *z*–score needs to be –1.175, however, since it indicates an area less than 50% to the left of the mean.

33. We need to find a *z*–score that corresponds to 0.6 – 0.5 = 0.1. Since 0.1 is one-fourth of the way between 0.099 and 0.103, the *z*–score would be 0.2525.

35. $z = \frac{87-80}{5} = 1.40$

37. $z = \frac{14-21}{4} = -1.75$

39. $z = \frac{48-38}{10.3} \approx 0.97$

41. $0.84 = \frac{x-60}{5}$
    $4.2 = x - 60$
    $64.2 = x$

43. $-0.45 = \frac{x-35}{3}$
    $-1.35 = x - 35$
    $33.65 = x$

45. $1.64 = \frac{x-28}{2.25}$
    $3.69 = x - 28$
    $31.69 = x$

47. a) 50%

    b) Since 7.5 is one standard deviation below the mean, we would expect $\frac{68\%}{2} = 34\%$ of the values to lie between 7.5 and 8. Therefore, 50% – 34% = 16% of the cups should have less that 7.5 ounces.

49. a) Since 202 is one standard deviation above the mean, we would expect $\frac{68\%}{2} = 34\%$ of the bags to have between 200 and 202 pieces in them.
    b) 50% – 34% = 16%

51. $z = \frac{140-120}{12} \approx 1.67$

53. 6 feet 8 inches corresponds to 80 inches and 7 feet corresponds to 84 inches. $z = \frac{84-80}{3} \approx 1.33$, and the percent of area that corresponds to players between 80 inches and 84 inches is 40.8%. Therefore, 50% – 40.8% = 9.2% have a height over 7 feet. Since $0.092 \cdot 324 = 29.808$, we would expect about 29 or 30 players out of the 324 to be over 7 feet.

55. $z = \frac{70-68}{4} = 0.50$, and this corresponds to an area of 19.2% + 50% = 69.2% for women with heart rates of less than 70 beats per minute. Since $0.692 \cdot 200 = 138.4$, we would expect 138 or 139 out of the 200 women to have heart rates of less that 70 beats per minute.

57. Since roughly 95% of data values are within 2 standard deviations of the mean, the standard deviation should be 9 since 250 is 18 below 268 and 286 is 18 above 268. We now need to find the *z*–score associated with 275 days. $z = \frac{275-268}{9} \approx 0.78$ and therefore, the percentage of pregnancies that would be expected to last at least 275 days would be 50% – 28.2% = 21.8%.

59. $1.5 = \frac{x-37}{11}$

    $16.5 = x - 37$

    $x = 53.5$

61. The top 10% of the class is equivalent to $A = 0.900 - 0.500 = 0.400$. From table 14.15, $z = 1.28$, so the cutoff for an A is $72 + 1.28(8) = 82.24$, which must be rounded up to 83 so that no more than the top 10% of the class will be given an A.

63. The top 20% of runners is equivalent to $A = 0.800 - 0.500 = 0.300$. From table 14.15, $z = 0.84$, so the cutoff time is $85 + 0.84(9) = 92.56$ minutes.

65. The bottom 95% of breaking strengths is equivalent to $A = 0.900 - 0.500 = 0.450$. From table 14.15, $z = 1.64$, so the cutoff weight is $150 + 1.64(8) = 163.12$ pounds.

67. $z = \frac{72-67}{10} = 0.50$, and this corresponds to an area of 19.2% + 50% = 69.2% of years with more than 72 inches of rain. So, we would expect to have $20(100\% - 69.2\%) = 20(30.8) \approx 6$ years.

69. Carew was more dominant.

    Jackie Robinson: $z = \frac{0.342-0.267}{0.0326} \approx 2.30$ above the mean

    Rod Carew: $z = \frac{0.350-0.261}{0.0317} \approx 2.81$ above the mean

71. The z–score that corresponds to 50% – 4%= 46% = 0.46 is 1.75. Moreover, this z–score should be – 1.75, since the accumulated area is less than 50%.

$-1.75=\dfrac{x-2,000}{800}\Rightarrow -1,400=x-2,000\Rightarrow x=600$ hours of play. Since a game is played for about 2 hours a day, the number of days the game is played is $\dfrac{600}{2}=300$. This implies that the warranty should be for about 10 months.

73. a) $z=\dfrac{9-7.8}{1.3}\approx 0.92$ and this corresponds to an area of 50% – 32.1% = 17.9%. So of the past 15 years, you would expect to receive at least 9% on the investments $0.179\cdot 15=2.685$ or 2 or 3 years.

b) $z=\dfrac{6-7.8}{1.3}\approx -1.38$, and this corresponds to an area of 50% – 41.6% = 8.4%. So of the past 15 years, you would expect to receive less than 6% on the investments $0.084\cdot 15=1.26$ or 1 or 2 years.

75. $z=\dfrac{480-500}{100}=-0.20$, and this corresponds to 50% – 7.9% = 42.1%.

77. a) The score is almost two standard deviations above the mean.
b) The score is almost one standard deviation below the mean.

79. By looking at the graph of the normal curve, there is clearly more area under the curve between 0.0 and 0.7 than there is between 1.3 and 2.0.

81. No solution provided.

83. The z–score that corresponds to 80% – 50% = 30% = 0.300 is 0.84.

85. The z–score that approximately corresponds to 75% – 50%= 25% = 0.250 is 0.673. $0.673=\dfrac{x-40}{4}\Rightarrow 2.692=x-40\Rightarrow x=42.692$

87. $z=\dfrac{75-68}{4}=1.75$ corresponds to an accumulated area of 50% + 46% = 96%. This implies the 96th percentile.

Section 14.5 Looking Deeper: Linear Correlation

1. positive correlation

3. a)

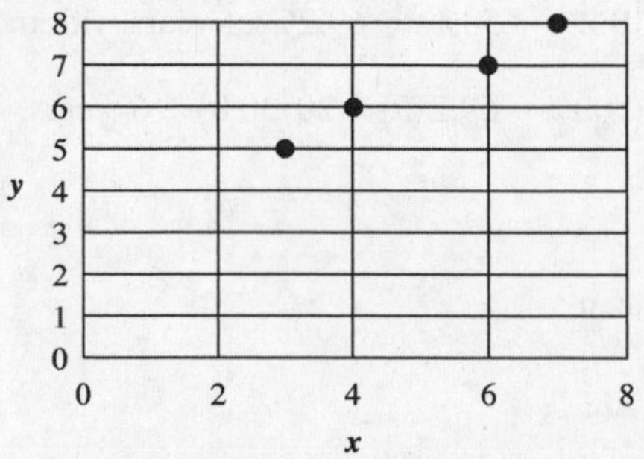

3. (continued)

b) No solution provided.

c)

x	y	x^2	y^2	xy
3	5	9	25	15
7	8	49	64	56
4	6	16	36	24
6	7	36	49	42
$\sum x = 20$	$\sum y = 26$	$\sum x^2 = 110$	$\sum y^2 = 174$	$\sum xy = 137$

$$r = \frac{n\sum xy - \left(\sum x\right)\left(\sum y\right)}{\sqrt{n\left(\sum x^2\right) - \left(\sum x\right)^2}\sqrt{n\left(\sum y^2\right) - \left(\sum y\right)^2}} = \frac{4 \cdot 137 - 20 \cdot 26}{\sqrt{4 \cdot 110 - 20^2}\sqrt{4 \cdot 174 - 26^2}}$$

$$= \frac{548 - 520}{\sqrt{440 - 400}\sqrt{696 - 676}} = \frac{28}{\sqrt{40}\sqrt{20}} = \frac{28}{\sqrt{800}} \approx 0.99$$

5. a)

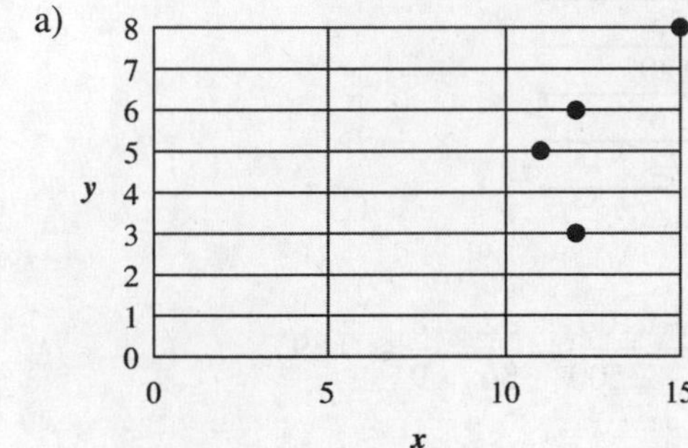

b) No solution provided.

c)

x	y	x^2	y^2	xy
11	5	121	25	55
15	8	225	64	120
12	3	144	9	36
12	6	144	36	72
$\sum x = 50$	$\sum y = 22$	$\sum x^2 = 634$	$\sum y^2 = 134$	$\sum xy = 283$

$$r = \frac{n\sum xy - \left(\sum x\right)\left(\sum y\right)}{\sqrt{n\left(\sum x^2\right) - \left(\sum x\right)^2}\sqrt{n\left(\sum y^2\right) - \left(\sum y\right)^2}} = \frac{4 \cdot 283 - 50 \cdot 22}{\sqrt{4 \cdot 634 - 50^2}\sqrt{4 \cdot 134 - 22^2}}$$

$$= \frac{1{,}132 - 1{,}100}{\sqrt{2{,}536 - 2{,}500}\sqrt{536 - 484}} = \frac{32}{\sqrt{36}\sqrt{52}} = \frac{32}{\sqrt{1{,}872}} \approx 0.74$$

7. We can be 95% confident, but not 99% confident.

9. neither

11. $y = 0.7x + 3$

x	y	x^2	xy
3	5	9	15
7	8	49	56
4	6	16	24
6	7	36	42
$\sum x = 20$	$\sum y = 26$	$\sum x^2 = 110$	$\sum xy = 137$

$$m = \frac{n\sum xy - \left(\sum x\right)\left(\sum y\right)}{n\left(\sum x^2\right) - \left(\sum x\right)^2} = \frac{4 \cdot 137 - 20 \cdot 26}{4 \cdot 110 - 20^2} = \frac{548 - 520}{440 - 400} = \frac{28}{40} = 0.7$$

$$b = \frac{\sum y - m\left(\sum x\right)}{n} = \frac{26 - 0.7 \cdot 20}{4} = \frac{26 - 14}{4} = \frac{12}{4} = 3$$

13. $y = 0.89x - 5.61$

x	y	x^2	xy
11	5	121	55
15	8	225	120
12	3	144	36
12	6	144	72
$\sum x = 50$	$\sum y = 22$	$\sum x^2 = 634$	$\sum xy = 283$

$$m = \frac{n\sum xy - \left(\sum x\right)\left(\sum y\right)}{n\left(\sum x^2\right) - \left(\sum x\right)^2} = \frac{4 \cdot 283 - 50 \cdot 22}{4 \cdot 634 - 50^2} = \frac{1{,}132 - 1{,}100}{2{,}536 - 2{,}500} = \frac{32}{36} = \frac{8}{9} \approx 0.89$$

$$b = \frac{\sum y - m\left(\sum x\right)}{n} = \frac{22 - \frac{8}{9} \cdot 50}{4} = \frac{198 - 8 \cdot 50}{36} = \frac{198 - 400}{36} = \frac{-202}{36} = -\frac{101}{18} \approx -5.61$$

15. We can be 99% confident that there is positive linear correlation.

x	y	x^2	y^2	xy
0	23	0	529	0
1	22	1	484	22
2	27	4	729	54
2	28	4	784	56
5	35	25	1,225	175
$\sum x = 10$	$\sum y = 135$	$\sum x^2 = 34$	$\sum y^2 = 3{,}751$	$\sum xy = 307$

$$r = \frac{n\sum xy - \left(\sum x\right)\left(\sum y\right)}{\sqrt{n\left(\sum x^2\right) - \left(\sum x\right)^2}\sqrt{n\left(\sum y^2\right) - \left(\sum y\right)^2}} = \frac{5 \cdot 307 - 10 \cdot 135}{\sqrt{5 \cdot 34 - 10^2}\sqrt{5 \cdot 3{,}751 - 135^2}}$$

$$= \frac{1{,}535 - 1{,}350}{\sqrt{170 - 100}\sqrt{18{,}755 - 18{,}225}} = \frac{185}{\sqrt{70}\sqrt{530}} = \frac{185}{\sqrt{37{,}100}} \approx 0.96$$

17. neither

x	y	x^2	y^2	xy
30	26	900	676	780
40	31	1,600	961	1,240
50	33	2,500	1,089	1,650
60	31	3,600	961	1,860
70	26	4,900	676	1,820
$\sum x = 250$	$\sum y = 147$	$\sum x^2 = 13,500$	$\sum y^2 = 4,363$	$\sum xy = 7,350$

$$r = \frac{n\sum xy - \left(\sum x\right)\left(\sum y\right)}{\sqrt{n\left(\sum x^2\right) - \left(\sum x\right)^2}\sqrt{n\left(\sum y^2\right) - \left(\sum y\right)^2}} = \frac{5 \cdot 7,350 - 250 \cdot 147}{\sqrt{5 \cdot 13,500 - 250^2}\sqrt{5 \cdot 4,363 - 147^2}}$$

$$= \frac{36,750 - 36,750}{\sqrt{67,500 - 62,500}\sqrt{21,815 - 21,609}} = \frac{0}{\sqrt{5,000}\sqrt{206}} = \frac{0}{\sqrt{1,030,000}} = 0.00$$

19. $y = 2.64x + 21.71$

x	y	x^2	xy
0	23	0	0
1	22	1	22
2	27	4	54
2	28	4	56
5	35	25	175
$\sum x = 10$	$\sum y = 135$	$\sum x^2 = 34$	$\sum xy = 307$

$$m = \frac{n\sum xy - \left(\sum x\right)\left(\sum y\right)}{n\left(\sum x^2\right) - \left(\sum x\right)^2} = \frac{5 \cdot 307 - 10 \cdot 135}{5 \cdot 34 - 10^2} = \frac{1,535 - 1,350}{170 - 100} = \frac{185}{70} = \frac{37}{14} \approx 2.64$$

$$b = \frac{\sum y - m\left(\sum x\right)}{n} = \frac{135 - \frac{37}{14} \cdot 10}{5} = \frac{1,890 - 370}{70} = \frac{1,520}{70} = \frac{152}{7} \approx 21.71$$

21. $y = 0 \cdot x + 29.4$

x	y	x^2	xy
30	26	900	780
40	31	1600	1240
50	33	2500	1650
60	31	3600	1860
70	26	4900	1820
$\sum x = 250$	$\sum y = 147$	$\sum x^2 = 13,500$	$\sum xy = 7,350$

$$m = \frac{n\sum xy - \left(\sum x\right)\left(\sum y\right)}{n\left(\sum x^2\right) - \left(\sum x\right)^2} = \frac{5 \cdot 7,350 - 250 \cdot 147}{5 \cdot 13,500 - 250^2} = \frac{36,750 - 36,750}{67,500 - 62,500} = \frac{0}{5,000} = 0$$

$$b = \frac{\sum y - m\left(\sum x\right)}{n} = \frac{147 - 0 \cdot 250}{5} = \frac{147 - 0}{5} = \frac{147}{5} = 29.4$$

23. We can be 99% confident that there is a significant linear correlation between the variables of car weight and mileage.

25. No solution provided.

27. 1

Chapter Review Exercises

1.

Number of Accidents	Frequency
4	1
5	3
6	6
7	2
8	10
9	4
10	5

2.

Number of Accidents	Relative Frequency
4	0.03
5	0.10
6	0.19
7	0.06
8	0.32
9	0.13
10	0.16

3.

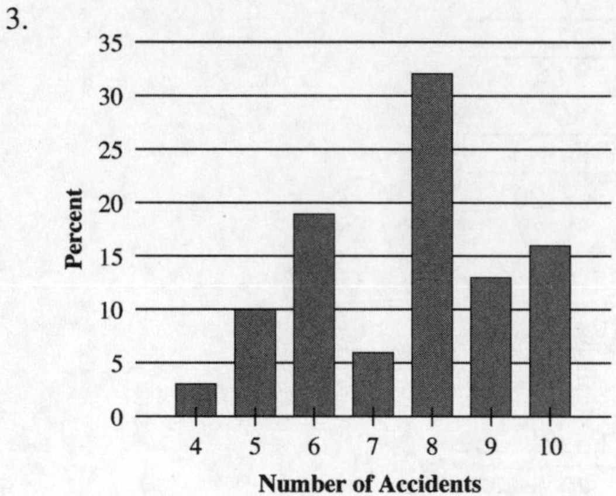

4. a) Four riders occurred three times.
   b) five riders
   c) $3 + 8 + 6 + 7 + 3 + 5 + 3 + 2 = 37$

5. a)

F		M
9 9 8 6 6 5 1	2	9
9 8 5 5 5 4 3 3 3 2 1 0 0	3	1 2 5 6 7 7 7 7 8 8 9
9 9 5 2 1	4	0 2 3 3 5 5 5 6 7
	5	0 1 2 2 4
1 1	6	0 1
4	7	6
0	8	

b) The actors generally seem older than the actresses.

6. Place the data in order.
   29, 32, 37, 42, 50, 50, 53, 54, 54, 54, 59, 64

   Mean: $\bar{x} = \dfrac{\sum x}{n} = \dfrac{578}{12} \approx 48.2$

   Median: $\dfrac{50+53}{2} = \dfrac{103}{2} = 51.5$

   Mode: 54

7. The mean is the arithmetic average; the median is the middle score; the mode is the most frequent score.

8. Place the data in order.
   3, 4, 6, 6, 6, 7, 7, 8, 9, 10, 11, 12, 13, 17, 20
   lower half: 3, 4, 6, 6, 6, 7, 7
   upper half: 9, 10, 11, 12, 13, 17, 20
   minimum: 3
   first quartile is the median of the lower half: $Q_1 = 6$
   median is the middle number: 8
   third quartile is the median of the upper half: $Q_3 = 12$
   maximum: 20

9. minimum: 3
   first quartile: $Q_1 = 6$
   median: 8
   third quartile: $Q_3 = 12$
   maximum: 20

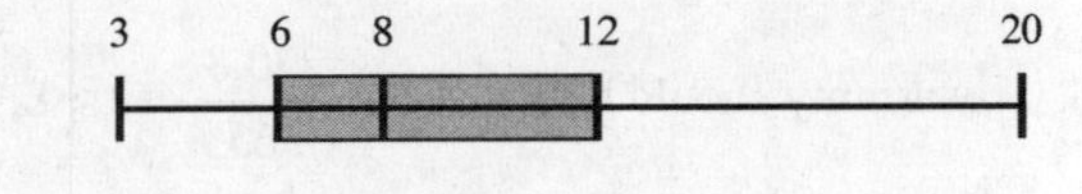

10. $\bar{x} = \dfrac{\sum x}{n} = \dfrac{40}{8} = 5$ and $s = \sqrt{\dfrac{\sum(x-\bar{x})^2}{n-1}} = \sqrt{\dfrac{12}{8-1}} = \sqrt{\dfrac{12}{7}} \approx 1.31$

Number x	Deviation $x-\bar{x}$	Deviation^2 $(x-\bar{x})^2$
4	4 − 5 = − 1	1
6	6 − 5 = 1	1
7	7 − 5 = 2	4
3	3 − 5 = − 2	4
5	5 − 5 = 0	0
6	6 − 5 = 1	1
4	4 − 5 = − 1	1
5	5 − 5 = 0	0
$\sum x = 40$		$\sum(x-\bar{x})^2 = 12$

11. the spread of the distribution

12. See box on page 739.

13. $z = \dfrac{85-80}{7} \approx 0.71$

14. $1.35 = \frac{x-60}{5}$

$6.75 = x - 60$

$x = 66.75$

15. For 75, $z = \frac{75-72}{4} = 0.75$, so 2.73% of the scores are between 75 and the mean of 72.

For 82 $z = \frac{82-72}{4} = 2.5$, so 4.94% of the scores are between the mean of 72 and 82.

So, 4.94% – 2.73% =2.21% of the scores are expected to lie between 75 and 82, so $1{,}000(.0.221) = 221$ scores should lie between 75 and 82.

16. The history exam score is relatively better, since it has a higher z–score.

History: $z = \frac{82-78}{3} \approx 1.33$ Anthropology: $z = \frac{84-79}{4} = 1.25$

17. The z–score that corresponds to 50% – 4%= 46% = 0.46 is 1.75. Moreover, this z–score should be –1.75, since the accumulated area is less than 50%.

$-1.75 = \frac{x-2{,}500}{500} \Rightarrow -875 = x - 2{,}500 \Rightarrow x = 1{,}625$ hours of use. Since a TV is used for about 6 hours a day, the number of days the iTouch is used is $\frac{1{,}625}{6} = 270.8$ days. This implies that the warranty should be for about $\frac{270.8}{365} \times 12 \approx 8.9$ months.

18. a) positive linear correlation b) no linear correlation

19. a) $r = 0.977$ b) $y = 2.5x - 2$

x	y	x^2	y^2	xy
3	5	9	25	15
4	9	16	81	36
5	10	25	100	50
6	13	36	169	78
$\sum x = 18$	$\sum y = 37$	$\sum x^2 = 86$	$\sum y^2 = 375$	$\sum xy = 179$

$$r = \frac{n\sum xy - \left(\sum x\right)\left(\sum y\right)}{\sqrt{n\left(\sum x^2\right) - \left(\sum x\right)^2}\sqrt{n\left(\sum y^2\right) - \left(\sum y\right)^2}} = \frac{4 \cdot 179 - 18 \cdot 37}{\sqrt{4 \cdot 86 - 18^2}\sqrt{4 \cdot 375 - 37^2}}$$

$$= \frac{716 - 666}{\sqrt{344 - 324}\sqrt{1{,}500 - 1{,}369}} = \frac{50}{\sqrt{20}\sqrt{131}} = \frac{50}{\sqrt{2{,}620}} \approx 0.977$$

$$m = \frac{n\sum xy - \left(\sum x\right)\left(\sum y\right)}{n\left(\sum x^2\right) - \left(\sum x\right)^2} = \frac{4 \cdot 179 - 18 \cdot 37}{4 \cdot 86 - 18^2} = \frac{716 - 666}{344 - 324} = \frac{50}{20} = 2.5$$

$$b = \frac{\sum y - m\left(\sum x\right)}{n} = \frac{37 - 2.5 \cdot 18}{4} = \frac{37 - 45}{4} = \frac{-8}{4} = -2$$

Chapter Test

1.

Number of Visits	Frequency
4	1
5	4
6	8
7	9
8	7
9	6
10	2
11	2
12	1

2.

Number of Visits	Relative Frequency
4	0.025
5	0.100
6	0.200
7	0.225
8	0.175
9	0.150
10	0.050
11	0.050
12	0.025

3.

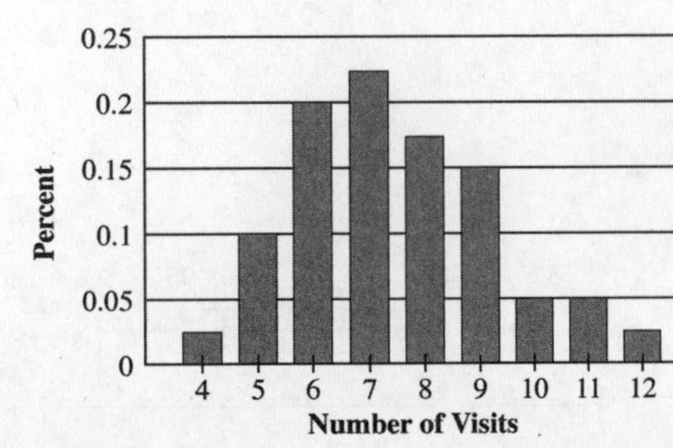

4. the spread of the distribution

5. $\bar{x} = \frac{\sum x}{n} = \frac{54}{9} = 6$ and $s = \sqrt{\frac{\sum (x-\bar{x})^2}{n-1}} = \sqrt{\frac{32}{9-1}} = \sqrt{\frac{32}{8}} = 2$

Number x	Deviation $x-\bar{x}$	Deviation^2 $(x-\bar{x})^2$
3	$3-2=-3$	9
4	$4-2=-2$	4
5	$5-2=-1$	1
6	$6-2=0$	0
5	$5-2=-1$	1
6	$6-2=0$	0
8	$8-2=2$	4
9	$9-2=3$	9
8	$8-2=2$	4
$\sum x = 54$		$\sum (x-\bar{x})^2 = 32$

6. See box on page 739.

7.

Ruth		Aaron
5 2	2	
5 4	3	0 2 4 4 8 9 9
9 7 6 6 6 1 1	4	0 0 4 4 4 4 5 7
9 4 4	5	
0	6	

8. a) Place the data in order.

82, 85, 86, 92, 93, 94, 95, 95, 96, 98

lower half: 82, 85, 86, 92, 93

upper half: 94, 95, 95, 96, 98

minimum: 82

first quartile is the median of the lower half: $Q_1 = 86$

median: $\dfrac{93+94}{2} = \dfrac{187}{2} = 93.5$

third quartile is the median of the upper half: $Q_3 = 95$

maximum: 98

b)

minimum: 82

first quartile: $Q_1 = 86$

median: 93.5

third quartile: $Q_3 = 95$

maximum: 98

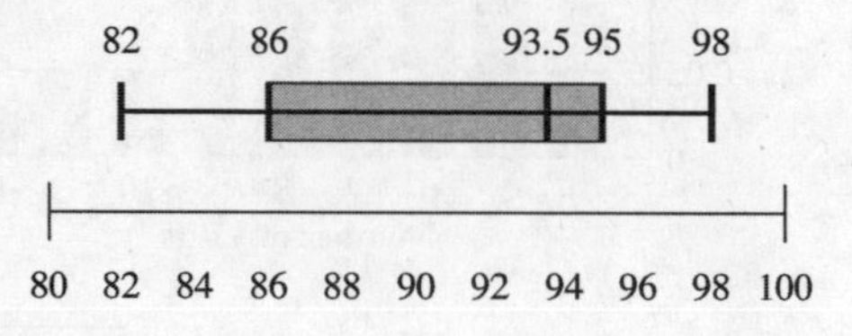

9. The statistics exam score is relatively better, since it has a higher z–score.

Statistics: $z = \dfrac{85-78}{4} = 1.75$

Sociology: $z = \dfrac{88-80}{5} = 1.60$

10. a) Three hits per minute occurred seven times.

b) four and six hits per minute

c) 7 + 8 + 7 + 8 + 5 + 3 = 38

11. a) no linear correlation

b) negative linear correlation

12. The mean is the arithmetic average; the median is the middle score; the mode is the most frequent score.

13. Place the data in order.

3, 5, 5, 6, 6, 6, 7, 7, 8, 8, 9, 9, 9, 9, 9, 9, 10, 10, 11, 11

Mean: $\bar{x} = \frac{\sum x}{n} = \frac{157}{20} = 7.85$

Median: $\frac{8+9}{2} = \frac{17}{2} = 8.5$

Mode: 9

14. $1.83 = \frac{x-50}{6}$

$10.98 = x - 50$

$x = 60.98 \approx 61$

15. $z = \frac{82-75}{5} = 1.4$

16. For 58, $z = \frac{58-54}{5} = 0.80$, so 28.8% of the scores are between 58 and the mean of 54.

So, 50% – 28.8% =21.2% of the scores are expected to lie above 58, so $1{,}000(.0.212) = 212$ scores should lie above 58.

17. The z–score that corresponds to 50% – 6%= 44% = 0.44 is 1.555, since 0.44 lies between 1.55 and 1.56. Moreover, this z–score should be –1.555, since the accumulated area is less than 50%.

$-1.555 = \frac{x-2{,}800}{400} \Rightarrow -622 = x - 2{,}800 \Rightarrow x = 2{,}178$ hours of use. Since the iTouch is used for about 4 hours a day, the number of days the iTouch is used is $\frac{2{,}178}{4} = 544.5$. This implies that the warranty should be for about 18 months.

18. a) $r = 0.832$ b) $y = 1.1x + 2.8$

x	y	x^2	y^2	xy
3	6	9	36	18
4	8	16	64	32
5	7	25	49	35
6	10	36	100	60
$\sum x = 18$	$\sum y = 31$	$\sum x^2 = 86$	$\sum y^2 = 249$	$\sum xy = 145$

$$r = \frac{n\sum xy - (\sum x)(\sum y)}{\sqrt{n(\sum x^2) - (\sum x)^2}\sqrt{n(\sum y^2) - (\sum y)^2}} = \frac{4 \cdot 145 - 18 \cdot 31}{\sqrt{4 \cdot 86 - 18^2}\sqrt{4 \cdot 249 - 31^2}}$$

$$= \frac{580 - 558}{\sqrt{344-324}\sqrt{996-961}} = \frac{22}{\sqrt{20}\sqrt{35}} = \frac{22}{\sqrt{700}} \approx 0.832$$

$$m = \frac{n\sum xy - (\sum x)(\sum y)}{n(\sum x^2) - (\sum x)^2} = \frac{4 \cdot 145 - 18 \cdot 31}{4 \cdot 86 - 18^2} = \frac{580 - 558}{344 - 324} = \frac{22}{20} = 1.1$$

$$b = \frac{\sum y - m(\sum x)}{n} = \frac{31 - 1.1 \cdot 18}{4} = \frac{31 - 19.8}{4} = \frac{11.2}{4} = 2.8$$